Chinese Environmental Contention

Chinese Environmental Contention

Linking Up against Waste Incineration

Maria Bondes

Routledge
Taylor & Francis Group
LONDON AND NEW YORK

First published in 2019 by Amsterdam University Press Ltd.

Published 2025 by Routledge
4 Park Square, Milton Park, Abingdon, Oxon OX14 4RN
605 Third Avenue, New York, NY 10158

Routledge is an imprint of the Taylor & Francis Group, an informa business

ISBN: 9789462988927 (hbk)
ISBN: 9781041176824 (pbk)
ISBN: 9781003692362 (ebk)
NUR 761

Cover illustration: Gaoantun incinerator, Beijing, November 2012
Source: Maria Bondes

Cover design: Coördesign, Leiden

DOI: 10.5117/9789462988927

Table of Contents

List of Tables and Figures

Tables

Figures

List of Abbreviations

BBS	Bulletin board system
BMAC	Beijing Municipal Administration Commission
BOT mode	Build-operate-transfer mode
CAMS	China Academy of Meteorological Sciences
CATS	Communities against Toxics
CCTV	China Central Television
CDM	Clean Development Mechanism
CLAPV	Center for the Legal Assistance to Pollution Victims
CPPCC	Chinese People's Political Consultative Conference
CWIN	China Waste Information Network
EIA	Environmental impact assessment
EPB	Environmental protection bureau
GAIA	Global Alliance of Incinerator Alternatives / Global Anti-Incinerator Alliance
IPEN	International POPs Elimination Network
KfW	Kreditanstalt für Wiederaufbau
LED	Light-emitting diode
MEP	Ministry of Environmental Protection
MSW	Municipal solid waste
MSWI	Municipal solid waste incinerator
NDRC	National Development and Reform Commission
Ng	Nanogram
NIABY	Not-In-Anybody's-Backyard
NIMBY	Not-In-My-Backyard
NPC	National People's Congress
PhD	Doctor of Philosophy
POPs	Persistent organic pollutants
PRC	People's Republic of China
PX	Paraxelene
RMB	Renminbi
SEE	Society of Entrepreneurs and Ecology
SEPA	State Environmental Protection Agency
TEQ	Toxic equivalents
UNFCCC	United Nations Framework Convention on Climate Change
WtE	Waste-to-Energy

Preface

When I came to China in 2011 to look for a specific issue field to study in my dissertation project on local environmental contention in China, the *Guardian*'s Jonathan Watts, whose environmental reporting about China I had followed and admired for a while, was kind enough to meet me and share his insights about the most recent developments in China's environmental sphere. He pointed me towards the then-newly founded environmental organization *Nature University,* whose staff member Chen Liwen urged me to take a closer look at the issue of waste and particularly waste incineration. Having spent time studying desertification and water issues in Western China for my master's research, at first I found the problem of waste, while certainly urgent, rather uninspiring – thinking mainly of the waste collection and separation projects I had come across in Western China. However, upon taking a closer look I found it to be a most fascinating issue field, one which, as a member of one environmental organization put it, encompasses the broader environmental problems and regulatory failures in China. Apart from the environmental and health risks associated with waste incineration, more general problems such as the lax local implementation of environmental laws and regulations, lack of public participation and transparency in the environmental sector, failure to guarantee the rights of pollution victims and affected communities, and local corruption issues are all reflected in the struggle against China's incineration policies and specific waste incinerator projects.

My original idea was to analyze the widening spectrum of contentious methods and strategies employed by affected communities in their fight for a clean and healthy living environment and the factors for the success or failure of individual campaigns. Apart from the difficulty of pinning down the meaning of 'success', particularly where different social groups are affected, I also made another observation that led me to shift the focus of my research. During my first interviews with the members of affected communities that had staged contentious action against waste incineration projects, I was surprised by the frequent references to other cases and reports of receiving assistance from supra-local environmental organizations, experts, lawyers, and activists from other localities. This did not seem to fit my assumption that (particularly rural) local communities were largely isolated from both each other and from the support of supra-local actors – as was the widespread opinion in the literature on social contention in China at the time. I therefore decided to study the linkages among the different actors in the issue field and their impacts on the emergence and development of contention at both the local and higher political levels. During my research I uncovered a dense network of contention spanning different sites and actor groups that had a major impact on both local campaigns and national-level advocacy activities, as described in this book.

The political climate in China has drastically chilled since the bulk of the data for this study was collected between 2011 and 2013. Under the reign of Xi Jinping, the maneuvering space for both supra-local actors and for local contention and collective action has significantly shrunk amidst the general tightening of political control. The political opportunity structures during the later Hu Jintao era (in the late 2000s and early 2010s) – facilitated by the then-booming spread of social media in China – that enabled the formation of networked contention as described in this study should be regarded as a window of political opportunity that has narrowed during the course of Xi Jinping's consolidation of power.

At the local level, affected communities across China continue to fight against waste incineration plants in their neighborhood. However, state reactions to local contention and particularly large-scale street protests, including those against waste incinerators, seem to have become fiercer in the years after the main observation period and have led to violent clashes between contenders and state forces in several anti-incinerator cases – such as in Hangzhou (2014), Wuhan (2015), Xintao (2016), and Liaoning (2018). In recent years the supra-local actor groups that are important nodes for networked contention have also experienced significant drawbacks that hamper their advocacy activities and engagement at the local level. Since 2017, a new law requiring the registration and strict oversight of international organizations limits their action range and ability to provide assistance and financial aid to Chinese organizations and local campaigns (Hsu and Teets 2016; Shieh 2018). This move to tighten control over social organizations has been regarded as part of an overall trend of 'shrinking [the] spaces' for civil society both in China and beyond (Hayes et al. 2017; Lang and Holbig 2018; Richter 2018). At the same time, the party-state's crackdown on *weiquan* ('rights protection') lawyers has made it more difficult for affected communities to find legal support and advice (Duggan 2015; Jacobs and Buckley 2015; Fu 2018). The Chinese media sphere that was characterized by an overall liberalization and 'greening' under Hu Jintao has also – together with the Chinese Internet – felt the clout of the central party-state (Economy 2018; Bandurski 2019).

In other words, the conditions for the formation of networked contention in China have become more adverse than described in this book.

Nonetheless, it is likely that the different actors in the environmental arena will continue to find ways to link up via the pathways observed in this study – particularly to exchange information via informal ties. The identified mechanisms through which such linkages may promote both local and higher-level contention remain the same under a more repressive framework, but it has become more difficult for such linkages to foster contentious action – or, in the terms of social movement theory, to translate into diffusion effects. On the other hand, networked contention as a loosely organized form of resistance that permits its actors to stay relatively under the radar of state attention compared to conventional social movement organizations or formal networks may hold even more merit under the present adverse political climate.

I would like to thank my two dissertation supervisors, Michael Friedrich from Hamburg University and Björn Alpermann from the University of Würzburg, for their supportive guidance of my research that helped me keep a critical eye throughout the course of the project. They always had an open ear and were excellent and compassionate mentors beyond the scope of this single research project.

I am also indebted to my colleagues at the GIGA German Institute of Global and Area Studies in Hamburg and in particular the colleagues at the GIGA Institute of Asian Studies (IAS) and the GIGA Research Team 'Persistence and Change of Nondemocratic Regimes' who offered a great environment for learning the rules of the game of academic life. My special thanks go to the IAS 'China community', namely Karsten Giese, Heike Holbig, Günter Schucher, Margot Schüller and Georg Strüver, for many inspiring discussions, helpful reading of my writings, and delightful lunches and cups of coffee. Thanks, Günter, for teaching me to take academia just seriously enough to not miss the fun part of it! I also thank our excellent head of library Uwe Kotzel for always being up to date on the relevant China literature and for more delightful lunches and cups of coffee. And I thank our IT team for equipping me with enough technological gimmicks to feel like I could keep my sources safe enough during field research while touching upon sensitive issues.

I am further indebted to my GIGA colleague Heike Holbig, who brought me to GIGA as a staff member of her third-party funded research project 'Ideological Change and Regime Legitimacy in China' but was generous enough to give me enough leeway to also conduct my own research on environmental contention. I thank the German Federal Ministry of Education and Research (BMBF), the German Aerospace Center/Project Management Agency, and in particular Rolf Geserick, who made my research stays in

China and numerous conference visits possible. The research project on ideological change (BMBF project 01UC1011D) was part of the BMBF-funded research network 'Governance in China' (2010-2016), and it was a real treasure to be part of this excellent research cluster. I am very grateful to the network's other heads of projects Björn Alpermann, Thomas Heberer, Sebastian Heilmann, and Gunter Schubert, as well as the other project staff, Anna Ahlers, Sandra Heep, Susanne Löhr, Elena Meyer-Clement, Baris Selcuk, Lea Shih, René Trappel, Eva Wieland, and Katja Yang. I also thank the members of the Association for Social Science Research on China (ASC). The annual meetings and conferences that brought together not only German-speaking China scholars but also renowned international scholars such as Ching Kwan Lee, Susan Whiting, Stig Thøgersen, Vivienne Shue, Anne-Marie Brady, Greg Mahoney, Frank Pieke, Patricia Thornton, Andrew Nathan, Carolyn Hsu, and Andrew Kipnis, to name only a few, were always an inspiration and an excellent platform to present our own research.

The research network 'Governance in China' also introduced me to Chinese colleagues from the China Center for Comparative Politics and Economics (CCCPE) in Beijing, who kindly opened their doors and hosted me as visiting scholar during my research stays. I thank the CCCPE colleagues for their heartening welcome and assistance. I also thank Greg Mahoney from East China Normal University for welcoming me whenever I passed through Shanghai. I further thank Thomas Johnson from the University of Sheffield for sharing his insights as an 'old hand' in the issue field of Chinese anti-incineration contention and for his great cooperation during our joint work on a topical special section in the *Journal of Contemporary China*. A further debt is owed to the editor of this series, Anna Lora-Wainwright, the series' editorial board, and an anonymous reviewer for their very helpful comments and suggestions on this manuscript, as well as to Amsterdam University Press editor Saskia Gieling for her patience.

My special thanks go to Chen Liwen, Mao Da, Feng Yongfeng, and the other staff members at Nature University, as well as the staff members of all the other environmental organizations, experts, and lawyers that populate this book. They were extremely helpful in making this research possible and spent hours and hours of their time explaining complex waste management and policy issues, establishing contacts, and letting me in on their work. I am full of admiration for their strenuous efforts to work towards a clean environment and a just society. I further thank the numerous unnamed community members in this study who stood up for their rights in spite of all hardships and let me take a glimpse into their lives and activities. I am also indebted to Heidrun Reimers, who shared her personal experiences in the struggle against a waste incineration plant in Ahrensburg, Germany, and provided me with community member interviews, protocols, and other documentary data that helped me better understand both the universality of grievances related to waste incineration projects and the specifics of Chinese anti-incineration contention. And I thank the dog that bit me during field research in Dagong village in 2013. The long-term side-effects from the following rabies shot in a small rural hospital outside Beijing knocked me out for a while, but also gave me time to ponder my priorities. Nevertheless, I can only advise other researchers to go to China fully vaccinated.

Last but not least, I thank my family for their unconditional support, my husband Georg, who set up a home base away from home in Beijing during my research stays, as well as our two sons Samuel and Joshua, who remind me of the importance of a clean environment and a just society every day.

1 An Emerging Network in China's Green Sphere

Towards an Environmental Movement?

Introduction

Environmental contention in China has undergone significant change in recent decades. Chinese environmental activism has long centred on the campaigns and activities of 'embedded' environmental actors (Ho 2007; Ho and Edmonds 2007) – environmental organizations with close ties to the Party-state that act on behalf of broader environmental and conservationist concerns – in collaboration with their journalist counterparts in a greening Chinese media sphere (Ho 2007; Ho and Edmonds 2007; Xie 2009; Yang and Calhoun 2007). In recent years, however, a plethora of new actors has entered China's environmental arena.

Victims of pollution and local communities facing the environmental drawbacks of China's rapid development are becoming increasingly outspoken in demanding their right to a clean and healthy living environment. They voice their grievances and concerns through a diverse claim-making repertoire ranging from legal actions, such as petitions and environmental litigation, to more disruptive activities like protests and sit-ins (Deng and Yang 2013; Herrold-Menzies 2010; Lora-Wainwright 2013b; Matsuzawa 2012; O'Brien and Deng 2015; Stern 2013; van Rooij 2010). Large-scale protests against hazardous construction projects such as paraxylene (PX) plants and waste facilities have become a frequent phenomenon that have spread beyond China's major cities to both smaller cities and rural areas (Ansfield 2013; Huang and Yip 2012; Johnson 2010, 2013b; Steinhardt and Wu 2015). With advances in Chinese environmental law and growing legal consciousness, both affected communities and environmental organizations are turning to (environmental) lawyers and legal associations such as the Beijing-based organization Center for Legal Assistance to Pollution Victims (CLAPV, 污染受害者法律帮助中心, *Wuran shouhaizhe falü bangzhu zhongxin*) for support (Stern 2011, 2013; Ying 2010). Moreover, against the backdrop of the knowledge-dependent and highly contested environmental risks that are typical of modern societies (Beck 1986; Yan 2012; Zhao and Ho 2005) and which generate a sense of risk and uncertainty that is amplified by the public perception of reliable information as unattainable, experts have also come to play an important role in Chinese environmental contention (Holdaway 2010; Lora-Wainwright 2013a, 2013b; van Rooij 2010). China's environmental arena is thus made up of an increasingly complex network of actors that go far beyond the environmental organizations and green journalists conventionally regarded as the core of Chinese environmental activism.

In Western societies, it is the local to national and (trans-)national linkages and networks between such different actors and social groups that make up and drive environmental movements (Hadden 2015; Keck and Sikkink 1998; Rootes 2004, 2013; Saunders 2013). Social movement scholars have shown how diffusion and learning processes between different contentious groups can contribute to the emergence of 'protest waves' and 'cycles of contention', and how disparate local struggles can 'scale up' to become regional, national, or transnational movements (McAdam 1995; McAdam, Tarrow, and Tilly 2001; Tarrow 1995; Traugott 1995b). Networks and alliances between different local communities, particularly those affected by similar issues, have often been the basis for both issue-specific movements, such as the anti-nuclear and anti-incineration movements, as well as for broader environmental and social justice movements in the United States and many European countries (McAdam and Boudet 2012; Rootes and Leonard 2013; Sherman 2011a; Walsh, Warland, and Smith 1997).

Grievance-driven local communities and supra-local environmentalists acting on behalf of broader environmental concerns have frequently joined together to undertake collaborative action despite the often-diverging goals and interests of these two groups (Keck and Sikkink 1998: McAdam and Boudet 2012; Rootes 2007; Walsh, Warland, and Smith 1997). Local groups have brought issues (back) onto the agenda of environmental movements and added weight to the claims and activities of environmental organizations. At the same time, environmentalists, experts, and lawyers have provided local communities with assistance that has played an important role in the emergence and development of local contention (Fischer 2000; McCormick 2009; Tesh 2000; Walsh, Warland, and Smith 1997). Despite the frequency of such phenomena, the dynamics of diffusion and the processes through which localized struggles scale up to higher levels remain poorly understood (Givan, Roberts, and Soule, 2010; McAdam and Boudet 2012: 134). This is particularly true in the study of restrictive political regimes – which are characterized by nondemocratic principles and practices such as restrictions on the freedom of expression, limited possibilities of participating in the political process, and weak rule of law – where the dynamics of contention tend to widely diverge from those observed in democratic contexts (Almeida 2003; Osa and Schock 2007; Soule 2004: 304).

The linkages and networks that connect different actors in the Chinese environmental sphere have received little academic attention, as the literature on environmental activism in China has long focused on the campaigns and activities of 'embedded' environmental organizations and journalists (Alpermann 2010b; Ho 2007; Ho and Edmonds 2007; Xie 2009; Yang and Calhoun 2007). In recent years, however, local environmental contention has also moved into the focus of academic research (Grano and Zhang 2016; Herrold-Menzies 2010; Jian and Chan 2016; Johnson 2010, 2013a, 2013b; Matsuzawa 2012; Steinhardt and Wu 2015; Tang forthcoming; van Rooij 2010; Wright 2018). In these studies, 'embedded' environmentalist activism and grievance-driven local contention have largely been investigated as two separate facets of China's green activism – partly because of the widespread assessment that Chinese environmental organizations tend to avoid becoming engaged at the local level or developing too close of ties with local contenders so as to avoid politicizing local campaigns and to ensure the survival of the their own organization (Ho 2007; Ho and Edmonds 2007; Spires 2011; Yang 2005). In this context, local communities have been described as largely 'isolated' from supra-local support (van Rooij 2010): for example, environmental organizations have reportedly shied away from getting involved in urban mass protests against hazardous construction projects despite being approached by local campaigners (Johnson 2010, 2013a; Matsuzawa 2012; Tang 2012; Zhao 2007).

Recent studies have pointed to the emergence of networking and its important role for Chinese social actors. For example, Wu (2013) and Peng and Wu (2018) have found that the development of country-spanning alliances of environmental and social organizations in China has greatly expanded their manoeuvring space and enhanced the survival of these organizations. In a similar vein, Sieckmann (2015) writes that the formation of a national network of Chinese environmental organizations focused on tackling climate change issues has significantly strengthened their national-level advocacy activities. However, these studies focus on inter-organizational connections and NGO-driven activism, leaving out grievance-driven social mobilization. The linkages between the 'two facets' of Chinese environmental activism remain understudied.

A notable exception is Mertha's (2008) study of contention against Chinese hydropower projects. He outlines the role of what he terms 'policy entrepreneurs' – primarily social organizations, media representatives, and disgruntled opponents both in- and outside of the government – who (at the time of his research) had entered the pluralizing policymaking process in China and played critical roles in the three cases presented in his study. Mertha argues that these new actors have found ways to impact the policy process by adopting strategies to manoeuvre within the framework of 'fragmented authoritarianism' – particularly by (re-)framing the issues at stake and providing 'neutral' information in a context where reliable information is limited and hence extremely valuable. Mertha's study does not, however, attempt to conceptualize the exact nature and role of the linkages between the different actors in his cases or to identify the broader network of actors working on the same issues.

The linkages between different affected communities have also not received much academic attention. Most studies of local environmental contention focus on individual cases without paying much attention to diffusion processes and linkages between different localities. In other words, local contention has largely been regarded as locally contained, parochial, and disconnected from others doing similar work (Cai 2010; Chen 2011; Hsing and Lee 2010; O'Brien and Li 2006). This can partially be explained by the fact that the pathways of diffusion – i.e., the channels along which information may travel between different localities and actors – have long been severely limited in China. It is only in some very recent studies that these issues have started to be addressed (Bondes and Johnson 2017; Steinhardt and Wu 2015; Sun, Huang, and Yip 2017; Zhu 2017).

During the government of Hu Jintao (2003 to 2013), the liberalizing media sphere (Mertha 2008; Shirk 2010), spread of the Internet and particularly social media, and diversifying range of activities engaged in by environmental organizations (Geall 2013) created new opportunities for communication across geographic spaces and between different actors in China's environmental scene. While political control has since tightened under Xi Jinping, who became President of the People's Republic of China in 2013, to understand the current developments in Chinese environmental activism it is necessary to investigate the linkages between environmental actors that emerged during the time of political opportunity under Hu Jintao, what they mean for the spread and development of local environmental contention, and the potential for scaling up local contention to higher levels or for the emergence of an issue-specific or broader environmental movement. The investigation of these questions also helps to shed light on the dynamics of diffusion and processes whereby local struggles can scale up to higher levels within a restrictive political setting.[1]

1 This touches upon several complex questions, including under what conditions communities oppose present or anticipated pollution and what factors lead to the success or failure of local campaigns. While these are interesting questions, they are not the main focus of this study and

Drawing on social movement theory, this book develops a comprehensive analytical framework for the systematic assessment of relationships between different kinds of environmental actors and the impact of these linkages on both local and higher-level environmental contention. I argue that *networked contention* among different types of environmental actors permits a diffusion of information and resources that plays a significant role in both the spread and development of local contention and the scaling up of local struggles to higher levels. The actors or network nodes that make up such networked contention encompass both the local communities that are directly affected by environmental grievances and the supra-local actors that act more out of environmentalist or social rights concerns. In this study, ties amongst affected communities are termed *horizontal linkages*, while relations between local communities and supra-local actors, i.e., between the 'two facets' of environmental contention, are termed *vertical linkages*.[2] Networked contention does not have to take the shape of a full-blown environmental movement. It can level out at a meso-level of contention that spans different sites and actor groups between fragmented activism and a full-grown movement. Particularly in the context of a restrictive political regime, such a loosely organized form of contention can hold significant advantages for contentious actors by strengthening supra-local policy advocacy and fostering local campaigns without drawing too much attention from state forces.

The empirical section of this book applies this analytical framework to the field of anti-incineration contention in China. One of the adverse side-effects of China's rapid economic development in recent decades has been a serious municipal waste problem brought on by the growing amounts of garbage generated by China's urban population. With China's major cities 'besieged by waste', during the last decade the Chinese government has proclaimed a national 'waste crisis' and promoted waste incineration as a space-efficient and environmentally friendly waste-treatment strategy. The government's push for incineration has not gone unimpeded, however. China's waste treatment policies and the growing number of

have been addressed more directly by other scholars for both China and other world regions. For China, see for instance Cai (2010); Deng and Yang (2013); Johnson (2013a); Li and O'Brien (2008); Lora-Wainwright (2017); and van Rooji (2010). For other world regions – particularly in the context of disputes over the location of polluting sites – see, among others, Boudet and Ortolano (2010); Hallman and Wandersman (1992); Kasperson (1988); Lober (1995); McAdam et al. (2010); McAdam and Boudet (2012); Sherman (2011a, 2011b); Walsh, Warland, and Smith (1997).

2 On the concept of 'networked contention' and an abbreviated version of this argument, see also Alpermann and Bondes (forthcoming).

incinerator projects mushrooming throughout the country have been met with fierce public resistance, similar to that seen in other countries and regions including the United States, many European countries, Japan, and Taiwan (Botetzagias and Karamichas 2009; Leonard, Fagan, and Doran 2009; McCauley 2009; Rootes 2009a, 2009b; Rootes and Leonard 2009; Shen and Yu 1997; Sherman 2011a; Walsh, Warland, and Smith 1997).

As in other countries, in China opposition against both the national waste policies and individual incinerator facilities has mounted from two sides and spurred a fierce public and media debate about incineration: first, a (trans-)national network of domestic and international experts, environmentalists, and (environmental) lawyers, which publicly criticize China's waste treatment strategy, related regulatory failures, and broader environmental problems reflected in the issue field – here described as the 'no burn' community; second, numerous local communities living near proposed or active incinerator sites, which have spoken out against and protested the use of these facilities. This contention has taken manifold forms from legal means like petitions and lawsuits to more disruptive means like sit-ins and large-scale street protests. It has produced a wave of local resistance against incinerator projects across the country, similar to the series of protests observed in opposition to other industrial and infrastructure projects such as PX plants.

Several large-scale protests against waste incinerators – like those in Beijing's (北京市, *Beijing shi*) Liulitun (六里屯, 2006/2007) and Asuwei (阿苏卫, 2009) neighbourhoods; Guangzhou's (广州市, *Guangzhou shi*) Panyu district (番禺区, *Panyu qu*, 2009); and Shanghai's (上海市, *Shanghai shi*) Songjiang district (松江区, *Songjiang qu*, 2012) – have attracted major public and academic attention and are frequently brought up as examples of Chinese environmental protests together with prominent large-scale street actions against other types of construction projects like those in the cities of Xiamen (厦门, 2003), Dalian (大连, 2011), Shifang (什邡, 2012), Qidong (启东, 2012), or Kunming (昆明, 2013). Along with these other mass mobilizations, anti-incinerator contention is often given as an example of what scholars have termed a newly emerging 'Chinese NIMBY (Not-In-My-Backyard) activism', similar to that seen in Western societies (e.g., Cui 2011; Huo 2013; Johnson 2010; Lang and Xu 2013; J. Liu 2013; Otsuka 2009; Tang 2013; Wasserstrom 2008; Xia 2014).[3]

3 'Not-In-My-Backyard' (NIMBY) activism is a term frequently used in Western scholarship and media debates to refer to local communities' resistance against construction projects in their neighbourhood. The term NIMBY (Chinese: 邻避, *linbi*, often used with the supplement

Since similar episodes of anti-incineration contention have emerged in many other world regions, comparisons with the Chinese context can provide an assessment of the specific characteristics of diffusion and scale shift within a restrictive political regime. While anti-incineration contention in China has received some academic attention in recent years (Huang and Yip 2012; Johnson 2010, 2013a, 2013b; Lang and Xu 2013; Steinhardt and Wu 2015; Wong 2016; K. Zhao 2011), these studies have largely focused on individual cases, mostly the homeowner campaigns in Beijing city and Guangzhou city's Panyu district, and not paid much attention to the role of linkages across different cases or between the different actors in the issue field. This reflects the broader literature on environmental contention in China.

In many ways, anti-incineration contention represents a most likely case for the emergence of linkages both between different sites of contention and amongst the different actors working on the issue in China. The close interrelation between waste incineration and broader waste policies permits the alignment of local grievances with broader claims and has attracted the attention of a large number of environmental organizations, experts, and other supra-local actors. Further, the presence of a vivid global anti-incineration movement and the close linkages between the Chinese and transnational 'no burn' communities have also promoted Chinese anti-incineration contention and the emergence of a national network that benefits from assisting local communities, because the local engagement of national actors expands their action range and strengthens their political claims. While the developments described in this study are likely more pronounced than in other issue fields, the findings from this study do point to a broader tendency of networking and cooperation amongst the different types of contentious environmental actors in China – at least between 2011 and 2013 – and demonstrate the dynamics and mechanisms of how diffusion processes and scale shift can occur in the context of a restrictive political setting.

Drawing on a total of eight months of fieldwork between 2011 and 2013 (September and October 2011; September to November 2012; and April to August 2013) and a wealth of material collected both during field research and via online sources, this book investigates – to varying depths – nine cases of local contention against incinerator projects in urban and rural China. To gain a comprehensive picture of the issue field and to select

运动, *yundong*, 'movement' or 'campaign') entered Chinese media and public debates after the large-scale opposition of urban residents to a paraxylene (PX) plant in Xiamen in 2007.

cases for in-depth case studies, I collected as many known cases of local anti-incinerator contention as possible via media analysis and during field research. By 2013, at least 39 cases of local contention against incinerator facilities had occurred, encompassing a variety of forms from legal actions to large-scale street protests. These cases were mainly clustered in the larger Beijing, Shanghai, and Guangzhou metropolitan areas. A first macro-level assessment of these struggles suggests that a 'protest wave' and diffusion processes were indeed at play. A systematic investigation of the linkages between different actors, their role in the local campaigns under study, and the potential for a scaling up of local contention has to be based on in-depth case studies, however. This book thus uses a comparative case study approach (George and Bennett 2005; Stake 2006; Yin 2002) and investigates nine cases – the most recent urban, peri-urban, and rural cases from each of the above cluster regions – drawing on ethnographic methods.

Based on these cases, the book finds that during the Hu Jintao era a complex network of ties from the local up to the (trans-)national level has emerged in China's waste realm, which has significant benefits for both local communities and the supra-local actors in the issue field. Such networked contention with the Chinese 'no burn' community at its core facilitates the spread and development of local environmental contention, on the one hand, and has fostered a national issue network dedicated to sustainable waste policies and to exposing broader regulatory failures, on the other.

Within this network, local communities affected by planned or operating facilities in their neighborhood are learning from their predecessors in other localities and significantly impact each other. The emerging linkages are not restricted to urban cases, but also connect rural and peri-urban contenders. The case studies show that particularly social media have played a crucial role as source of less strictly censored and critical information. Also members of the Chinese 'no burn' community have made deliberate efforts to link up affected communities. Despite such opportunities for personal contact, the relations between the local campaigners in this study remained largely restricted to nonrelational ties. In most cases, the local actors greatly benefitted from the information about or provided by other communities via the internet, mass media or brokers. Nonetheless, they refrained from establishing direct personal relations. This can partly be attributed to the restrictive political setting in which closer ties across different localities still pose a significant risk. Particularly rural and peri-urban communities also showed little interest in the grievances of other groups or broader waste and environmental issues. Since sustained or broader action would have to be based on a shift from a Not-In-My-Backyard (NIMBY) to

Not-In-Anybody's-Backyard (NIABY) attitude, this significantly limits the prospects of a broader movement based on horizontal alliances.

Despite these limitations of horizontal collaborations, the findings from this book show that vertical linkages between local campaigners and the Chinese 'no burn' community have left their imprint also at higher political levels. In contrast to the widespread assessment in the literature that Chinese environmentalists are shunning direct local engagement, the case studies demonstrate that the environmental organizations, experts and lawyers in this study have been increasingly active at the local level and fostered individual campaigns. At the same time, they have aggregated the disparate local grievances and transformed the mostly short-lived local struggles into more sustained policy advocacy – both for more sustainable waste policies and for exposing broader regulatory failures such as the lax local implementation of environmental laws and standards; weaknesses in China's environmental litigation system; or lacking public participation and transparency.

It is unlikely that these developments will consolidate into a full-blown issue-specific or environmental movement in the near future. Particularly under the current restrictive political climate, anti-incineration contention in China will likely remain at a meso-level stage of networked contention that permits its actors to stay small and loosely organized enough not to trigger a crackdown on the network and its members. However, within the Chinese political context, the existence of networked contention as a meso-level social phenomenon is a significant development in its own right.

Outline of the Book

The rest of this book is organized as follows: The next section introduces the analytical framework for assessing horizontal and vertical linkages and their role for local and higher-level contention. The chapter also outlines the methods and data used for this study. Chapter Two introduces China's incineration policies during the last decade and the two main social forces challenging these policies – China's national 'no burn' community, on the one hand; and the growing number of contentious local communities directly affected by incinerator projects, on the other. The chapter also discusses first macro-level patterns that point towards a 'protest wave' and diffusion processes at play.

Chapters Three to Five present three in-depth case studies of local contention against planned incineration facilities in the larger Beijing metropolitan area. Chapter Three takes a closer look at the urban case of a homeowner campaign against a planned incinerator in Beijing city's Asuwei area, which was temporarily halted due to public pressure. The case shows how urban activists from different localities have significantly influenced and learned from each other. Chapter Four investigates the rural case of a villager contention against a planned incinerator in Hebei Province's (河北省, *Hebei sheng*) Panguanying village (潘官营村, *Panguanying cun*), which was obstructed by the local community. The case demonstrates how different local and supra-local actor groups in China's environmental sphere can join forces to their mutual benefit. Chapter Five outlines the peri-urban case of a failed struggle against a planned – and by now completed – facility in Beijing's Dagong village (大工村, *Dagong cun*), showing how even the strongest linkages fail to yield effects if not rooted in sustained local contention.

Chapter Six compares the findings from the above case studies and from the broader case spectrum, outlining the network of contention that has emerged in the waste realm. The chapter also discusses the specifics of the issue field and how the study's findings can be transferred to other areas. It then moves beyond China and discusses what the findings can tell us about the dynamics of diffusion and the processes of upward scale shift in a restrictive political setting.

Networked Contention: Horizontal and Vertical Linkages and the Diffusion of Contention

The importance of networking is widely acknowledged by the literature on contentious politics. This is particularly true in the environmental arena, where the issues that motivate these disputes tend to be highly contested and require a great deal of information and expertise to trigger collective action. Networks of environmental actors have played an important role for both the spread and development of local contention and the emergence of regional or (trans-)national movements (Diani and Donati 1999; Hadden 2015; Keck and Sikkink 1998; McAdam et al. 2010; McAdam and Boudet 2012; Saunders 2007, 2013; Sherman 2011a; Rootes 2004, 2007; Rootes and Leonard 2013; Tarrow 2005; Walsh, Warland, and Smith 1997; Wu 2013; Peng and Wu 2018).

This book is interested in a systematic analysis of exactly how the various types of contentious actors in the environmental arena link up with each other, and how these linkages impact both local struggles and higher-level contention. I argue that environmental contention is fostered by a network of ties amongst different contentious actors. The actors or network nodes that make up such *networked contention* encompass grievance-driven local communities directly affected by environmental degradation and supra-local actors in the issue field that act out of environmentalist or social rights concerns – termed 'intermediaries' by van Rooij (2010). Linkages between different local communities are here termed *horizontal linkages*; relations between local communities and supra-local actors are referred to as *vertical linkages*.[4]

On a horizontal scale, networks and alliances amongst local communities, particularly those affected by similar issues, have been the basis for issue-specific or broader environmental movements in many world regions (McAdam and Boudet 2012; Rootes and Leonard 2013; Sherman 2011a; Walsh et al. 1997). On a vertical scale, grievance-driven local communities and environmentalists have frequently entered a 'symbiotic relationship' (Rootes 2007: 725) with each other and closed ranks to undertake collaborative action despite their often-diverging goals and priorities (Keck and Sikkink 1998: 121, 140; McAdam and Boudet 2012: 173; Michaud, Carlisle, and Smith 2008), which can render these relationships somewhat 'uneasy alliance[s]' (McAdam and Boudet 2012: 135). Still, environmental organizations, experts, lawyers, and other individuals have been credited with providing multifaceted assistance to contentious local communities, from the provision of information to more active local engagement, while affected communities have brought issues (back) onto the agenda of higher-level movements and added weight to the claims and activities of environmental organizations (Fischer 2000; Herrold-Menzies 2010; Keck and Sikkink 1998: 121-163; Lora-Wainwright et al. 2012; Matsuzawa 2012; McCormick 2009; Mertha 2008; Tesh 2000; van Rooij 2010; Walsh, Warland, and Smith 1997).

Much of the literature on environmental networks focuses on inter-organizational alliances and does not take into account the role of local communities (Hadden 2015; Saunders 2013; Sieckmann 2015; Tarrow 2005; Wu 2013; Peng and Wu 2018). Some theoretical conceptions, such as Keck and Sikkink's (1998) 'transnational advocacy networks', do include different kinds of actors, but they tend to focus primarily on policy impacts at higher political levels and do not tackle the role of such networks for local

4 Compare Alpermann and Bondes (forthcoming). The term 'networked contention' has also been used in the context of digitally enabled social activism, among others by Bennett and Segerberg (2012, 2014). A similar conceptualization of horizontal and vertical relations is used by Wu (2013), although her investigation is focused on inter-organizational connections rather than the linkages between different types of environmental actors. Her study uses the term 'vertical relations' in reference to hierarchically structured connections between a social organization and its peer organizations outside the province, while 'horizontal relations' is used to refer to more equally structured connections between a social organization and its peer organizations within the province.

contention.[5] Mertha's concept of 'policy entrepreneurs' captures several of the main functions played by supra-local actors in the environmental realm – (re)framing the issues at stake, providing valuable information, and initiating media and policy advocacy campaigns – but does not attempt to conceptualize the full breadth of functions played by such supra-local players, particularly for local contention, or to identify the broader network of actors in the issue field.

Social movement scholars have long been intrigued by how contention spreads from one site or social group to another. It is widely recognized that contentious struggles are rarely isolated instances that develop independently; instead, contentious actors in different localities and over time influence and learn from each other, especially when faced with similar grievances (Soule 2004: 295; McAdam and Boudet 2012: 132-169). In social movement theory, the geographic spread of local contention and the influence exerted by different social actors are captured in the theory of the 'diffusion of contention'. However, this branch of the social movement literature has failed to acknowledge some of the unique aspects of environmental contention, such as linkages between local communities and supra-local intermediaries and the central importance of information and expertise for often highly technical environmental disputes.[6]

Such linkages permit a diffusion of information and other resources that can play a significant role for both the spread and development of local contention and the scaling up of local struggles to higher policy arenas.[7] To do this, networked contention does not have to take the shape of a full-blown environmental movement, but can instead level out at a meso stage between

5 In a similar vein, while the notion of epistemic communities includes scientists, most scholars of epistemic communities explicitly exclude activist groups from their definition.

6 The literature on the diffusion of contention has largely centered on the spread of specific tactics or movement frames across social movements or geographic spaces. Scholars tend to focus on tracing out the communication channels along which these tactics and ideas have travelled and how they have been adapted or 'emulated' (Givan, Roberts, and Soule 2010; McAdam and Rucht 1993; Soule 2004; Tarrow 2010). This does not capture the specific role of the relationships between different types of actors as seen in the environmental sphere, in particular supra-local intermediaries, or the role of other types of information and resources that may be shared amongst contentious actors.

7 For social scientists, 'diffusion' very broadly means 'the flow of social practices among actors within some larger system', in which 'social practices' refers to anything from agricultural practices to organizational forms, specific policies, or forms of contention such as riots or sit-ins (Soule 2004: 295). Diffusion processes are acknowledged as a central force for the emergence and development of contention which can initiate 'protest waves' or 'cycles of contention' and a scaling up of local struggles to higher political arenas (McAdam 1995; McAdam and Rucht 1993; Soule 2004; Tarrow 1995, 2010).

fragmented activism and a full-grown movement. In a restrictive political context, such meso-level networked contention can have significant benefits for contentious actors by fostering both local campaigns and national-level policy advocacy without attracting too much attention.

The study of the diffusion of contention generally encompasses three aspects (Givan et al. 2010; Soule 2004): the channels of diffusion, i.e., the pathways via which the transmission takes place; the contents of diffusion, i.e., what is being diffused; and the effects of diffusion, either for the adopting social movement or group or as broader effects, including the scaling up of localized struggles to higher levels. Drawing on this approach, the remainder of this section outlines the nature of the horizontal and vertical linkages that make up networked contention, the content that can be transmitted or 'diffused' via these ties, and how exactly this diffusion impacts local and higher-level forms of contention.

The Nature of Horizontal and Vertical Linkages: Relational and Nonrelational Ties

The diffusion literature describes two types of channels through which information can be transmitted (Givan, Roberts, and Soule 2010; McAdam and Rucht 1993; Oliver and Myers 2003; Soule 2004):[8] relational and nonrelational channels. 'Relational channels' refers to the transmission of content based on node-to-node interpersonal ties or networks. Such processes of interpersonal communication can occur in multiple forms, from direct face-to-face interactions to more indirect communication across space via the telephone or online media like email or online networks. 'Nonrelational channels' refers to the transmission of content in the absence of relational ties between the transmitter and the adopter. Such nonrelational diffusion is largely based on broadcast forms of communication, including the mass media and the Internet.[9] In recent years the Internet has significantly widened the scope of information flow.

Of the two, direct relational ties are more likely to produce diffusion effects due to higher levels of trust in and identification with the transmitter

8 As a third pathway of diffusion, the literature outlines 'mediated diffusion' or brokerage, i.e., diffusion based on the engagement of brokers defined as third parties that deliberately 'connect people who would not have otherwise met' (Tarrow 2010: 209). In this study, 'mediated diffusion' is not conceptualized as a separate pathway of diffusion, but rather as an intermediary mechanism, as outlined below.

9 By functioning as broadcast media, online pathways of relational diffusion, such as online networks, blogs, or micro-blogs, may also enable nonrelational diffusion.

of the information, while nonrelational channels are more effective in breaching geographical and social distances (Givan, Roberts, and Soule 2010; McAdam and Rucht 1993; Oliver and Myers 2003). When aided by the new information and communication technologies, relational diffusion processes can also span large geographical distances, since communication no longer relies on the co-presence of transmitter and adopter in time and space (Earl and Kimport 2011; Myers 2010: 312). Nonetheless, direct face-to-face interactions are likely to remain more effective in building relations of trust and influence, particularly in the context of restrictive political regimes.

Since many of the pathways of diffusion identified above are significantly limited in authoritarian settings such as China, authors have highlighted the important role of informal networks that have to 'substitute both for organizations and the mass media' in the dissemination of uncensored information (Osa 2003: 78). Much attention has also been paid to the role of information and communication technologies, particularly social media, in the broadcast dissemination of information and relational diffusion processes across geographical space in restrictive political settings (Bennett and Segerberg 2012; Diamond 2010; Diamond and Plattner 2012; Earl and Kimport 2011; Segerberg and Bennett 2011). However, since the majority of research on the diffusion of contention has been conducted in the context of democratic regimes, the channels that drive diffusion processes in authoritarian settings remain understudied (Soule 2004: 304).

The Contents of Diffusion: Information Flows and More Manifest Resources

The horizontal and vertical ties described above permit the diffusion of information and other resources. Such transmissions play a particularly pivotal role for environmental contention because of its heavy reliance on information, knowledge, and expertise. The literature suggests that the contents that can be disseminated via networked contention include information about issue interpretations, (movement) frames, and strategies employed by other movements or social actors, which may then be 'emulated' by the adopting actors (Givan, Roberts, and Soule 2010; McAdam 1995; McAdam and Rucht 1993; Soule 2004; Tarrow 1995, 2010).[10] In the case of environmental contention, this includes (but is not limited to) issue

10 With regards to the contents of diffusion, movement scholars agree that the actual 'thing' that is transmitted is information (Oliver and Myers 2003). The literature on the diffusion of contention mainly focuses on the transmission of specific tactics and movement frames.

interpretations, frames,[11] and environmental concepts employed by regional or (trans-)national actors or environmental movements that can be adapted to the local context (Burningham and O'Brien 1994; Heiman 1990; Johnson 2013b).

As with all forms of contention, grievances have to first be perceived by the affected communities or individuals before they can be transformed into collective claims based on a sense of injustice and a clear attribution of blame (Felstiner, Abel, and Sarat 1980; Futrell 2003; McAdam and Boudet 2012; Sherman 2011a; van Rooij 2010).[12] The risks around which environmental contention tends to evolve are generally deeply contested, highly technical in nature, and characterized by an inherent difficulty in demonstrating causal links with their mostly long-term effects (Beck 1986). Thus, 'the issues that ultimately motivate these [environmental] disputes are initially far more ambiguous and uncertain than is true for rights movements', where the issues are generally rather clear-cut (McAdam and Boudet 2012: 97; cp. also Walsh, Warland, and Smith 1997: 45).

For these reasons, the interpretive processes by which affected communities and individuals try to make sense of their environmental situation are heavily dependent on information (Brown 2007; Epstein 2009; Gould 1993; Kasperson et al. 1988; Keck and Sikkink 1998: 137-138; McCormick 2009; Tesh 2000). While the affected tend to actively search for reliable information once an initial sense of threat prevails – often becoming lay experts on the risks and hazards they face (Brown 2007; Fischer 2000; Lora-Wainwright 2013b; McCormick 2009) – they frequently face high levels of uncertainty and a lack of reliable, or contradictory, information, which Futrell (2003: 378) terms 'information haze'. This makes it difficult to define clear empirical grounds on which claims can be staked. Especially in the initial stage of ambiguity, a clear public perception of risk is dependent on trusted information and interpretive frames, or 'cognitive cues', that provide a cognitive justification for resistance (Futrell 2003). Walsh, Warland, and Smith (1997: xvi), for instance, attribute a major role to the growing accessibility of critical

11 Snow and Benford (1992, 137) define frames as "interpretative schemata that simplif[y] and condens[e] [...] the 'world out there' by selectively punctuating and encoding objects, situations, events, experiences, and sequences of actions," thus allowing individuals "to locate, perceive, identify and label" events within their life space or the world at large. As a strategic device on the part of contentious actors, movement or collective action frames have the same interpretative function as frames, but "in ways that are intended to mobilize potential adherents and constituents, to garner bystander support and to demobilize antagonists" (Benford and Snow 2000, 614).

12 This awareness process is known as 'cognitive liberation' in the contention literature (McAdam 1999; McAdam, Tarrow, and Tilly 2001).

technical information about incineration in the emergence of opposition to US waste incinerators in the mid-1980s.

Social actors engaging in environmental contention also rely heavily on technical and legal expertise. Scholars of environmental contention both in China and beyond have described the pivotal role of environmental organizations, environmentalists, experts, and lawyers in providing local communities with the knowledge and expertise that is needed to take action, including evidence of environmental pollution, health impacts, or legal failures (Futrell 2003; McAdam and Boudet 2012; Mertha 2008; Tesh 2000; Walsh, Warland, and Smith 1997).

Diffusion processes can play a particularly crucial role for the dissemination of trusted information, evidence, and data in the context of authoritarian settings where uncertainty tends to peak against a backdrop of restricted access to information paired with broader crises of trust and credibility (Lora-Wainwright 2013a; Mertha 2008). Providing this trusted information is one of the core functions of the 'policy entrepreneurs' described by Mertha (2008). In a political context where information is suspiciously guarded, the leaking and sharing of information becomes a powerful political tool which is strategically employed by the officials, journalists, organizations, and experts discussed in Mertha's study. Moreover, as demonstrated in the literature on contentious politics in China, in the context of restricted political opportunity structures contentious actors also have to rely on knowledge about the legal and political system and administrative structures. This includes knowledge of their rights, the boundaries of the permissible, effective tactics and strategies, loopholes provided within the power structures, and suitable targets and responsible institutions (Cai 2008, 2010; Chen 2011; Hsing and Lee 2010; O'Brien and Li 2006; Stern 2011). In such a context, relevant action-related information also includes news about contentious action in other localities and the authorities' reaction to these struggles, especially information that (specific forms of) contention might be tolerated by the regime. Such information flows can alter the adopting actors' perceptions of what is politically feasible and within the boundaries of the permissible, thus encouraging or discouraging (specific modes of) action (Lohmann 1994; Myers 2010: 311; Osa 2003; Osa and Schock 2007).[13]

Beyond the mere provision of information, intermediaries also support local communities with more manifest resources that can be transmitted via

13 This aspect prominently features in the 'information cascades' described by Lohmann (1994) in the context of the Monday demonstrations in East Germany's Leipzig in 1989-1991, and has garnered new attention in the events of the Arab Spring.

relational linkages. This includes financial or material resources (McCormick 2009: 63-69), as well as active engagement related to the mobilization and organization of action, such as encouraging local communities to initiate contention and providing assistance in and training on mobilizing opposition, sometimes in the form of comprehensive 'organizing manuals' (Walsh, Warland, and Smith 1997: 15-16), or by taking on leadership functions from the outside (McAdam and Boudet 2012: 120). Intermediaries have also been reported to play a pivotal role in facilitating local access to justice by providing or conveying legal assistance or representation and helping obtain the necessary evidence. In a similar vein, intermediaries can also facilitate or provide access to the media, experts, or other intermediaries (Herrold-Menzies 2010; Matsuzawa 2012; Mertha 2008; van Rooij 2010; Xie 2011; G. Yang 2010). Other studies have found intermediaries to be functioning as 'mouthpieces' or representatives of the affected populations, disseminating and advocating their claims and issue frames vis-à-vis the responsible party-state institutions, targeted companies, or the public (Matsuzawa 2012; McAdam and Boudet 2012; McCormick 2009). Framing issues at higher political levels is another of the central strategies of the 'policy entrepreneurs' described by Mertha (2008), and is used to impact the policymaking process while staying within the boundaries of the politically permissible. In doing so, intermediaries influence the interpretation of an issue both by the public – to mobilize widespread social support – and by affected communities – to promote local contention. While providing other forms of support is usually seen as the role of supra-local actors, it can also be undertaken by local communities on a sharing basis, as this study shows.

To sum up, the content transmitted through the horizontal and vertical linkages that make up networked contention may include information and 'cognitive cues' about: prevalent or future risks; the specific issues; the expertise needed to take action; the legal and political system; and contentious action elsewhere and responses by the state. Beyond information, more manifest resources and active engagement can also be disseminated and often significantly impact the course of environmental struggles.

Impacts on Local Environmental Contention

The information and resources transmitted via the linkages behind networked contention can impact local environmental struggles through five core mechanisms, which alter the local dynamics of contention and either lead to the emergence of new contentious action or impact the course of ongoing action.[14] While derived from the literatures on the diffusion of contention and environmental contention, these five mechanisms have not yet been systematically drawn together into an analytical framework that can explain how the linkages between different social actors foster local struggles.[15]

First, external information and 'cognitive cues' about prevalent or future risks may *initiate or foster the awareness process* by facilitating the emergence of grievances and the attribution of blame, and by providing the cognitive justification for resistance, thus enabling the formation of the 'contentious consciousness' that is a prerequisite for contentious action. Resource-poor communities tend to be particularly dependent on external cognitive cues for the formation of a collective interpretation of the situation, a clear attribution of blame, and a feeling of injustice and entitlement (Felstiner, Abel, and Sarat 1980; Futrell 2003; van Rooij 2010).

Second, the literature on environmental contention has found that intermediaries, particularly experts and scientists, have played a central role in many environmental and health-related struggles by not only providing and disseminating, but also *certifying* information (Brown 2007; Diani 2003b; Fischer 2000; McCormick 2009; Sherman 2011a; Tesh 2000; Walsh, Warland, and Smith 1997). Particularly in a context of uncertainty and distrust, information needs to be validated in order to attain credibility for not only the affected communities, but also the broader public. Environmental and health-related contention, which often centres on suddenly imposed or unfamiliar threats, is thus dependent on 'neutral experts' or influential public figures who interpret the issues at stake and achieve credibility (Broadbent 2003: 225; Epstein 2009;

14 Mechanisms are a 'delimited class of events that alter relations among specified sets of elements in identical or closely similar ways over a variety of situations', and which 'recur in different combinations with different aggregate consequences in varying historical settings' (McAdam, Tarrow, and Tilly 2001: 24). Located at a level of analysis 'in-between pure description and story-telling, on the one hand, and universal social laws, on the other' (Hedström and Swedberg 1996: 281), the focus on mechanisms affords 'mid-range generalizations about regularized patterns of interaction that allow for contingency and contextual specificity at both local and larger-scale levels' (Mische 2003: 265-266).

15 In the context of the broader discussion about the study of mechanisms in the social sciences (see Hedström and Swedberg 1998; Hedström and Ylikoski 2010; Mayntz 2004; Stinchcombe 2005: 149-182) and in the research on contentious politics more specifically (McAdam, Tarrow, and Tilly 2001, 2008; McAdam and Tarrow 2011), it has been widely recognized that it is necessary to identify and trace the dynamics and mechanisms that make up the process of the diffusion of contention (Givan, Roberts, Soule 2010: 6; Oliver and Myers 2003: 175; Tarrow 2010: 218). Due to the rather narrow conceptualization of the diffusion of contention in the literature, however, the study of the effects of diffusion on local contentious struggles has been largely limited to what Tarrow (2010: 204) terms the 'fact of diffusion', i.e. the observation and tracing of the geographical or social spread of the tactics or frames under investigation.

Futrell 2003; McAdam and Boudet 2012: 178; Mertha 2008). Certification has two aspects: the *'internal' certification* of information towards other members of the local community, which is closely linked with the awareness process; and the *'external' certification* of information, which is directed at party-state organs, targeted companies, and the broader public, and can therefore lend credibility to the claims and actions of the contentious community. While the literature largely describes the process of certification as a responsibility of intermediaries, this can also happen through the sharing of information provided by or about other affected communities.

Third, as outlined in the literature on the process of diffusion in authoritarian contexts, information about contentious action in other localities and the authorities' reaction may *change the adopting actors' perception of threat and opportunity*, as was the case during the Monday demonstrations in 1989 in Leipzig in Eastern Germany (Lohmann 1994; Mueller 1999) or the uprisings in the Arab region, which started in 2010 (Allagui and Kuebler 2011; Lynch 2011; Michael-Matsas 2011). 'Perceptions of threat' here refers not to the risks posed by environmental issues or construction projects, but to the risks of conducting contentious action. Scholars of contentious politics have identified the perception of opportunities or threats as a crucial activating mechanism for the mobilization of 'previously inert populations' (McAdam, Tarrow, and Tilly 2001: 43-47). This includes the contentious actors' belief in the likelihood of success, which is pivotal for the decision to take up action (Cai 2005).

Fourth, diffusion scholars have pointed to the role of information flows for *identity formation*, i.e., the formation of a shared identity that motivates collective action. Information flows may contribute to the construction or 'borrowing' of a collective identity either within the local context or as part of a broader contentious struggle or movement (McAdam and Rucht 1993: 64; Osa 2003: 78-79; Soule 2004: 296; Tarrow 1995: 13). McAdam, for example, refers to the spread of 'identity frames' that 'offer the group an altered [...] collective vision of itself' (1995: 228). Similar to the formation of a 'contentious consciousness', the construction of such a collective identity is regarded as a crucial mechanism for the mobilization of contention (McAdam, Tarrow, and Tilly 2001: 132-159). In this context, intermediaries can play an important role as diffusion nodes or 'translators' between regional, national, and transnational actors and movements, on the one hand, and local communities, on the other, by providing broader identity frames that can be adopted to the local context or by permitting local activists to feel as part of a larger movement (Burningham and O'Brien 1994; Heiman 1990; Johnson 2013b; G. Yang 2010).

Fifth, the content transmitted via horizontal and vertical linkages may have a significant *impact on the resource structure* of the adopting community, thus enabling or facilitating the mobilization and organization of local contention (cp. McAdam et al. 2010). This includes the 'emulation', i.e., adaptation, of information about specific (effective) tactics, strategies, and organizational forms that have been applied in other sites of contention, which is the core focus of the diffusion of contention literature. However, it also encompasses information about the structure of and opportunities provided within the legal and political system, as well as issue-specific information and expertise that is critical to often highly technical environmental disputes. Beyond information, more manifest resources provided by and active engagement of supra-local actors or the members of other contentious communities can also significantly change the resource structure of the adopting community. Networked contention can thus help to fill resource gaps, particularly for resource-poor communities.

Brokerage as an Intermediary Mechanism

Intermediaries have also been credited with an important role in establishing ties between otherwise unconnected communities, thus fostering horizontal linkages. This linking of contentious groups across different localities, termed '*brokerage*', is regarded by the literature on contentious politics as an important mechanism that drives diffusion processes. It has also been found to significantly impact the onset and development of local environmental contention (Diani 2003a; McAdam and Boudet 2012; McAdam and Rucht 1993; McCormick 2009; Sherman 2011a; Walsh et al. 1997). As defined in the literature, 'brokers' are nodes that facilitate communication and mutual recognition between otherwise largely isolated groups (Diani 2003a; McAdam 2003: 294; McAdam, Tarrow, and Tilly 2001; Vasi 2011). Apart from fostering horizontal communication between different local communities, brokers can also function as nodes for transnational diffusion processes, i.e., the flow of information and other resources across national boundaries (Della Porta and Tarrow 2005; Givan, Roberts, and Soule 2010: 13; McAdam and Rucht 1993). While brokerage is generally attributed to intermediaries, it can also be provided by members of local communities that link two or more otherwise unconnected sites or social groups, including by introducing other contentious communities to intermediaries. In this study, brokerage is regarded as an intermediary mechanism that may facilitate the formation of both horizontal and vertical linkages and impact local environmental contention via the five mechanisms outlined above.

Beyond Localized Struggles: Upward Scale Shift and Impacts on Higher-Level Contention

The literature on contentious politics has shown great interest in the processes that enable the 'scaling up' of local contention to broader regional or (trans-)national movements (Givan, Roberts, and Soule 2010: 13-14; McAdam, Tarrow, and Tilly 2001: 331-340; McAdam and Boudet 2012; Tarrow 2010). Linkages and diffusion processes between different social actors are regarded as an integral part of such *'upward' scale shifts*.[16]

On a horizontal scale, linkages between local communities may spawn new actors or sites of contention and thus create incentives for those actors to 'begin to coordinate with one another, or create new representative or coordinating bodies to articulate their claims on larger political arenas' (Givan, Roberts, and Soule 2010: 3). Local environmental contention has frequently birthed issue-specific organizations, networks, and alliances that are active beyond local settings, link individual struggles, and tackle higher-level policy issues, thus widening the scope of the contention beyond local claims and grievances (McAdam et al. 2010; Rootes and Leonard 2013; Sherman 2011a; Walsh, Warland, and Smith 1997). Accordingly, the signs of scale shift through horizontal linkages include: collaborative action across different sites; the formation of claims and issue interpretations that reach beyond localized grievances' the emergence of broader collective identities; and the establishment of new organizational forms and alliances that are active beyond individual local settings.

On a vertical scale, the cooperation of local communities and intermediaries can foster higher-level contention by strengthening the political efforts of supra-local contentious actors who aggregate localized grievances into broader public demands and long-term advocacy campaigns. This can not only lift disparate local claims to the national level, but can also transform the mostly short-lived local struggles into more sustained policy advocacy. In the literature on contentious politics, local contenders have been reported to have closed ranks with supra-local environmentalists by deliberately broadening the scope of their claims and aligning their interpretations of the issues with regional or (trans-)national environmental struggles (Burningham and O'Brien 1994; Heiman 1990; Johnson 2013b).

16 Scale shift may also move in a downward direction, i.e., from (trans-)national to regional and local levels (McAdam, Tarrow, and Tilly 2001; Tarrow 2010). This includes, for instance, the local adaptation of frames, claims, and strategies employed by (trans-)national actors and movements. Since these aspects are captured by the above mechanisms, the rather broad concept of 'downward scale shift' is not analytically employed in this study.

Local campaigns have also been credited with bringing issues (back) to the agenda of institutionalized environmental organizations (McCormick 2009; Rootes 2007) and facilitating higher-level policy changes by increasing the pressure on policymakers (Johnson 2010, 2013a). Local environmental struggles have thus spurred or fed into broader environmental movements, such as the environmental justice and health movements in the United States and movements against specific types of infrastructure such as nuclear plants, dams, and waste facilities in different world regions (McAdam and Boudet 2012; Sherman 2011a; Walsh, Warland, and Smith 1997). The linkages among the different contentious actors in the environmental arena can also contribute to an upward scale-shift or foster higher-level contention without culminating in a full-blown movement.

Factors Mediating the Emergence of Networked Contention

There are several factors at the national, regional, and local level that mediate whether and what kinds of linkages will emerge amongst contentious actors. Such contextual factors also impact the role that linkages, once established, play in local or higher-level contention. In restrictive political settings, for example, demonstrated risks linked to the adaptation of certain strategies or claims may inhibit their use by other adopting actors (McAdam 2003: 295; Oliver and Myers 2003: 175; Tarrow 2010: 215), while resources such as information and expertise may play a particularly vital role as access to such resources is limited. The specific impact of contextual factors on diffusion effects depends strongly on the individual case, and is here kept open for analysis. This section focuses instead on the factors that impact whether and what kinds of linkages emerge between contentious actors.

At a national level, it is the political and institutional context and political opportunity structures that determine the dynamics of not only contention, but also diffusion. Social movement scholars point out that contentious action always mirrors the institutions it challenges, because it is shaped by the existing opportunity structures (Almeida 2003; Inclán 2009; Osa 2003; Tarrow 2010; Tilly 1986; Tilly and Wood 2003; Traugott 1995a). In a similar vein, the ties between social actors are also shaped by their institutional and political context and the constraints it imposes, since this determines what kinds of channels are available and what content may be transmitted (Oliver and Myers 2003; Osa 2003; Osa and Schock 2007).

Restrictive political settings can severely limit the formation of linkages. Since political elites tend to be aware of the power of networks and communication among contentious actors, linkages between them are often actively impeded and actor networks are faced with severe constraints or repression. In such a context, relational linkages in particular can be associated with significant personal risks and thus less prone to develop. Restrictive political contexts are also often characterized by a closed media environment, which severely curtails the available information channels, hampering both nonrelational information flows and the relational ties that might potentially be established based on media-transmitted information. In such contexts, scholars have pointed to the central role of informal networks and the Internet to 'substitute both for organizations and the mass media' in the dissemination of uncensored information (Bennett and Segerberg 2012; Diamond 2010; Diamond and Plattner 2012; Earl and Kimport 2011; Osa 2003: 78; Segerberg and Bennett 2011).

It is important to note, however, that neither political and institutional contexts nor political opportunity structures are static entities, but are instead constantly in flux (McAdam, Tarrow, and Tilly 2001; Steinhardt and Wu 2015) – as demonstrated by the political closure experienced under the government of Xi Jinping. Apart from changes over time that are unrelated to contention, prior cases of contention may also have an important impact through learning on the part of both contentious and state actors, such as the strategic siting of contested facilities (McAdam and Boudet 2012: 56). Moreover, growing repression or coercion of spreading contention on the part of the government may lead to more contentious action, stifle contention, or impact the mode of action that is employed (Almeida 2003; Inclán 2009; McAdam 2003: 295; Oliver and Myers 2003: 175; Osa 2003; Tilly and Wood 2003). This also impacts the kinds of linkages that will develop.

Regional and local factors within the political regime also impact what kinds of linkages will emerge between contentious actors. At a regional level, these are primarily the political and media environment and the presence of intermediaries in the region. Local communities in regions with a relatively more closed political and media environment are expected to develop fewer linkages with other affected communities and intermediaries, and more nonrelational than relational ties. Where relational linkages develop, they are expected to be based on Internet communication rather than face-to-face personal ties. The presence of intermediaries in a region is also expected to increase the likelihood of linkages. Intermediaries are more likely to become involved in cases within their own neighbourhood, particularly where their engagement is based on their own initiative. The presence of intermediaries in an area also facilitates local communities' knowledge about and access to those intermediaries. This significantly increases the likelihood of the development of (particularly relational) vertical ties. Since they can function as brokers between different local communities, a high density of intermediaries in a region may also increase the likelihood of horizontal linkages forming there.

At a local level, the resource structure of the affected community and the proximity to other cases can be regarded as central factors that impact the emergence and nature of linkages.[17] In reference to local communities' resource structure, Givan, Roberts, and Soule note that 'some actors are receptive to outside influences, whereas others are more closed or resistant' (2010: 3). This willingness to accept outside help is influenced by several factors: the affected communities' socio-economic structure, including their educational background and level of income; their organizational capacity and prior experiences with contentious action, including the population density in the region as potential basis for (or audience of) contentious action; the density of network ties outside the community; and their access to channels for the transmission of information, such as access to the Internet and different mass media (Givan, Roberts, and Soule 2010: 3; McAdam and Boudet 2012: 56). These factors impact the likeliness of not only contentious action, but also the establishment of linkages with other actors. Communities with limited access to transmission channels, particularly the Internet as a central tool for the collection and dissemination of (uncensored) information, are expected to be less likely to develop linkages with intermediaries or other communities. The same holds for communities that have limited network ties to the outside world. At the same time, resource-poor actor groups, particularly those with limited educational backgrounds, are more likely to depend on external information and resources in both the initial phase of ambiguity and the mobilization of collective action.

In addition, social movement scholars have outlined the importance of geographical proximity to other contentious struggles, particularly those covering similar issues. Several studies have demonstrated that prior or simultaneous contentious action undertaken nearby significantly increases the likelihood of contention in a given locality (Biggs and Andrews 2010: 188; Givan, Roberts, and Soule 2010: 11; Soule 2004: 296-297). This can partly be attributed to the increased likelihood that relational ties will emerge across short geographical distances. Local communities located in close proximity to other cases of contention are thus more likely to develop (particularly

17 Various studies have further outlined the role of a local economic dependence on polluters or the (expected) benefits from an industry or proposed facility, which reduces the likelihood of local contention (Deng and Yang 2013; McAdam and Boudet 2012). This also impacts whether local communities seek contact with or assistance from other actors. However, since local economic dependence is rarely the case with MSWIs (and did not occur in any of the cases investigated for this study), this factor is not included in my discussion.

direct relational) linkages or cooperative action with other contentious communities. Vertical ties with intermediaries can also be regarded as more likely in these cases, since intermediaries who are active in one locality are more prone to pay attention to nearby cases and affected communities can learn about intermediaries via nearby communities. Further, nonrelational information flows are more likely to breach short distances, such as via the local media. Overall, the geographical proximity to other (similar) cases is expected to increase the likelihood of linkages with other actors, particularly with those based or active in the same region.

Methods and Data

Since the nature of the linkages between different social actors, their exact roles in local contention, and the potential for 'scaling up' to higher levels can only be captured through detailed case analyses, this study takes a comparative case study approach (George and Bennett 2005; Stake 2006; Yin 2002). The case selection process was completed in two steps. In the first step, as many known cases of local anti-incinerator contention as possible were collected via media analysis, which was later complemented with information collected during fieldwork. In the second step, a theoretically-based sample of nine cases as outlined later in this section was selected for which a comprehensive data collection was conducted. Of these nine cases, three were chosen for the comprehensive case studies presented in Chapters Three through Five. The results of a more focused analysis that includes the full spectrum of cases are presented in Chapter Six.

Comparative Case Study Approach and Case Sampling

Following the logic of theory-building comparative case study analyses, this study aims to capture as wide of a variety of the phenomenon under study as possible to maximize analytical variance while permitting a structured analysis of the role of the factors underlying that variance (Baxter and Jack 2008; George and Bennett 2005; Hurst 2010; King, Keohane, and Verba 1994; Seawright and Gerring 2008; Stake 2006). The case sampling was based on the factors that were expected to impact the emergence and role of linkages, as discussed above. The case selection matrix is therefore organized along two lines: cross-regional variance with regards to the factors of the openness of the political and media environment and presence of intermediaries; and intra-regional variance at the local level with regards to the factors of community

Figure 1.1 Case sampling matrix

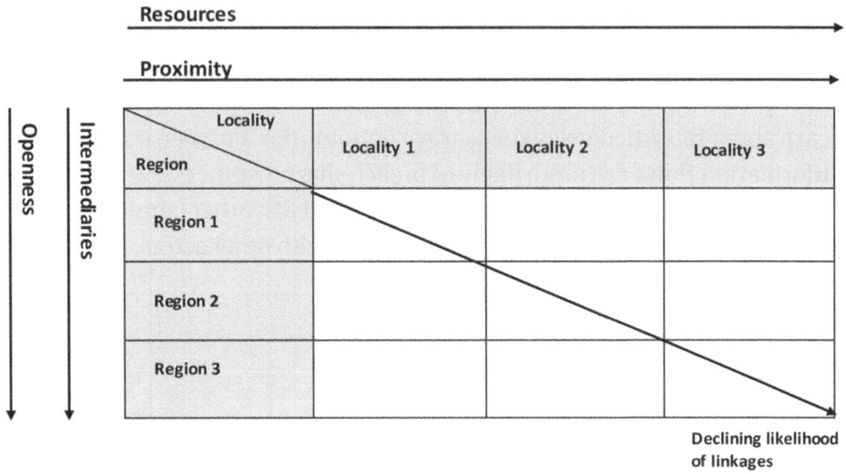

resources and proximity to other cases. As described above, the likelihood of linkages is expected to decline with decreasing openness of the political and media environment, lower presence of intermediaries in the region, increasingly resource-poor communities, and declining proximity to other cases.

Case Selection Process

For the case selection, all available information about local activism against waste incinerators in China was collected via a systematic analysis of major Chinese and international news media, using the material provided in their online archives from between April 2003 and April 2013.[18] This includes contention against waste incinerators in all stages (planned, under construction, or in operation) and in manifold forms, from legal means like petitions and lawsuits to more disruptive means like sit-ins and large-scale protests.

18 Due to the sensitivity of many cases and the very limited Chinese-language media reporting about contention in China, English-language news reports in China's major news media, which often feature more sensitive topics than the Chinese-language press, were regarded as the most useful source of systematic information. Chinese media included in the analysis are the party-led English-language newspapers *Global Times* (环球时报, *Hanqiu shibao*) and *China Daily* (中国日报, *Zhongguo ribao*), as well as the English editions of the Party-affiliated newspaper *People's Daily* (人民日报, *Renmin ribao*) and the main state news agency *Xinhua News Agency* (新华通讯社, *Xinhua tongxunshe*). As more liberal news outlets, the English-language edition of the Hong Kong-based independent newspaper *South China Morning Post* (南华早报, *Nanhua zaobao*), two major international newspapers, *The New York Times* and *The Guardian*, and *Radio Free Asia* were included. Beyond the traditional news media, archive searches were also conducted for three online media: *China Dialogue*, *China Digital Times*, and *China Media Project.*

The media analysis produced a total of 27 cases. To avoid biasing the case selection towards the engagement of intermediaries or linkages between local activists (when choosing cases that were brought to my attention through intermediaries or other community members), the selection was based solely on this media analysis. During my field research, the media analysis was complemented with information from different sources including interviews, field visits, documents, Chinese media articles, and information found online. This raised the total number of cases that had occurred by the summer of 2013 to 39. Basic information about all of the collected cases – including location, onset, important dates, engaged actors, major claims and strategies, and outcomes – was collected in an internal database (see Appendices III and IV). Individual cases were delimited as contentious action against one incinerator facility, which may include action at different points in time or taken by different local actors. Incinerators that have been successfully obstructed and moved to a different site are treated as a different case regardless of whether the new site is close to the original one.

When plotting all of the collected cases, three main cluster regions along the East coast emerged, with most cases located around the larger Beijing metropolitan area, including Hebei Province (7); the Shanghai metropolitan area, including the Jiangsu and Zhejiang Provinces (江苏省, *Jiangsu sheng;* 浙江省, *Zhejiang sheng*) (12); and the Guangzhou metropolitan area, including Guangdong Province and Hong Kong (14) (see Figures 1.2 to 1.5).[19] This geographical distribution is not surprising, since the cluster regions are major urban agglomerations and not only have a large number of urban residents with good access to resources and information, but were also the first regions to experience a high density of municipal solid waste incinerator (MSWI) projects. A total of 6 outlier cases were located outside these cluster regions.

According to the sampling logic presented above, these three regions were analysed in terms of the openness of the political and media environment and the presence of intermediaries. Since it is hard to substantiate the political restrictiveness of the regions, my calculation of the openness of the political and media environment was primarily based on the regional media situation. Overall, the Guangzhou cluster region seems most conducive to the formation of linkages due to its close proximity to Hong Kong, which has more liberal news reporting, and the relatively liberal media environment in

19 The Beijing cluster region covers an area of about 200,000 square kilometres; the Shanghai cluster region covers about 210,000 square kilometres; and the Guangzhou cluster region about 180,000 square kilometres.

Figure 1.2 Cases of local anti-incinerator contention

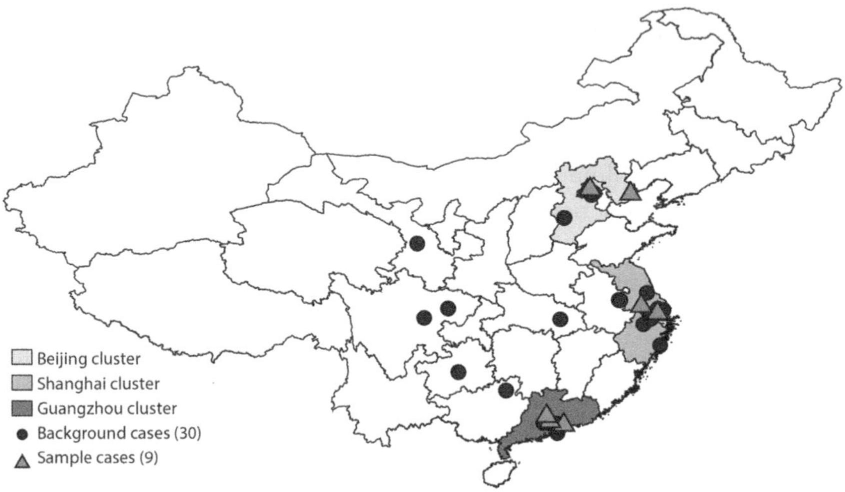

Source: Own compilation using the Stata module spmap by Maurizio Pisati. The China administrative maps were downloaded from: https://data.humdata.org/dataset/china-administrative-boundaries (27 April 2019). I thank Georg Strüver for help with generating these maps.

Guangdong Province itself. Guangdong Province is also the base of numerous social and environmental organizations, suggesting a high presence of intermediaries who may choose to become involved in local cases or function as brokers (cp. Wu 2013). Moreover, while hard to substantiate and hence only a secondary factor, several of my interview partners regarded the political framework in Guangzhou as relatively permissive of social mobilization compared to other parts of China. The Beijing cluster region is also regarded as quite conducive to the formation of linkages. While it lacks proximity to Hong Kong, Beijing is the headquarters of many news media institutions that regularly cover environmental issues and of several independent media houses. Beijing is also home to numerous social and environmental organizations and to many of the main intermediaries involved in resistance to waste incinerators (including environmental organizations, experts, and lawyers). The Shanghai cluster region is regarded as more adverse to the formation of linkages in comparison to the other two regions, because the media environment is more closed and there are only a limited number of social and environmental organizations. While again hard to substantiate and not regarded as a main factor, several of my interview partners also portrayed Shanghai's local political framework as relatively restrictive.

Since most of the collected cases were located in the three cluster regions – and since most MSWIs are located in these areas – this research is focused on cases located within these regions, following a 'typical case study' approach

Figure 1.3 Beijing cluster

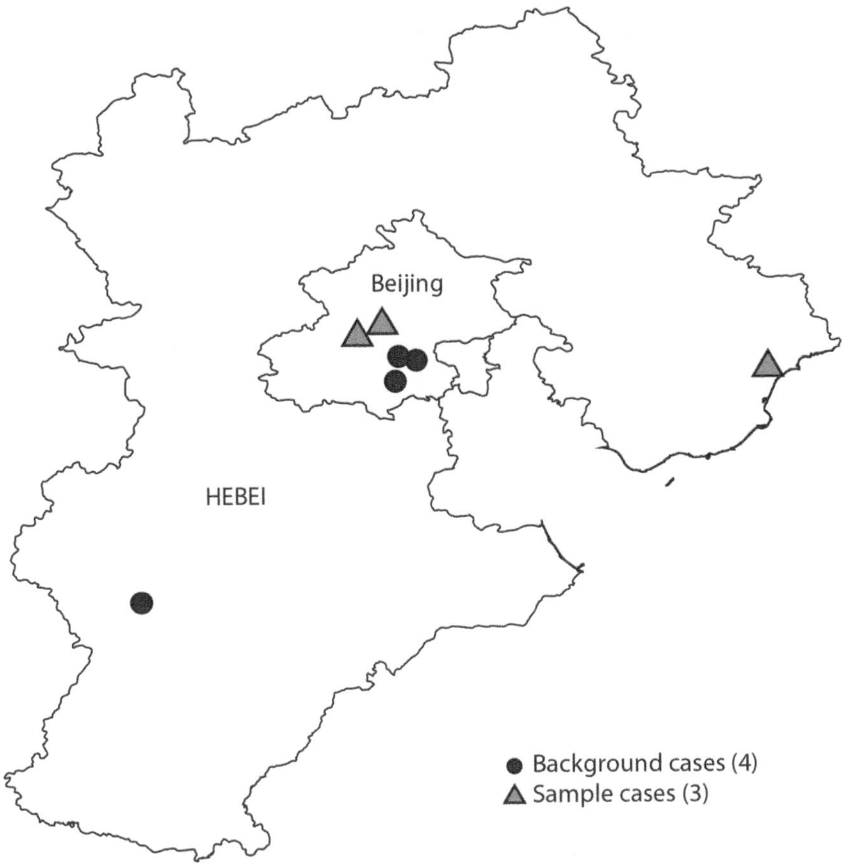

Source: Own compilation using the Stata module spmap by Maurizio Pisati. The China administrative maps were downloaded from: https://data.humdata.org/dataset/china-administrative-boundaries (27 April 2019). I thank Georg Strüver for help with generating these maps.

(Hurst 2010). However, it is important to note that the spatial focus on these three cluster regions limits the scope of findings. While linkages are more likely to emerge within the cluster regions due to the proximity of other cases and the agglomeration of intermediaries, it is unlikely that the emerging network of linkages is limited to these regions, since the information flows and opportunities for communication provided by information and communication technologies span larger geographical distances. More importantly, the 'no burn' community deliberately reaches out to affected communities in more distant parts of the country. How and to what extent cases outside the three cluster regions are linked to the network of actors working on incineration and whether the diffusion processes described in this study reach more disparate localities would provide an interesting basis for future research.

Figure 1.4 Shanghai cluster

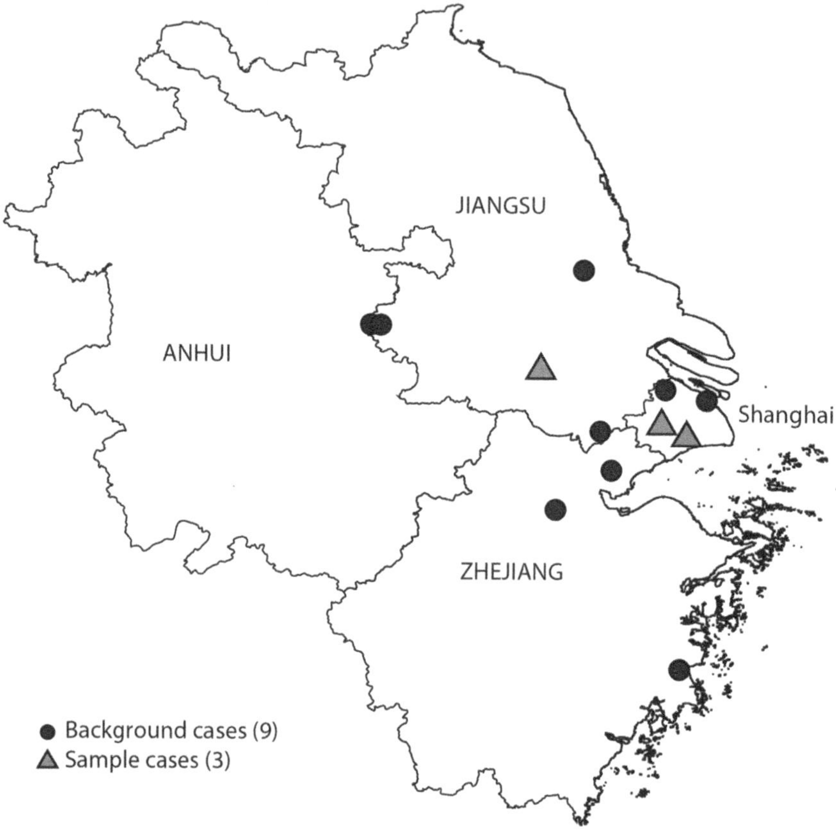

Source: Own compilation using the Stata module spmap by Maurizio Pisati. The China administrative maps were downloaded from: https://data.humdata.org/dataset/china-administrative-boundaries (27 April 2019). I thank Georg Strüver for help with generating these maps.

Within each of the three cluster regions, three localities were selected for in-depth case studies (George and Bennett 2005; Herriott and Firestone 1983; Stake 2006; Yin 1993, 2002). The cases were organized into three categories: urban, peri-urban, and rural. These different settings are expected to vary in conduciveness to the formation of linkages since they differ in the resource structure of the affected communities and their proximity to other cases. Urban cases, defined as cases of contentious action staged primarily by urban residents (mainly homeowners) within the main cluster metropolitan area (i.e., Beijing, Guangzhou, or Shanghai) are regarded as resource-rich communities with a close proximity to other cases, since many cases in the cluster regions are located in or near the main metropolitan area. Peri-urban cases, defined as cases of contention conducted primarily by villagers within the main cluster municipality, are regarded as (relatively) resource-poor communities, but

Figure 1.5 Guangzhou cluster

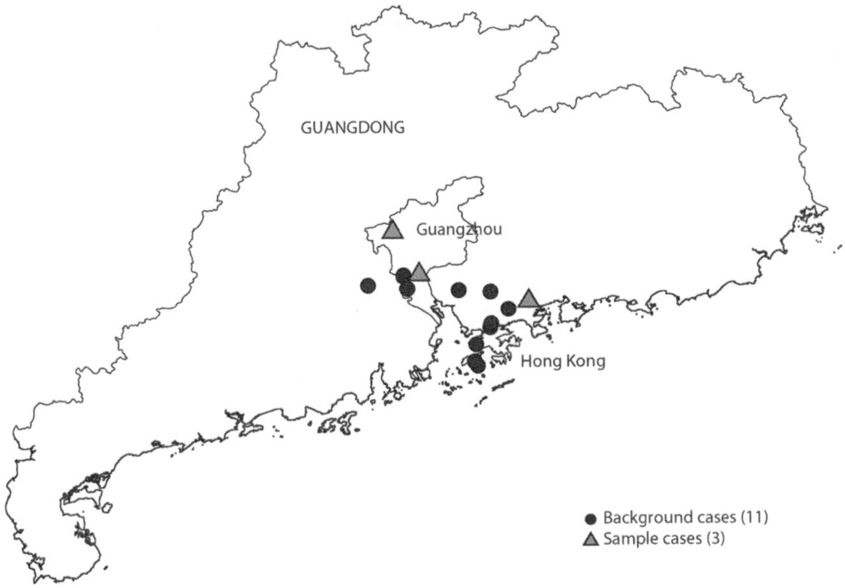

Source: Own compilation using the Stata module spmap by Maurizio Pisati. The China administrative maps were downloaded from: https://data.humdata.org/dataset/china-administrative-boundaries (27 April 2019). I thank Georg Strüver for help with generating these maps.

with a close proximity to other cases.[20] Since urban and peri-urban cases are located within the same municipality and thus within the same political-administrative setting, they allow a closer investigation of the role of resources in the formation of linkages. Rural cases, defined as cases of contention staged by villagers in rural areas within the cluster region but outside of the main metropolitan area, are regarded as (relatively) resource-poor communities with larger geographic distance separating them from other cases. The likelihood of forming linkages is expected to decline across the three categories from urban to peri-urban to rural, while dependence on external information and resources is expected to increase for resource-poorer communities.

The most recently reported urban, peri-urban, and rural case in each cluster region were selected for in-depth case studies based on the findings of the media analysis. Choosing the latest reported case provided maximum

20 Here, a mere geographic differentiation (i.e., urban cases closer to the city center and in areas mainly filled with apartments, peri-urban cases in the periphery) would not have been sufficient, since wealthy apartment owners with non-agricultural residency permits (户口, *hukou*) were the main activists in various cases despite the facilities being located in small villages along the outskirts of the major metropolitan areas.

Table 1.1 Sample and background cases

Cluster \ Setting	**Urban** – Resource-rich – Close proximity to other cases	**Peri-Urban** – Resource-poor – Close proximity to other cases	**Rural** – Resource-poor – Larger distance to other cases
Guangzhou – High presence of intermediaries – Open media environment – (Relative political openness)	**Huadu, Guangzhou** Panyu, Guangzhou	**Likeng, Guangzhou**	**Huiyang, Huizhou, Guangdong**
Beijing – High presence of intermediaries – Relatively open media environment – (Relative political openness)	**Asuwei, Beijing** Liulitun, Beijing Gaoantun, Beijing	**Dagong, Beijing** Nangong, Beijing	**Panguanying, Qinhuangdao, Hebei**
Shanghai – Closed media environment – Low presence of intermediaries – (Relative political restrictiveness)	**Songjiang, Shanghai** Caolu, Shanghai	**Fengqiao, Shanghai**	**Huangtutang, Wuxi, Jiangsu** Hai'an, Nantong, Jiangsu Taizhou, Zhejiang Binjiang, Hangzhou, Zhejiang

Declining likelihood of linkages

Notes: Cases in bold letters indicate sample cases. The case of Yongxing village in Guangzhou is generally referred to as "Likeng" case in the media and by the actors in this study, following the name of the facility. This is adopted in this book.

potential of linkages having developed with other cases in the cluster region and maximized the number of cases in other regions that could function as communication nodes. Since the time of onset (e.g., first complaints) was hard to discern from the media analysis, the selection was based on the first year of reported major events including mass petitions, protests, or litigation (see Appendix III).

While case selection according to the above logic was possible without further constraints for the Beijing cluster, difficulties emerged for the other two cluster regions. For the Shanghai cluster, the media analysis produced no peri-urban cases. The selection of the peri-urban Shanghai cluster case was therefore based on additional information obtained at a conference for affected communities organized by several intermediaries in Shanghai in June 2013, leading to the possibility that the sampling of this case may be biased towards the occurrence of vertical linkages. Selection of the rural Guangzhou cluster case also proved difficult because the region is so densely populated and all reported cases outside Guangzhou were based on the joint action of villagers and urban residents or homeowners (mostly of apartment buildings in the area). It is therefore debatable whether this case can count as truly 'rural'. Apart from these difficulties, the case selection occurred according to the sampling logic, which permits a systematic case comparison across the three cluster regions and three types of settings. The sample cases and further background cases for which data was collected are listed in Table 1.1. For more information on the cases, see Appendices III and IV.

Data Collection and Ethnographic Fieldwork

For the nine sample cases, a comprehensive collection of as wide a variety of data as possible was conducted to allow the triangulation of results. The primary data collection was conducted during a total of eight months of field research between 2011 and 2013 (in September and October 2011; September to November 2012; and April to August 2013), drawing on ethnographic methods including semi-structured interviews, participant observation, and the collection of documents. This data was complemented with online documents and media reports about both the cases and the activities of intermediaries who are involved in the issue field.

Due to the large number of cases under study and the sensitivity of conducting fieldwork on contention, particularly in rural China, the field research was conducted more like what could be termed an 'intermediary ethnography'. This meant having a base in the main municipality of each cluster region and spending time in participant observation with the local 'no burn' community members and urban communities while paying shorter site visits to the peri-urban and rural localities – just like the intermediaries in the issue field do. Wherever possible, site visits included visits to the respective incinerator or construction site, as well as to the other main waste facilities such as landfills. Apart from the sample cases, detailed data was collected for several other cases, either because they were necessary for an understanding of the sampled case or because the opportunity was given. Overall, site visits were made to 11 localities, interviews were conducted with members of 17 affected communities, and primary documents were collected for 17 cases (see Table 1.2).

During field research, a total of 104 interviews were conducted. These include interviews with the main domestic and international intermediaries active in the issue field and at least two core members of each of the contentious communities under study. Interviews were also held with a few officials and MSWI operators, as well as several experts on waste and environmental issues (a list of interviews is provided in Appendix I). The number of interviews

Table 1.2 Data collected during field work

Case	Field Visits	Interviews with Community Members	Primary Document Collection
Beijing Asuwei (BJ/U)	17/10/2012 27/05/2013	√	√
Beijing Dagong (BJ/PU)	01/06/2013 22/07/2013 23/07/2013	√	√
Qinhuangdao Panguanying (BJ/R)	04-06/11/2012 27-29/07/2013	√	√
Shanghai Songjiang (SH/U)	17/05/2013 18/05/2013 20/05/2013 23/05/2013 25/05/2013	√	√
Shanghai Fengqiao (SH/PR)	14/07/2013	√	√
Wuxi Huangtutang (SH/R)	×[1]	√	√
Guangzhou Huadu (GZ/U)	07/07/2013 08/07/2013	√	√
Guangzhou Likeng (GZ/PU)	04/07/2013 09/07/2013	√	√
Huizhou Huiyang (GZ/R)	11-12/07/2013	√	√
Beijing Liulitun (x/BJ/U)	×	×	√
Beijing Gaoantun (x/BJ/U)	01/11/2012 09/11/2012	√	√
Beijing Nangong (x/BJ/PU)	×	NA[2]	√
Shanghai Caolu (x/SH/U)	22/05/2013	√	√
Shanghai Yuqiao (x/SH/PU)	15/05/2013	√	NA[3]
Nantong Hai'an (x/SH/R)	×	√	√
Taizhou (x/SH/R)	×	√	×
Hangzhou Binjiang (x/SH/R)	×	√	√
Guangzhou Panyu (x/GZ/U)	01/07/2013	√	√
Wuhan Yongfeng (x/O/U)	×	√	√

Notes: The letters in brackets indicate the cluster region (BJ-Beijing, SH-Shanghai, GZ-Guangzhou, O-Outlier) and the setting (U-urban, PU-peri-urban, R-rural). (1) Site visit not possible due to sensitivity of case. Local community members were interviewed in Beijing and Shanghai. (2) Contention limited to "no burn" community activities. (3) Interviewed villagers deny the occurrence of contention at the time of field visit despite respective information from "no burn" community members.

with local officials was limited, since the first interviews with local cadres did not produce significant new information relevant to the research questions and seeking access to local party-state officials would have attracted attention to my research, which I particularly tried to avoid in more sensitive cases and rural areas. All interviews were held in Chinese and most interviews lasted between one and three hours. Interviews in rural (including peri-urban) areas often lasted throughout the day and were conducted in the presence of several villagers. With the permission of the interview partners, most interviews were recorded and fully transcribed. Interviews with community members followed a three-step approach. In the first step, an open narrative about the events was collected. In the second, the interview partners were guided to recall the events in more detail and in chronological order. Throughout these steps, no special interest in linkages was mentioned in order not to distort the narratives. In a third step, the interview partners were asked more focused questions about the existence and role of linkages if they had not already been tackled during the first two stages.

For all sample cases, extensive document collections were compiled, mostly drawing on documents provided by members of the local communities and engaged intermediaries. In several cases, community members provided comprehensive data collections (for a list of cited documents see Appendix II). These include materials written by community members (such as complaints, petitions, legal documents, and correspondence with state and party organs), official documents (such as public notices, official letters, project reports, or environmental impact reports), documents written by intermediaries (such as correspondence with state and party organs, documents written on behalf of the local communities, or petitions or public letters related to the case), legal documents, medical documents in health-related cases, videos and photographs taken by community members or intermediaries (including footage of contentious activities), and Chinese or English-language media reports. The data also encompassed the collection of a wealth of information about the broader activities of the main intermediaries, including both documents and participant observation of numerous conferences and workshops, in order to get a comprehensive understanding of the intermediaries' range of actions. Both during and between field research periods, this data was complemented with material available online – including the live observation of several microblogs during relevant events (such as during a village election, in the rural Beijing cluster case, or a road blockade, in the urban Beijing case), the continuous observation of mailing lists and social media run by intermediaries and local community members, and online Chinese and English-language media reports about the cases and activities of supra-local actors.

Where possible, field access to the selected sample cases was attempted via the Chinese 'no burn' community. For most cases, an intermediary had already established contact with a local community member or could find a contact online for me (e.g., via social media). Since several of the cases were still ongoing at the time of field research, many local community members were hesitant to talk about sensitive issues with a stranger, particularly when they had been the target of contention-related repercussions. In cases where major large-scale protests had occurred, it took some effort to gain access to core community members since they were concerned about retaliation if their identities were revealed. Establishing first contact via intermediaries helped to gain the communities' trust. Once trust was established, many local community members were glad to share their story in the hopes of receiving (international) attention for their cause. Site visits were not possible for the rural Shanghai cluster case due to its high sensitivity and interview partners had to be met outside of the locality. Similarly, the two site visits for the rural Beijing cluster case were limited in time due to the tense local political situation at the time of field research. While the villagers were grateful for the external attention and the potential protection it could provide – even explicitly inviting me for another visit – stays in the village were kept to a maximum of three days to avoid causing trouble for all sides. Visits to the peri-urban Guangzhou locality also had to be limited in number due to the high political sensitivity at the time of research, when a second incinerator had just opened in the vicinity. After I was interviewed by local public security during the first site visit and warned against further visits, only one careful follow-up visit was conducted.

While I tried to retain a neutral position as an observant researcher throughout my field research, this was at times difficult when I encountered hopes for assistance from aggrieved local community members faced with severe threats to their livelihoods. Particularly in the rural cases, I was often perceived as more of an expert on incineration and environmental or health issues than as a researcher with a social science background and no expertise in the technical or health-related aspects of incineration and waste management. In some of the cases I thus became a node in the network of information flows myself, providing contact information to potentially helpful intermediaries before leaving the villages. Throughout the months of field research and as news about my research activities spread through the anti-incineration network, I became regarded as a (temporary) member of the 'anti-incineration' community, which greatly facilitated field access to the individual cases.

Data Analysis and Case Studies

Due to the large amount of collected data, data analyses for the sample cases were conducted at different levels. Based on its 'middle' position between the Guangzhou and Shanghai cluster cases with regards to the expected role of linkages (see the case sample in Table 1.1) – again following a 'typical case' logic – and the sampling difficulties experienced in the other two cluster regions, the three Beijing cluster cases were selected for holistic case studies based on systematic events data analyses (presented in Chapters Three to Five). For the other sample cases, a more focused analysis concentrating on the nature and role of linkages was conducted, the results of which are presented in Chapter Six.

Following comparative case study methods (Stake 2006), comprehensive case protocols were compiled for all cases, which permitted the systematic comparison of different cases. A case protocol template is provided in Appendix V. The protocols include information about the respective case setting and incinerator project, prehistories such as prior contention, as well as the main actors and their self-perception, framing, claims, and strategies. For all sample cases, the case protocols further include a systematic analysis of the main research questions, i.e., the nature and role of horizontal and vertical linkages and signs of scale shift. For the three Beijing cluster cases, a detailed chronology of events was also compiled based on all available information, which served as template for the case study narratives provided in Chapters Three to Five. In the other cases for which data was collected during fieldwork (i.e., the sample and background cases), the analysis focused on the research questions about linkages and diffusion processes, so the chronology of events listed in the case protocols was limited to central dates and events.[21]

All recorded interviews were fully transcribed with the help of two Chinese student assistants. The interview transcripts and other collected data were analysed using the qualitative data analysis software MAXQDA, which was an invaluable tool for systematizing the large volume of data. It helped to organize and code the data according to the above case protocol logic. Throughout the study, no identities of interview partners are revealed according to standard practice in social science research on China, except for those featuring prominently in the media. All photographs and figures used in this study were taken or compiled by the author unless otherwise indicated.

21 I thank Carolin Kautz for a first pre-sorting of the Shanghai and Guangzhou cluster case data, which greatly facilitated the analysis.

2 A Burning Issue

Waste Incineration in China

During the last decade, waste incineration has drawn major public attention and become a hotly debated issue in the Chinese media. This is due to the massive scale of the Chinese party-state's waste incineration strategy and the fierce public opposition to its implementation. This chapter introduces China's waste treatment policies and rush for incineration before turning to the two main groups of incineration opponents: the (trans-)national 'no burn' community, and the affected local communities choosing to stand against incinerator projects in their neighbourhoods.

A 'Golden Age' of Chinese Waste Incineration: Policies and Economic Incentives

Waste treatment and disposal first started to become an issue in China in the 1980s. With the reform and opening-up period beginning in 1978, both the overall amount of municipal solid waste (MSW)[22] and its ratio of inorganic components such as plastics started to grow. Waste could no longer be dumped on farmland as organic fertilizer. This led to the question of alternative disposal methods (Balkan 2012; Dong et al. 2010; Huang et al. 2006; Wang et al. 2011). By the early 1990s, most urban waste was being pitched into ponds, scrublands, and basic garbage dumps around the outskirts of cities without any further treatment, occupying growing stretches of farmland (ibid.).[23] This practice soon led to significant soil, air, and water pollution, as well as several accidents when waste gas at the dumps ignited and exploded (Wang and Nie 2001a, 2001b).

22 According to Chinese law, solid waste is classified into three types: industrial solid waste, municipal solid waste, and hazardous waste. Municipal solid waste is composed of residential waste, municipal services waste (from street and park cleaning), and institutional and commercial waste. Of those types, residential waste has the largest share and the most complex composition, which varies strongly across China due to the imbalance in economic development. 'Solid waste' is defined as any solid material intentionally discarded for disposal. Much of this waste, such as recyclables, has value for someone else and can be extracted from the waste stream (Huang et al. 2006; Wang and Nie 2011a; World Bank 2005).

23 Between 1980 and 2000, the amount of waste increased between 3 and 10 percent annually; at the beginning of the 1990s, less than 2 percent of waste was treated before dumping (Dong et al. 2010; Huang et al. 2006; Wang and Nie 2001a, 2001b; Wang et al. 2011).

A Mounting Waste Crisis

In the 1990s, the mounting waste problem started to draw government attention from the local to the national level. The Ministry of Housing and Urban-Rural Development (中华人民共和国住房和城乡建设部, *Zhonghua renmin gongheguo zhufang he chengxiang jianshe bu*) is China's main national authority for MSW management. Under its jurisdiction, the municipal Environmental Sanitation Bureaus (环境卫生局, *Huanjing weisheng ju*) are responsible for the collection, transportation, and disposal of municipal solid waste. At the same time, the Ministry of Environmental Protection (MEP, 中华人民共和国环境保护部, *Zhonghua renmin gongheguo huanjing baohu bu*) and its local Environmental Protection Bureaus (环境保护局, *huanjing baohu ju*)*Huanjing baohu ju*, EPB) are responsible for the regulation and implementation of waste-related pollution control. The National Development and Reform Commission (NDRC, 国家发展和改革委员会, *Guojia fazhan he gaige weiyuanhui*), the Ministry of Commerce (商务部, *Shangwu bu*), and various government institutions at the provincial and municipal levels are also engaged in the formulation of waste-related laws, regulations, and standards (Dong et al. 2010; FON 2011d; Normura 2011; Pöyry n.d.; Wang and Nie 2011b). This fragmentation and lack of a clearly divided administrative and legislative structure or consistent policy and strategic planning is frequently raised in critiques of the Chinese waste management system (Dong et al. 2010; FON 2011d; Wang and Nie 2001b; World Bank 2005; interviews Zhao 8-11-12, NGO Boell 25-10-12, NGO FON1 19-10-12, NGO GAIA 23-6-13, NGO GVB 18-10-11, NGO NU1 11-10-11, NGO WEC 23-4-13).

In the 1990s, several large Chinese cities began building a system of waste collection centres and the first sanitary landfills to address the growing waste pollution issue. In the late 1990s, municipal governments made several large-scale investments in MSW management projects in the hope of increasing the ratio of waste treated in an environmentally sound manner (Dong et al. 2010; FON 2011d; Huang et al. 2006; Wang et al. 2011; World Bank 2005).

National legislation also started to address the problem. In 1995, the National People's Congress (NPC, 全国人民代表大会, *Quanguo renmin daibiao dahui*) issued the first law on solid waste management, the 'Law of the People's Republic of China on the Prevention and Control of Environmental Pollution by Solid Waste' (中华人民共和国固体废弃物污染环境防治法, *Zhonghua renmin gongheguo guti feiqiwu wuran huanjing fangzhi fa;* amended in 2004), often referred to as the Solid Waste Act (MEP 2010; National People's Congress 1995). While the Solid Waste Act calls for reducing waste at the source, recycling, and the environmentally sound treatment and disposal of waste, its implementation proved difficult due to a lack of corresponding rules and regulations (Pöyry n.d.; Huang et al. 2006; Nomura 2011; Wang and Nie 2001a, 2001b; World Bank 2005). Despite increasing efforts to treat and dispose of waste in a more environmentally friendly manner, municipal governments could not keep up with the rapidly growing amount of generated garbage. By the year 2000, the ratio of sufficiently treated urban waste had only reached 59 percent – 90 percent of which was landfilled – with the rest continuing to be dumped without prior treatment (Nie and Wang 2011b).[24] Moreover, since many sanitary landfills were not well operated, they continued to produce significant pollution.

Apart from pollution issues, municipal waste – which continued to grow at a rate of 8 to 10 percent annually – soon started to also experience a problem of space. In 2004, China surpassed the United States as the largest producer of MSW, producing 190 million tons of urban solid waste – nearly one quarter of the world's recorded waste that year (Balkan 2012; Dong et al. 2010; Lau 2010; Wang et al. 2011; World Bank 2005). As the World Bank pointed out, no country had ever experienced as large, or as rapid, an increase in waste generation (World Bank 2005). With rapidly growing waste volumes, many of the facilities installed in the 1990s started to reach their capacity limits in the 2000s. At the same time, the small percentage of waste treated in composting facilities generated only low-quality compost because the garbage entered the facilities without sufficient prior separation. As a result, the generated compost was unsuitable for use as agricultural fertilizer, could not be sold, and instead piled up at the composting facilities (ibid.; Dong et al. 2010; Huang et al. 2006; Nie and Wang 2001a; Wang et al. 2011). Residents' complaints about the odour and pollution generated by landfills and composting facilities had also become more frequent. Although they had been opened on the outskirts of cities, the facilities were soon engulfed by sprawling residential areas and now caused discontent among the surrounding residential communities. By the early 2000s, Chinese waste management had become one of China's most urgent environmental and urban construction problems (Balkan 2012; Chen et al. 2010; Dong et al. 2010; Pöyry n.d.; Huang et al. 2006; Ma 2009; Nie and Wang 2001a, 2001b; Nomura 2011; Wang et al. 2011; World Bank 2005; *Xinhua* 2010b). With increasing real estate prices and land becoming an ever-scarcer resource, municipal governments started to look for more space-efficient ways to treat the growing piles of garbage.

24 The situation is even worse in rural areas, where the majority of local authorities at the township and village level still fail to provide solid waste treatment services (Wang et al. 2011; Zheng, Niu, and Zhao 2015).

The Search for Alternative Waste Treatment Strategies

Around this time, the State Environmental Protection Agency (SEPA, since 2008 the Ministry of Environmental Protection) commissioned a group of researchers, including professor Nie Yongfeng from Tsinghua University's Department of Environmental Science and Engineering – who later became one of the chief advocates of incineration technology in China – to conduct a comprehensive study on the situation of and problems with China's waste management system (interview Nie 7-6-13; Wang and Nie 2001a, 2001b). The group of researchers suggested two main strategies for solving China's waste crisis that mirrored the recommendations also given by other experts. First, they called for comprehensive measures to promote waste segregation and recycling to reduce the amount of waste requiring treatment and disposal. Second, they suggested that the government promote waste incineration, i.e., the combustion of solid waste at high temperatures, as practiced in many developed countries (Wang and Nie 2001a, 2001b).

Calls for the inclusion of better waste segregation and recycling measures in China's waste management system have also been frequently requested by other waste experts and environmentalists (interviews Nie 7-11-12, NGO FON1 19-10-12, NGO GAIA 23-6-13, NGO GVB 18-10-11, NGO NU1 11-10-11, NGO WEC 23-4-13, Zhao 8-11-12; see also Balkan 2012; Bradsher 2009; *China Daily* 2011; Dong et al. 2010; FON 2011d; Huang et al. 2006; Huo 2009; Ma 2009; Meng 2010c; Wang and Nie 2001b; Wang et al. 2011; C. Yang 2011). However, apart from individual projects introduced in some of China's major cities in the late 2000s, there were no government recycling initiatives or systematic source-separated collection before 2010, and recycling was not officially included in the MSW management system. Instead, most waste segregation in China was and still is conducted by the large informal sector of private waste pickers, who make their living by collecting and selling materials with an economic value – such as paper, metals, or plastics – at different points in the waste collection process (Balkan 2012; Dong et al. 2010; *Economist* 2015; Pöyry n.d.; Nomura 2011; Wang and Nie 2001a).[25] The World Bank (2005) estimated that around 2.5 million people were making a living in the informal waste sector in 2005, while only 1.3 million were employed by local governments or businesses in the formal urban waste collection system.

25 Chinese households have a habit of collecting valuable materials separately and selling them to waste collectors in their neighborhoods, who resell them to companies. Waste pickers also sort recyclables from residential garbage directly at the collection bins in the residential units, at waste collection and transfer stations, and at landfills and garbage dumps.

Despite this large informal waste sector, recycling rates have long reported to be surprisingly low in China – a trend that is seeing a decreasing tendency since low material prices provide diminishing incentives for households to pre-sort and sell their recyclables as incomes rise (ibid.; Dong et al. 2010; Huang et al. 2006; *Xinhua* 2006). According to waste experts and environmentalists, official inclusion of waste sorting and recycling in the MSW management system would increase recycling rates by facilitating MSW regulation and management and also improve the precarious living situation of waste pickers (interviews BMAC 7-11-12, Zhao 8-11-12, Nie 7-11-12, NGO FON1 19-10-12, NGO GAIA 23-6-13, NGO GVB 18-10-11, NGO NU1 11-10-11, NGO WEC 23-4-13; Balkan 2012; FON 2011d; Ma 2009; World Bank 2005). Another waste management strategy often raised in this context is waste reduction by the consumer – another of the 'three Rs' of sustainable waste management (reduction, reuse, recycling). This includes the introduction of a volume-related household waste service fee system to increase households' incentive to reduce residential waste, as well as regulations for companies that reward the reduction of consumer waste such as packaging during the production process (FON 2011d; Ma 2009; Wang and Nie 2001a, 2001b; World Bank 2005).

After 2010 individual municipalities have stepped up their efforts to increase waste segregation and recycling; urban separated waste collection and recycling targets were finally introduced in the 12th Five Year Plan (2011-2015) in 2011 (*China Daily* 2011, 2012; Cui 2010c, 2010d; L. Li 2010b; Xie and Du 2011; Xie and Zhang 2011; *Xinhua* 2011; Zhai 2010). Still, experts and environmentalists continue to regard the government incentives as half-hearted and the Chinese recycling system as ineffective. In addition to the lack of public awareness and participation, this is largely attributed to the fragmented responsibility structures and weak regulations of the MSW sector (interviews ASW 17-10-12, BMAC 7-11-12, Zhao 8-11-12, Nie 7-11-12, NGO FON1 19-10-12, NGO GAIA 23-6-13, NGO GVB 18-10-11, NGO NU1 11-10-11, NGO WEC 23-4-13; World Bank 2013).

A 'Great Leap Forward' in Waste Incineration

The second suggestion of the SEPA-commissioned researchers, to promote waste incineration technology in China, quickly built up steam. By significantly reducing the volume and mass of solid waste while easily controlling odours, waste incineration seemed like the logical answer to China's waste problems. Both the national and municipal governments soon started to promote incineration at a massive scale, with the support of many Chinese researchers and experts. However, the immense initial cost for the construction and operation of waste incinerator facilities meant that municipal governments were dependent on external funding and investment – which, in China, was long regarded as a government responsibility. The fragmented regulatory, administrative, and operational responsibilities therefore impeded the development of a market mechanism in the MSW management sector.

Starting in the early 2000s, several national-level government institutions started to issue policies promoting investment in the municipal solid waste incineration sector in China. This heralded what is often called the 'golden age of waste incineration' or 'Great Leap Forward in waste incineration' (e.g., interview Nie 7-11-12; Biswas and Zhang 2014; Meng 2009; C. Yang 2013; Yu 2012). Due to these policies, companies investing in and operating municipal solid waste incinerators in China became eligible for a variety of tax rebates and subsidies as well as prioritized commercial bank loans. In addition, electricity produced by waste-to-energy plants was listed as a form of renewable energy in China's 11th Five Year Plan (2006-2011). This makes Waste-to-Energy (WtE) plant operators eligible for substantial tax cuts and favourable credit conditions. During the period of operation, a subsidy is further provided for every ton of waste treated – with the local government responsible for the collection and transportation of MSW to the facility as well as providing a sufficient supply of garbage (Balkan 2012; Hook 2012; Hu 2015; National People's Congress 2008; Nomura 2011; Pöyry n.d.; *Waste Management World* 2012; World Bank 2005; Xie 2010a; C. Yang 2011; Yu 2011, 2012). Moreover, since most waste incineration plants in China are Waste-to-Energy facilities, i.e., they use the heat from the combustion process to generate electricity, operators are eligible for subsidies for every kilowatt hour fed into the state energy grid; with WtE-generated electricity is also dispatched with priority by the power grid (National People's Congress 2008; Nomura 2011; Pöyry n.d.; *Waste Management World* 2012; C. Yang 2011). Finally, due to their designation as renewable energy projects, WtE plants can apply to be listed as Clean Development Mechanism (CDM) projects with the United Nations Framework Convention on Climate Change (UNFCCC), which makes them eligible for additional subsidies (World Bank 2005).

In China, most waste incinerators are built on a Build-Operate-Transfer (BOT) model. Based on joint investment by the responsible municipal government and a company selected through a public tendering process, the facilities are designed, constructed, and successively operated by the companies for a 20- to 30-year licensing period. After this period, the facility returns to the responsible government. The industry estimates that it takes between eight and twelve years to earn back the cost of constructing an electricity-generating waste incinerator. This means that investing companies can enjoy up to 22 years of profit from these deals. Overall, the waste incineration sector in China promises extremely high profit margins with high initial investment followed by low operating costs and large and stable profits (Nomura 2011; Pöyry n.d.; *Waste Management World* 2012).

Together with large-scale government investment in the technology, these high profit margins have led to a major boom in the Chinese incineration sector during the last decade. Many domestic companies quickly started to push into the incineration market – in some cases submitting loss-making tenders just to get a foothold (Balkan 2012; Hook 2012). Many Chinese incineration companies, often state-owned enterprises, moved into the sector from different backgrounds. This has led to major concerns about unqualified plant operators, as raised by experts and environmentalists. Foreign companies and producers of waste incineration technology have also been pushing onto the Chinese market – a move enhanced by the international trend of reducing or banning incineration in many other countries due to its related pollution harms. Even development banks such as the Asian Development Bank and German Development Bank (Kreditanstalt für Wiederaufbau, KfW) have made substantial investments in the Chinese waste incineration sector (interviews NGO GAIA 23-6-13, NGO IPEN1 22-6-13, NGO NU1 11-10-11; FON 2011d; Hook 2012; NEEC 2009; Nomura 2011; Pöyry n.d.).

The real peak or 'golden' years of Chinese waste incineration started during the second half of the 11th Five Year Plan (2006-2011) and continued throughout the 12th Five Year Plan period (2012-2016). In 2008, a report by the Ministry of Housing and Urban-Rural Development declared that more than one third of China's cities were facing a 'waste crisis' (垃圾危机, *laji weiji*) – a term rapidly picked up by the media (*Global Times* 2009; Hook 2012; Huo 2009; Feng 2011; Meng 2010c, 2010d; Wang 2009; Wang and Jing 2009; Xie and Zhang 2011; *Xinhua* 2009b; Xu 2010). Following this report, the 11th and 12th Five Year Plans officially promoted waste incineration as the primary solution for this problem and both government investment in the waste incineration sector and official targets for the ratio of waste treated via incineration skyrocketed.

While the location of China's first modern waste incinerators was restricted to large cities along the Eastern coastline – the first one opened as early as 1988 in Shenzhen – with the new government backing waste incinerator projects started to spread to China's second- and third-tier cities, with numerous project plans announced in 2009. For the 11th Five Year Plan period, the national government planned the construction of 82 new waste incinerators and, according to official statistics, the number of operating plants had increased to 138 by 2012 (Hu 2015). By the end of the 12th Five Year Plan period in 2015, the number of operating facilities was slated to increase to 300 and the national incineration ratio was to reach 30 percent – with the targeted ratio as high as 48 percent for the well-developed eastern areas (Balkan 2012; Cheng 2011a; X. Li 2011 Wang 2014; Yu 2011; Zhai 2010). Total government investment in the solid waste treatment industry was expected to reach up to 280 billion Renminbi (RMB)[26] for the period between 2011 and 2015, with about 100 billion RMB spent on waste incineration (Balkan 2012; X. Li 2011; Nomura 2011; Wang 2014; Yu 2011; Zhai 2010). A 2009 study by the Standard Chartered banking group estimated that over one-half of global orders for new MSWI facilities came from China (Balkan 2012; Biswas and Zhang 2014). Techsci Research, a consultancy firm, reportedly expected the Chinese market for incinerators to further double in size between 2015 and 2018 (*Economist* 2015). Incinerator operators, mostly state-owned enterprises or Sino-foreign joint ventures, therefore have high financial stakes in the unobstructed construction of waste incineration projects.

Implementation at the Local Level: Government Incentives, Legislative Framework, and Procedures for MSWI Projects

At the local level, the national policies promoting incineration are reflected in the municipal solid waste treatment plans announced by the municipal governments for every Five Year Plan period. For many municipal governments, WtE facilities are attractive projects, not only as a way to rid their cities of growing amounts of garbage, but also as large-scale economic investments that have the benefit of simultaneously being classified as 'environmentally friendly' – and thus helping boost local governments' green records. Moreover, like the operators, municipal governments have financial incentives for the construction of incineration facilities not only due to their function as investors, but also because they can expect to receive high profits after the end of the facilities' licensing period. The viewpoint of lower-level local, particularly village, governments is often less clear. They might have financial, relational, or political incentives to promote the projects, especially when corruption comes into play. Local politics, transparency, and corruption are often major issues in anti-incineration disputes. In several cases, village committees have also sided with the

26 One RMB is equivalent to approximately 0.13 euros.

opposing local population, however. This can be for different reasons, such as the village committee members themselves living in close proximity to the plants or their complex (kin) relations with the other villagers.

MSWI projects must first be listed in the relevant municipal plans and be approved by several departments at the municipal or city level. After a site for the project is selected within the municipality's jurisdiction, the project plans have to be approved by the appropriate departments at the district, county, township, town, or village levels of government, depending on the planned site. With increasing public opposition to MSWI projects in recent years, slated project sites have moved towards districts along the outskirts of cities or to rural areas within municipalities' administrative divisions. Through a public bidding process, a company is then selected as a construction and operation unit (interviews BMAC 7-11-12, Nie 7-11-12, Zhao 8-11-12, NGO FON1 19-10-12, NGO NU2 14-10-12; Nomura 2011; Pöyry n.d.).

As large-scale construction projects, a large part of the remaining procedures is organized by the 'Environmental Impact Assessment (EIA) Law of the PRC' (环境影响评价法, *Huanjing yingxiang pingjia fa*), issued by the National People's Congress in 2002, as well as several related regulations stipulated by SEPA in 2004 and the 'Regulations on Environmental Impact Assessment for Plans' (规划环境影响评价条例, *Guihua huanjing yingxiang pingjia tiaoli*) issued by the State Council (国务院, *Guowuyuan*) in 2009 (National People's Congress 2009; Wang 2011). According to the EIA Law and related regulations, the construction and operation unit of the construction project (in this case the incinerator) commissions an Environmental Impact Assessment unit[27] to conduct an environmental impact assessment study, which will evaluate the impact of the project on the surrounding environment and residents. The results are summarized in an EIA report, of which – since September 2012 – an abridged version has to be published. The EIA procedures encompass public participation measures, including a twofold public announcement period for the project and the distribution of questionnaires soliciting the opinion of residents and other affected units within two kilometres of the project site. After the EIA report is completed and shows that the project will not have a significant impact on the environment or human health and that it has acquired the consent of the surrounding residents and other affected units, the project is ratified

27 EIA units in China are appointed by the MEP. They can obtain two levels of qualification. Units categorized as grade-A can function as EIA units for all construction projects. Units categorized as grade-B are only permitted to evaluate small- and medium-sized projects. The environmental evaluation of incinerator projects requires a grade-A qualification.

by the responsible Environmental Protection Bureau and construction can commence (interviews BMAC 7-11-12, Nie 7-11-12, Zhao 8-11-12, NGO NU2 14-10-12; MEP 2012; Nomura 2011; Pöyry n.d.).

During plant operation, the local EPB is responsible for the monitoring and control of the created pollution. In the early 2000s, different national-level institutions issued standards and codes for the control of pollution from waste incineration, which were complemented by some local regulations and standards. Long criticized as outdated and insufficient, some of these standards have been revised in recent years (interviews BMAC 7-11-12, Zhao 8-11-12, NGO Boell 25-10-12, NGO FON1 19-10-12, NGO GAIA 23-6-13, NGO IPEN1 22-6-13, NGO NU1 11-10-11, NGO WEC 23-4-13; Ni et al. 2009; Pöyry n.d.; World Bank 2005; C. Yang 2011). These standards include air pollution emission limits for waste incinerators. Among others, emission limits for dioxins, a group of highly carcinogen toxic pollutants and the main products used by critics to oppose waste incineration, were long set at 1 nanogram Toxic Equivalents per cubic meter (1 ng TEQ / m^3) – ten times higher than permitted by EU emission standards. In July 2014, Chinese standards were lowered to meet EU standards due to major public complaints and in compliance with the Stockholm Convention on Persistent Organic Pollutants (POPs), an international environmental treaty ratified by China in 2004.[28] According to the new standards, new garbage incinerators are required to meet emission standards of 0.1 ng TEQ / m^3, whereas plants already in operation were required to meet the new standards by 2016 (D_IPEN1; Shuang Li 2010d; Wang 2010; Wang 2014; *Xinhua* 2010d).

Other regulations include a technical standard that requires the combustion of waste in MSWI facilities to be done at a minimum of 850 degrees Celsius to minimize toxic dioxin emissions,[29] as well as a 300-metre protection zone around incinerator sites that requires the relocation of all inhabitants within this area. However, this buffer zone is regarded as highly insufficient by Chinese environmentalists and some experts. Overall, critics point out that the main problem with China's regulatory framework for incineration is not the lack of standards and regulations, but their weak

28 The Stockholm Convention, initiated by the UN Environment Program, was signed by China in 2001 and went into effect in 2004 with the aim of reducing persistent organic pollutants (POPs). The declaration is regarded as a major challenge for Chinese waste incineration policies by experts and environmentalists (interviews NGO IPEN1 22-6-13, NGO IPEN2 23-6-13, NGO GAIA 23-6-13, 4-7-13; D_IPEN1, D_IPEN2).

29 According to conventional research on waste incineration, the amount of toxic dioxins that are produced during the combustion process is drastically reduced above temperatures of 850 degrees Celsius.

enforcement and lack of implementation at the local level (interviews Zhao 8-11-12, NGO Boell 25-10-12, NGO CLAPV 18-10-12, NGO FON1 19-10-12, NGO GAIA 23-6-13, NGO NU1 11-10-11, NGO WEC 23-4-13, Xia 30-7-13; Meng 2010d; Pöyry n.d.; Xu 2010; Yu 2012).

Critical Voices: The Chinese 'No Burn' Community

While the government was promoting waste incineration as a panacea for China's waste crisis, critics of the process – including domestic and international experts, environmental organizations, and environmentalists – soon started to publicly challenge China's waste treatment approach, which they considered unsustainable. They regarded China's push for incineration as a dangerous one-way street guided by vested economic interests at the cost of both the environment and human health. Their critique soon widened beyond the issue of waste management to tackle broader regulatory failures and environmental concerns reflected in the issue field. These experts and environmentalists were soon joined by (environmental) lawyers and legal associations, who were less concerned with the issue of incineration, but rather with the overall advancement of environmental litigation and the rights of communities living near planned or operating plants.

While constituting a heterogeneous group with partially diverging goals and interests, these different actors soon started to cooperate in joint activities both at the national level and in individual local disputes. This more or less loosely connected supra-local network of actors, which resembles the transnational advocacy networks described by Keck and Sikkink (1998), is here termed the 'no burn' community since its members generally are opponents of incineration technology at large, of China's incineration policies more specifically, or of individual incineration facilities. 'No burn' is a reference to the Chinese term 'anti burn faction' (反烧派, *fan shao pai,* of which 烧, *shao* refers to 焚烧, *fenshao,* the Chinese term for incineration), which is frequently applied to opponents of incineration and was also used self-referentially by several of the actors. This is in contrast to the Chinese term 'pro burn faction' (主烧派, *zhu shao pai*), which is frequently used for advocates of incineration (e.g., Sipan Li 2010; Xu 2010). 'No burn' and 'pro burn' are also common terms used by the international anti-incineration community.

The following sub-sections introduce the three main groups that constitute the Chinese 'no burn' community: experts, environmental organizations, and environmental lawyers and legal associations. These three actor groups have further drawn on their personal networks with and received crucial support from journalists working for various media institutions across the country. Journalists are not included in this description of the 'no burn' community because their reporting is not necessarily targeted at impeding China's incineration boom or obstructing individual incinerators, but is often part of their broader reporting on environmental issues.

Panacea versus Health Hazard: Experts and the Public Debate about Incineration in China

In China, as in other countries, experts specialized in waste or related environmental and health issues have played a major role not only in the promotion of waste incineration, but also in its public contestation. By shaping the public debate about the risks of incineration and adequate mitigation strategies, and acting as a source of issue-specific knowledge and expertise at the local level, experts critical of China's incineration path constitute the first important group within the Chinese 'no burn' community. Similar roles are also played by the experts described in Mertha's (2008) study of contention against hydropower projects, particularly the campaign against the Nu River Project.

Due to the highly technical nature of incineration, a public understanding of the technology and its related risks is reliant on issue-specific information, knowledge, and expertise. This is particularly the case as the risks associated with waste incineration resemble those that are typical of modern 'risk societies' (Beck 1986): not clearly discernible, deeply contested, and characterized by an inherent difficulty of demonstrating causal links with their mostly long-term effects, particularly those related to human health. In such a context, the very definition of risk, the decision about what amount of risk the society should bear, and sufficient mitigation strategies are all a matter of contestation. This translates into related policies, standards, and regulations. Experts come to play a crucial role not only in concrete policy formulation, but also in the constitution of the larger issue field and the public debate about the matter (Fischer 2000; Kasperson et al. 1988; McCormick 2009; Mertha 2008; Tesh 2000; Walsh, Warland, and Smith 1997). At the local level, the nature of incineration and its related risks make affected communities dependent on expert knowledge, information, and expertise for both their awareness process and taking action as outlined in the previous chapter. Experts can be a crucial source of information (Brown 2007; Fischer 2000; Gould 1993; Sherman 2011a).

After the Chinese government started to promote waste incineration in the mid-2000s, a fierce public and media debate about incineration technology and its risks unfolded, mainly revolving around two expert camps (interviews BMAC 7-11-12, Nie 7-6-13; Balkan 2012; Lan 2010; Sipan Li 2010; Meng 2009, 2011; Meng 2010d; Qiu 2010; Wang 2014; Xie 2010a, 2010b; Xu 2010; Yu 2012). These two expert camps are often termed the 'pro burn faction' and 'anti burn faction', both in the Chinese media and by different actors in the issue field. The 'pro burn faction', which is in favour of incineration technology becoming China's primary waste treatment strategy, included the majority of Chinese experts on waste or related environmental issues employed at Chinese state research institutions. Two of its most prominent members are Nie Yongfeng and Xu Haiyun. Nie is a professor in Tsinghua University's Department of Environmental Science and Engineering and was among the group of researchers commissioned by SEPA to find alternative waste treatment strategies in the late 1990s. He also functions as a government advisor and expert in various positions. Xu is the chief engineer for environmental sanitation at the China Urban Construction Design and Research Institute Co., Ltd. (中国城市建设研究院有限公司, *Zhongguo chengshi jianshe yanjiuyuan youxian gongsi*). This institute was established in 1985 under the jurisdiction of the Ministry of Construction and is now subordinate to the China Construction Technology Consulting Group Co. Ltd. (中国建设科技集团股份有限公司, *Zhongguo Jianshe keji jituan gufen youxian gongsi*), a state-owned enterprise under the jurisdiction of the State Council. According to its self-description, the institute has a major impact on the formulation of standards and regulations and is actively involved in the drafting of China's Five Year Plans.

The 'pro burn faction' of experts has supported the government's push for waste incineration through academic articles and lectures, media articles, and public talks. As government advisors, the advocates have played a significant role in the shift of the country's waste policies towards incineration and in the formulation of related standards and regulations. Moreover, at the local level they have also functioned as government experts for the siting of individual facilities, both during the EIA process and in the case of disputes between local governments or companies and surrounding residents (interviews BMAC 7-11-12, Nie 7-6-13; Balkan 2012; Cui 2009; *Global Times* 2009; Lan 2010; Sipan Li 2010; Meng 2009, 2011; Meng 2010d; Qiu 2010; Wang 2009; Wang 2014; Xie 2010a, 2010b; Xinhua 2009b; Xu 2010; Yu 2012).

The 'pro burn faction's' arguments mirror the official stance on waste incineration. They praise incineration as a space-efficient and environmentally friendly solution to China's urgent waste crisis. In addition to long-term strategies such as recycling or reduction, incineration can instantly help rid China's cities of its overflowing landfills, they argue. It is important to note that recycling and reduction are regarded as necessary complementary long-term strategies by most advocates of incineration. The risks of incineration such as the production of dioxins can be sufficiently controlled, proponents claim, without posing a major harm to the environment or human health. This has been demonstrated in many developed countries such as Japan, the United States, and several European countries, they point out, where incineration technology has been used as a modern waste treatment strategy for years. For these reasons, advocates dismiss the public's fear of dioxin as emotive and based on misleading information from sensational media reports and irresponsible experts from the 'anti burn faction' (interviews BMAC 8-11-12, Nie 7-6-13; Cui 2009; *Global Times* 2009; Lan 2010; Sipan Li 2010; Qiu 2010; Wang 2009; *Xinhua* 2009b; Xie 2010a, 2010b; Yu 2012).

Journalists and members of the Chinese 'no burn' community have suggested, however, that several of the Chinese experts (and officials) advocating waste incineration are guided by vested economic interests – that there is 'expert-industry collusion' (Xie 2010a). Expert opinions, they claim, are presented to the public as neutral and based on objective research without disclosing the experts' entangled interests with the industry (interviews NGO NU 1, 7-11-12, NGO NU2 29-10-12, NGO NU3 25-7-13, NGO FON1 19-10-12, NGO FON2 21-10-11, Zhao 8-11-12, ASW1 31-7-13; Balkan 2012; Sipan Li 2010; Meng 2009, 2011; Xie 2010a; Xu 2010; Zhai 2010). One of these critics is professor Zhao Zhangyuan, a retired researcher at the Chinese Research Academy of Environmental Sciences (中国环境科学研究院, *Zhongguo huanjing keji yanjiuyuan*) and former deputy director and general-secretary of the Environment Special Committee of the Chinese Geophysical Society (中国地球物理学会环境专业委员会, *Zhongguo diqiu wuli xuehui huanjing zhuanye weiyuanhui*). While not trained as a waste expert, Zhao became engaged in the field of waste incineration through government-commissioned research on the pollution emanating from Beijing's landfills in 2001, which found significant water pollution and disease rates around the landfills (see Chapter Three). Disappointed by the government's reaction to his findings, Zhao started to immerse himself in personal research about waste treatment strategies and incineration and soon became one of China's fiercest critics of incineration technology (interview Zhao 8-11-12; China Daily 2011; D_ASW1 to D_ASW3; *China Speech* n.d.; Huo 2009; Sipan Li 2010; Meng 2011; Wang 2009; Xu 2010; Zhao 2009, 2010, 2011a, 2011b, 2012b).

After his engagement as an anti-incineration expert in several local anti-siting struggles, Zhao reached national public and media prominence when he was the only expert openly opposing incineration at an expert symposium discussing the contested Panyu incinerator and China's waste incineration strategy, held in Guangzhou city in February 2010 (cp. Chapter Three). The symposium, which brought together the leading 32 waste experts across the country – including incineration advocates Nie and Xu – attracted major nation-wide media attention, with most articles explicitly naming Zhao as the sole expert opposing incineration (interview Zhao 8-11-12; *Beijing News* 2010a; *Guangzhou Daily* 2010; Qiu 2010; *Southern Daily* 2010; *Southern Metropolis Daily* 2010; Xu 2010; Ye 2010). After this event, Zhao became the most prominent representative of China's 'anti burn faction' in the ongoing expert debate, with Nie and Zhao often considered as main adversaries in the debate and the quarrel between them becoming quite personal at times (interviews Nie 7-6-13, Zhao 8-11-12; Qiu 2010; Xu 2010).[30]

Zhao, who fundamentally rejects incineration technology, particularly in the Chinese context, has published numerous articles about his viewpoints in academic journals, in the media, and on the internet. His online articles can be found primarily on his personal Sohu blog and Sina microblog accounts, but also on various forums such as the well-frequented Tianya forum and as a guest author on many other blogs and microblogs. After incineration started to become a major public issue in 2009, Zhao was cited in most media reports on the topic as an expert opposing incineration and is frequently portrayed as a 'representative of the "anti burn faction"' ('反烧派'代表, *'fanshao pai' daibiao*) (Bradsher 2012; *China Daily* 2011; Cui 2009; Lan 2010; Meng 2009; Meng 2010d; Wang and Jing 2009; Wang 2014; Wei 2011; Xie 2010; Xu 2010; J. Yang 2010; Yu 2012). Zhao was also featured as an anti-incineration expert in an influential half-hour special feature about incineration and dioxins that was broadcast on the state-owned China Central Television (中国中央电视台, *Zhongguo zhongyang dianshi tai*, CCTV) in the context of the International Dioxin Symposium held in Beijing in August 2009, which was extensively covered in state media and had a major impact on the public debate about incineration (cp. Chapters Three and Four).

Mirroring the main arguments of incineration opponents both in China and abroad, Zhao argues that waste incineration is a waste treatment strategy that only tackles the symptoms of China's waste problems rather than its causes. Instead of attempting to reduce the rapidly growing amounts of municipal solid waste through policies such as waste disposal fees or better garbage sorting and recycling policies, waste incineration only transfers the

30 A similar expert debate related to the Nu River Project is also described by Mertha (2008) for the field of hydropower policies.

pollution problem into the air. Moreover, Zhao and other environmentalists argue that once chosen as primary waste treatment strategy, incineration becomes a one-way-street: due to the high initial and operating cost, incineration goes against more sustainable waste treatment strategies such as reduction and recycling, since the facilities need to be continuously fed large amounts of residue for combustion in order to cover their expenses (interview Zhao 8-11-12; D_ASW1 to D_ASW3; *China Daily* 2011; *China Speech* n.d.; Huo 2009; Sipan Li 2010; Meng 2011; Wang 2009; Xu 2010; Zhao 2009, 2010, 2011a, 2011b, 2012b). According to Zhao and other opponents, incineration causes serious harm to the environment and human health. Regardless of pollution standards, incineration can never be toxin-free, they argue, referring to several international studies by researchers and organizations such as Global Alliance of Incinerator Alternatives/Global Anti-Incineration Alliance (GAIA),[31] the World Bank, the World Health Organization, and Greenpeace. According to all of these studies, incinerators are one of the main sources of dioxin, a highly toxic carcinogen, which accumulates in the human body and causes severe health effects (Allsopp, Costner, and Johnston 2001; Tangri 2003; World Bank 1999).

In light of these general concerns, incineration is particularly unsuited for the Chinese context, caution Zhao and other members of the Chinese 'no burn' community. The frequent use of inferior technology, insufficient standards and regulations, the lack of pollution and emission controls, and the failure to enforce environmental regulations at the local level render incineration irresponsible in China, they warn. In addition, the lack of prior waste sorting and the high content of moist organic materials in Chinese waste – which often necessitates the addition of fuels that further hike emissions or make it altogether impossible to reach the 'safer' temperatures above 850 degrees Celsius – make Chinese waste unsuited for incineration and increases the danger of toxic emissions (interview Zhao 8-11-12; D_ASW1 to D_ASW3; *China Daily* 2011; *China Speech* n.d.; Huo 2009; Sipan Li 2010; Meng 2011; Wang 2009; Xu 2010; Zhao 2009, 2010, 2011a, 2011b, 2012b).[32] For support, Zhao and other opponents point to research that has found high levels of toxic emissions produced by Chinese incinerators (e.g. Ni et al. 2009; Themelis and Zhang 2010).

31 The web address of GAIA is www.no-burn.org (24.03.2016). GAIA is the largest transnational network against incineration, which according to their self-description is a 'worldwide alliance of more than 650 grassroots groups, non-governmental organizations, and individuals in over 90 countries whose ultimate vision is a just, toxic-free world without incineration'.

32 These arguments or parts thereof are also supported by other researchers and environmentalists, including Hu (2015), Ni et al. (2009), Wang and Nie (2001b), and World Bank (2005).

Like a classic policy entrepreneur, Zhao is determined to promote a change in China's waste policies and the creation of stricter standards and regulations, to alert the Chinese public to the dangers of incineration, and to help affected communities fight for their rights and against the siting of incinerators. To these ends, Zhao has in recent years joined ranks with environmentalists and environmental organizations active in the issue field and become an active member of the Chinese 'no burn' community. He has frequently participated in their public outreach activities, held lectures at their forums and symposiums, written articles for their publication outlets, provided expert certification for their work and claims, and joined in their activities at the local level, as outlined in more detail in the following case studies. Understanding himself to be part of the transnational 'no burn' community and having personal ties to several important actors in the global anti-incineration movement, Zhao has functioned as an important diffusion node between the transnational and Chinese 'no burn' communities. In his writings, he has repeatedly placed Chinese anti-incineration contention in the context of transnational 'no burn' activism and has cited and translated parts of the studies that have been influential in the global anti-incineration movement on his Sohu blog (interview Zhao 8-11-12; Zhao 2009, 2010, 2011b, 2012a, 2012b).

Apart from Zhao, other Chinese experts from different backgrounds, including environmental hygiene and medical studies, have also expressed critiques of incineration technology and the extent of its use in China via academic and media articles, and have played an active role in individual struggles at the local level. However, none of these experts have reached a similar level of prominence and influence as Zhao Zhangyuan. Their opposition and engagement has remained largely sporadic or limited to individual cases and the local level. Zhao has thus retained a singular position as the nationally renowned Chinese anti-incineration expert. As such, he resembles United States biology professor Barry Commoner and chemistry and toxicology professor Paul Connett, who have played central roles in the anti-incineration and environmental justice movements in the United States (Sherman 2011a; Walsh, Warland, and Smith 1997). Indeed, Professor Zhao is personally acquainted with Paul Connett, who has visited China on several occasions. In 2011, Zhao and his wife visited the Connetts during a trip to the United States during which they also met with staff from the Global Alliance of Incinerator Alternatives (GAIA), the largest transnational network of environmental organizations and individuals against incineration (interviews Zhao 8-11-12, NGO GAIA 4-7-13).

A (Trans-)National Issue Network: Environmental Organizations in the Issue Field

The second main group within the Chinese 'no burn' community consists of environmental organizations that are active in the field of incineration. While most of the organizations that were actively working on incineration issues early on were based in Beijing, a (trans-) national issue network dedicated to China's waste policies including incineration has developed during the past few years. In the early 2000s, several Chinese environmental organizations started working on waste issues, focusing mainly on waste separation, recycling, and hazardous waste. In China's major cities, primarily in Beijing, these environmental organizations started community waste sorting projects and hoped to promote sustainable and environmentally friendly waste treatment measures (interviews NGO Aifen 15-5-13, NGO ECO1/PY1 1-7-13, NGO GVB 18-10-11, NGO FON1 19-10-12, NGO FON2 21-10-11, NGO Boell 25-10-12, NGO IPE 13-10-11, NGO NU1 11-10-11, NGO NU2 18-10-11, NGO NU3 25-7-13, NGO PacEnv 27-10-11, NGO SEE 14-10-11, ASW 17-10-12).

Incineration was not an issue tackled by environmental organizations until the first local campaigns against planned incinerators – beginning in 2006/2007 with the first major anti-incinerator campaign staged by the residents of Beijing's Liulitun neighborhood – alerted the organizations to the issue. While several Beijing-based environmental organizations were approached by campaigners of Beijing's urban anti-incinerator campaigns (Liulitun, Gaoantun, and Asuwei), the organizations initially declined to engage in the local struggles for reasons of self-protection and to avoid increasing the risk for the campaigners by getting involved as organizations (cp. Chapter Three). When the number of local struggles against incinerator plants started to increase significantly in 2009, in combination with transnational influences, some environmental organizations started to pay attention to the issue and soon became actively involved in the issue field.

Individual Chinese environmentalists established the first linkages with the transnational anti-incineration community in 2007 and brought these influences to China. One of them was the young environmentalist Mao Da, who at the time was working on a project about chemicals for the Beijing-based environmental organization Beijing Global Village Environmental Education Center (北京地球村环境教育中心, *Beijing diqiu cun huanjing jiaoyu zhongxin*, hereafter referred to by its common English name Global Village of Beijing). During a research stay in the United States, he came into contact with several transnational networks and organizations active in the field of waste incineration, including GAIA and the International POPs Elimination Network (IPEN), a transnational network of organizations dedicated to establishing and implementing safe chemical policies and practices around the world. Through these contacts he learned about the problem of incineration and the main arguments of the transnational anti-incineration community. The Liulitun campaign alerted him to the issue in the Chinese context. When Liulitun and later Asuwei campaigners contacted Global Village of Beijing, he sent them information and materials from the United States (cp. Chapter Three) (interviews NGO NU3 25-7-13, NGO NU3 26-3-14; Mao 2010, 2011, 2012, 2013; Mao and Zhao 2010).

After his return to China in 2009, Mao took up doctoral studies in environmental history in Beijing and decided to research and write about China's waste problem, delving deeper into the issue and soon gaining expert-level knowledge about the topic. During this time, local campaigns against incinerator projects started to emerge more frequently and incineration started to enter the public discourse. Through his engagement with both the issue of waste in China and the history of incineration abroad, Mao became more and more critical of China's waste policies and the country's push for incineration, as well as the weak regulatory framework and lacking implementation of environmental standards and regulations. While not fundamentally an opponent of incineration technology, his stance that incineration at the scale envisioned by the Chinese government is unsuitable for the Chinese context resembles that of professor Zhao (interviews NGO NU3 25-7-13, NGO NU3 26-3-14; Mao 2010, 2011, 2012, 2013; Mao and Zhao 2010).

Also in 2009, environmentalist and *Guangming Daily* reporter Feng Yongfeng founded the environmental organization Green Beagle Environment Institute (达尔问环境研究所, *Daerwen huanjing yanjiusuo*, hereafter referred to as Green Beagle) in Beijing. Mao Da started to work for this organization as a waste expert, where he was soon joined by another colleague dedicated to China's waste problems, Chen Liwen. While also working on other issues, Green Beagle soon became the first environmental organization to have a special focus on incineration. Since providing assistance to affected local communities was at the core of the organization's mission, Green Beagle was also the first organization to become actively involved in anti-incinerator struggles at the local level. When part of the organization's staff broke away from Green Beagle and founded the organization Nature University (自然大学, *Ziran daxue*) in 2011, Mao and Chen were among the staff members who moved to the new organization and took over the lead of the organization's 'School of Waste' (垃圾学院, *Laji xueyuan*),[33] which specialized in dealing with waste issues including incineration. While the name of the central organization active in the issue field changed in 2011, the main staff members remained the same. Since Nature University was not yet established in the early phases of the case studies outlined in the following chapters, the staff members are referred to as Green Beagle staff throughout the book in order to avoid confusion.

Like Mao Da, the other Green Beagle/Nature University staff members were not fundamentally opposed to incineration, but instead concerned about its safe implementation in China. Rather than seeing incineration as a primary waste treatment strategy, they favoured more sustainable or 'zero waste' policies. 'Zero waste' policies are based mainly on the 'three Rs' of reduction, reuse, and recycling, and aim to leave no trash to be sent to landfills and incinerators. The organization members regarded the issue of waste incineration as an epitome of the broader environmental problems and regulatory failures in China, which the believed could be tackled via their engagement in the field of incineration. This included the lacking local implementation of environmental laws and regulations (such as the EIA or information disclosure laws and the control of emission standards and pollution), lacking transparency and public participation in the environmental arena, and the failure to guarantee the rights of pollution victims and communities affected by environmental problems (interviews NGO NU1 11-10-11, 7-11-12, NGO NU2 29-10-12, NGO NU3 25-7-13, 26-3-14). In the words of one organization staff member:

> Waste incineration is a public service and incinerators are public projects. So apart from the environmental and health risks particular to waste incineration, the other problems are the same as with all public projects. Land problems, regulatory problems, corruption problems, they are all there. So a lot of successful opposition against incinerators has actually not produced evidence of its environmental or health problems, but of typical other public regulatory problems. (Interview NGO NU3 25-7-13)

The organizations' activities were not aimed only at pushing for more sustainable waste policies, but also at disclosing broader environmental

33 Nature University is composed of different 'schools' specialized on various issues. Apart from the 'School of Waste', there are also the 'School of Birds and Beasts' (鸟兽学院, *Niaoshou xueyuan*), 'School of Native Soil' (乡土学院, *Xiangtu xueyuan*), 'School of Health' (健康学院, *Jiankang xueyuan*), 'School of Vegetation' (草木学院, *Caomu xueyuan*), 'School of Mountains and Rivers' (山川学校, *Shanchuan xueyuan*), and 'School of Parks and Forests' (园林学院, *Yuanlin xueyuan*).

problems and regulatory failures. By assisting local communities in their fight against planned or operating incinerators, they hoped to point out broader regulatory failures and environmental problems that were typical across different cases and to increase pressure on the government to retreat from their massive push for incineration (interviews NGO NU1 11-10-11, 7-11-12, NGO NU2 29-10-12, NGO NU3 25-7-13, 26-3-14; Zuo 2013) – again typical aims of policy entrepreneurs as described by Mertha (2008).

Through Mao's personal relations with GAIA and IPEN, Green Beagle/Nature University also established a close cooperation with the two transnational networks, including joint conferences, workshops, and projects. Among others, IPEN and Green Beagle were partners in the European Union-funded 'China Chemical Safety Project', which had the aim of 'strengthening the capacity of pollution victims and civil society organizations to increase chemical safety in China' and dealt with the local anti-incineration campaign against the Likeng incinerator in Guangzhou's Yongxing village (永兴村, *Yongxing cun*) (IPEN and Green Beagle 2015). During several of the organizations' activities, members of GAIA and IPEN have functioned as international experts certifying the Chinese organizations' claims and assessments, such as regarding the implications of the Stockholm Convention for waste incineration. Moreover, some of the organization's activities have also received funding from the transnational networks (ibid.; interviews NGO GAIA 23-6-13, 4-7-13, NGO IPEN1 22-6-13, NGO IPEN 2 23-6-13; D_IPEN1, D_IPEN2).

Green Beagle/Nature University was soon joined by another Beijing-based environmental organization in the waste and incineration realm. While not specifically focused on waste incineration, Friends of Nature (自然之友, *Ziran zhiyou*), one of China's oldest environmental organizations which had been working on waste issues for several years, joined forces with Green Beagle/Nature University and Global Village of Beijing for some incineration-related activities that touched upon broader waste problems and policies or more general environmental problems. This included their engagement in several local anti-incinerator struggles, particularly those located in and around Beijing. Friends of Nature also joined various national-level campaigns for more sustainable waste treatment strategies and for heightened attention to regulatory failures such as inadequate implementation of the EIA or information disclosure laws. Apart from their close cooperation, overlaps in personnel also happened, among other instances when Mao Da temporarily joined Friends of Nature for some project-related work. Like Green Beagle/Nature University, Friends of Nature does not entirely reject incineration, but instead opposes use of the technology under Chinese conditions and at the scale promoted by the Chinese government. Like the other organizations, Friends of Nature regards the issue as exemplary of broader environmental problems and regulatory failures. A central focus of the organization's work is transparency and the attainability of reliable environmental information and pollution data (interviews NGO FON1 19-10-12, 6-6-13, NGO FON2 21-10-11, NGO FON3 27-9-11, FON4 25-10-12, NGO NU1 11-10-11, NGO NU2 29-10-12, NGO NU3 25-7-13; FON 2011a, 2011b, 2011c, 2011d). As a staff member summarized their standpoint:

> We don't completely oppose incineration. But under the current conditions in China, with no waste separation and a kitchen waste rate of over 60 percent, with loose regulations and without supervision, under these conditions we oppose incineration. Moreover, such blind and large-scale construction, we also think that's not acceptable, because at a certain level it hinders the development of waste separation. But the most important thing at the moment is information disclosure. We hope to urge the government for better supervision by pushing for information disclosure. And via making information public, we hope to let the public understand [...] the real situation of incineration and to slow down the current trend of constructing incinerators in China at a massive scale and by all means. (Interview NGO FON1 19-10-12)

In 2009, the community was joined by another newly founded organization, which specialized in waste, particularly incineration, issues and consolidated the linkages with the transnational anti-incineration community. In October 2008, the environmental organization Wuhu Ecology Center (芜湖生态中心, *Wuhu shengtai zhongxin*) was founded in Wuhu city (芜湖市, *Wuhu shi*), Anhui Province (安徽省, *Anhui sheng*), with the help and under the lead of GAIA, who was introduced to the new organization through a Pacific Environment staff member. In 2009, the newly founded organization received some funds from GAIA to establish the 'China Waste Information Network' (CWIN, 中国垃圾信息工作网络, *Zhongguo laji xinxi wangluo*), a project with the aim of information sharing and capacity building for Chinese environmental organizations working in the waste realm. The project included the regular collection of news on waste-related issues and its distribution to interested environmental organizations and individuals via an email group. The organization thus played a crucial role in promoting the waste issue among Chinese environmental organizations and in fostering a loose network of the organizations that were active in the issue field (interviews NGO WEC 23-4-12, 23-4-13, NGO FON1[34] 19-10-12, 6-6-13, NGO GAIA 4-7-13, NGO PacEnv27-10-11, NGO Green Anhui 27-9-12, 23-10-12).

Under the influence of GAIA, who trained the new organization's staff members during work visits in China and abroad, the Wuhu Ecology Center soon started to focus more on incineration issues. In 2011 – and again typical of the strategies employed by policy entrepreneurs – the organization launched the 'Municipal Solid Waste Incineration Information Platform' (生活垃圾焚烧信息平台, *Shenghuo laji fenshao xinxi pingtai*), an online platform providing extensive information on China's incineration situation, including information on Chinese MSWIs, related laws and regulations, NGO activities, and local contention against incinerators across the country.[35] Acknowledging the complex and contested nature of incineration and the lack of publicly attainable independent information about the issue, the aim of this platform is to provide both Chinese environmental organizations and the Chinese public with accessible information about the situation and harms of waste incineration in China (interviews NGO WEC 23-4-12, 23-4-13, NGO FON1 19-10-12, 6-6-13). Since 2011, the organization has also been involved in several local anti-incinerator struggles and led or participated in many activities in cooperation with other organizations active in the issue field, including several national-level issue campaigns. Among others, Wuhu Ecology Center joined Friends of Nature to apply for the disclosure of emissions data from 122 waste incineration plants and 103 local environmental protection bureaus. The disclosed information showed that many plants' emissions exceeded national standards, and the organizations used these results to pressure the Ministry of Environmental Protection for stricter standards and a better enforcement of existing regulations (D_WEC1, D_WEC2; WEC 2011, 2012, 2013; Zhang 2015).

Through their close contacts with GAIA, Wuhu Ecology Center has also functioned as an important communication node between the transnational and Chinese 'no burn' communities. The organization has invited GAIA staff to China on several occasions, including for presentations of their work and stance on incineration at workshops and conferences that were well-visited by Chinese environmental organizations and individuals active in the waste realm, including affected community members. This has helped raise the domestic awareness of incineration and its related

34 Since one core Wuhu Ecology Center staff member moved to Friends of Nature in 2011, she is here cited with reference to her current organization's affiliation.

35 The platform can be found at: http://www.waste-cwin.org/ (28.04.2016). By April 2016, it had more than 2.6 million visits.

harms. Like Nature University and Green Beagle, Wuhu Ecology Center also disseminates Chinese translations of materials produced by the transnational anti-incineration community, such as GAIA's influential study 'Waste Incineration: A Dying Technology' (Tangri 2003).

In late 2011, the Chinese organizations took a further step to consolidate the national 'no burn' community by founding an official network dedicated to waste issues. At a conference in Beijing, several Chinese organizations including the core group of Wuhu Ecology Center, Green Beagle, and Friends of Nature decided that the complex Chinese waste problem could only be solved if tackled in a more comprehensive manner. While many Chinese environmental organizations were engaged in the issue field, they specialized in different parts of the waste stream and each focused on their individual activities without 'seeing the big picture' or combining their expertise and activities into a comprehensive approach. In December 2011, the organizations came together for another conference in Guangzhou and – supported by some funds from the German NGO Asia House Foundation – founded the Chinese 'Zero Waste Alliance' (零废弃联盟, *Ling feiqi lianmeng*). The alliance started out with about 20 members, all Chinese environmental organizations from across the country that were active in the waste realm. The alliance's goal is to promote 'zero waste' policies in China by bringing together the organizations' individual expertise and integrating their activities into a broader perspective or strategy, including joint action (interviews NGO WEC 23-10-12, NGO NU1 7-11-12, NGO NU2 8-5-13, NGO NU3 25-7-13, NGO FON1 19-10-12, NGO GAIA 23-6-13, NGO ECO1 1-7-13, NGO ECO2 17-6-13, NGO GCB 5-10-12; D_FON1).

With the foundation of the Zero Waste Alliance and its dedication to 'zero waste' policies, the Chinese organizations also linked up with a transnational 'zero waste' community, which was closely related to the transnational 'no burn' community, whose members are generally in favour of 'zero waste' policies. GAIA, for instance, officially supports and actively promotes 'zero waste' approaches and has also actively supported the Chinese Zero Waste Alliance. The 'zero waste' concept was established in the 1970s in the United States and has since moved to the core of sustainable waste approaches. Today, Zero Waste Alliances exist in many countries, united under the umbrella of the Zero Waste International Alliance.[36] In China, it was again

36 At a transnational level, zero waste alliances or groups include the Zero Waste International Alliance (http://zwia.org/) and Zero Waste Europe (https://www.zerowasteeurope.eu/). At a national level, zero waste alliances or groups exist in Germany, the UK, Ireland, France, and Brazil. In the United States, several cities have their own groups or alliances.

the young environmentalist Mao Da with his close transnational ties who promoted the idea among Chinese organizations and suggested the founding of a Chinese Zero Waste Alliance (interviews NGO WEC 23-10-12, NGO NU1 7-11-12, NGO NU2 8-5-13, NGO NU3 25-7-13, NGO FON1 19-10-12).

By 2013, the number of member organizations in the Chinese Zero Waste Alliance had grown to 53, networked among others through a very active email list. While not specialized in incineration issues, a central aspect of the 'zero waste' approach – and thus the alliance – is finding alternatives to waste incineration and opposing incineration, either fundamentally or when used on a large scale. While most of the member organizations do not specialize in incineration issues, they nonetheless provide a valuable network for the organizations that specifically deal with incineration matters. At a national level, these are primarily the above-mentioned environmental organizations with Green Beagle/Nature University, Friends of Nature, and Wuhu Ecology Center at the core. Members of the Zero Waste Alliance have joined forces for several incineration-related national issue campaigns, as outlined in more detail in the following chapters. The network has also been a valuable resource for the primarily Beijing-based organizations that are active in the incineration field. When dealing with local communities affected by incinerators in other parts of the country, the core organizations frequently point the local residents to alliance member organizations in the region for assistance. The Beijing-based organizations can then provide support via these local or regional partners, as has been the case in several of the Shanghai and Guangzhou cases investigated for this study. To facilitate this practice, Wuhu Ecology Center has compiled an interactive map indexing and providing brief information about all Chinese Zero Waste Alliance member organizations. The map was published on their 'Municipal Solid Waste Incineration Information Platform' to point local incinerator-affected communities looking for support to the closest organization dealing with waste issues. As a Friends of Nature staff member described the role of the Zero Waste Alliance:

> Since we have established the Zero Waste Alliance, we can better help our companions in different [local] cases. [...] If [affected communities in] other places have problems, we can encourage local environmental organizations to go and see them. If we go everywhere, then the costs are too high, so there are a lot of places where we can't come to their help. [...] But since we founded the Zero Waste Alliance in 2011, our approach in this regard [of assisting local communities] is to particularly focus on this aspect [of capacity building and encouraging local NGOs to get involved].

> If no organizations are there, or if they don't dare to get involved, then we still go. But we really urge the closest organization to go there. That way the extent of how we can help local residents has really increased tremendously. (Interview NGO FON1 7-11-12)

While the number of Chinese organizations specialized in waste incineration remains limited – particularly those who have gained national relevance in the issue field – they are now supported by a vast country-spanning network of Chinese organizations dealing with broader waste issues. This includes newly founded organizations that have emerged from local anti-incinerator campaigns (cp. Chapter Three). Moreover, the Chinese 'no burn' community has experienced major influence and support from the transnational 'no burn' and 'zero waste' communities and is well embedded in the transnational anti-incineration movement that has shaped the development of anti-incineration activism in China. (Sub-)national anti-incineration movements in other countries and regions, particularly those in Japan and Taiwan, have had a major impact on the Chinese 'no burn' community. While no close personal ties have been established with activists from these movements, the core Chinese organizations have benefitted from their experiences via shared information and materials. Among other things, many of the Chinese-language materials distributed to affected communities by Chinese organizations were written or translated by Taiwanese anti-incineration activists and have proven to be valuable resources for the Chinese 'no burn' community.

Fighting for Environmental Rights: (Environmental) Lawyers and Legal Associations

A third group within the Chinese 'no burn' community consists of (environmental) lawyers and legal associations. While concerned less with incineration or waste issues per se but instead with the advancement of environmental litigation and the rights of affected communities and pollution victims, they have become important allies for the experts and environmental organizations in several local anti-incineration disputes. The major legal organization that has been engaged in incineration-related cases is the Beijing-based organization Center for the Legal Assistance to Pollution Victims (CLAPV). The organization is dedicated to supporting pollution victims who cannot afford other means of legal representation and was founded in 1998 by staff from the law department at Beijing's China University of Political Science and Law (中国政法大学, *Zhongguo zhengfa daxue*). CLAPV has taken over the legal representation of or provided legal assistance to affected communities in several incineration-related cases.

A handful of individual Beijing-based lawyers have also become engaged in numerous incineration-related cases. Based on their successful engagement in some well-known early anti-incineration struggles, such as the Liulitun campaign in Beijing, they have been frequently contacted for legal assistance by other local communities across the country. Through their growing reputation as 'anti-incinerator' lawyers and via their joint engagement in several local cases, the lawyers and CLAPV started to develop close relations with environmental organizations and experts that were active in the issue field. They are now frequently called upon by other members of the 'no burn' community for both legal assistance in individual cases and as legal experts regarding environmental and waste-related laws and regulation. The lawyers and CLAPV members have also participated in many of the environmental organizations' activities, including functioning as legal experts for some of the organizations' campaigns, participating in and giving lectures at their workshops and conferences, and writing articles in their publication outlets. Despite their diverging goals and interests, they have thus become an active part of the Chinese 'no burn' community.

A Wave of Local Resistance: Local Community Contention Against Waste Incinerators

Together with the opposition to the Chinese government's incineration policies from the side of the 'no burn' community, resistance to Chinese incineration plants has – like in many other countries – also emerged at the local level. Local communities living in the surroundings of planned or operating facilities soon started to stand against the plants at their doorsteps. While these communities are often primarily guided by individual grievances rather than broader anti-incineration or environmental concerns, they have nonetheless become a significant counter-force to the Chinese government's incineration plans.

The first major local anti-incinerator campaign was staged by residents of Beijing's Liulitun neighbourhood in 2006/2007, which lead to a relocation of the project (cp. Chapter Three). After this campaign, which received nationwide public and media attention, a wave of local struggles against incineration plants started to spread across the country, reaching a first peak in 2009 and 2010 when a large number of new incinerator projects were announced. The systematic collection of cases via media analysis and field research has shown that at least 39 cases of urban and rural opposition against planned and operating waste incinerators had occurred by mid-2013 (cp. Appendices III and IV). This contention has taken manifold forms – from legal means like petitions and lawsuits to more disruptive means like sit-ins and large-scale street protests.

The first local anti-incinerator struggles were primarily staged by urban residents against planned facilities in China's major cities along the East coast such as Beijing, Shanghai, Nanjing (南京市, *Nanjing shi*), and Shenzhen. This can party be attributed to the fact that China's first incinerators were largely planned in or around these cities. With the spread of incinerator projects across the country, urban conflicts over the siting of incinerator projects have since reached smaller cities and the inland. The often-well-educated home-owners behind such urban campaigns are frequently regarded as members of the Chinese 'middle class' by both the media and academic observers.

This kind of contention often centres on concerns about property values and anticipated health effects. Moreover, as with other cases of environmental contention, most of these struggles also touch upon common regulatory failures such as the lack of transparency, lax EIA implementation, failed communication between residents and the government, or local corruption issues. Conflicts about waste incinerators are often also linked with ongoing struggles about prior waste facilities in the vicinity, such as landfills. In these cases, existing grievances about the pollution or stench from the earlier facilities tends to facilitate mobilization against yet another – potentially more harmful – waste facility. While the repertoire of actions used in the urban campaigns is diverse, peaceful 'strolls' (散步, *sanbu*) that bring hundreds or even thousands of residents to the streets have been a common part of the repertoire (cp. Chapter Three).

The urban anti-incinerator struggles have attracted major public and academic attention and are often cited as examples of a new 'Chinese NIMBY activism' similar to that seen in Western societies (e.g., Cui 2011; Huang and Yip 2012; Huo 2013; Johnson 2010; Lang and Xu 2013; J. Liu 2013; Otsuka 2009; Tang 2013; Wasserstrom 2008; Xia 2014). The Liulitun (2006/2007) and Asuwei (2009) campaigns against planned MSWI projects in Beijing, and the struggles against planned incinerators in Guangzhou's Panyu district (2009) and Shanghai's Songjiang district (2012) – all of which lead to a (temporary) project halt or relocation – tend to be brought up as prominent examples of a new kind of local Chinese environmental contention. As such, they are often compared to other large-scale anti-siting conflicts such as those in Xiamen (2003), Dalian (2011), Shifang (2012), Qidong (2012), and Kunming (2013), and have also received significant attention in the Chinese media.

While reporting on the Songjiang case – regarded as the largest public protest in Shanghai since the protest against the extension of the city's magnetic levitation train in 2008 – was mainly limited to English-language and international media outlets, in Chinese-language state media the Liulitun, Asuwei, and Panyu cases have been hailed as good examples of successful communication between an aggrieved public and the government (cp. Chapter Three). This media reporting has significantly facilitated the dissemination of information about these cases.[37]

Local resistance against waste incinerators has not been restricted to urban areas, however. While it receives less media attention, local opposition against planned MSWI plants has also mounted in the countryside (cp. Appendix III). The growing number of rural (and peri-urban) cases can in part be attributed to a shift in the siting of projects. Partially in response to growing urban opposition, governments and operators have in recent years moved the sites of new projects away from the city centres towards smaller villages in the outskirts, since less resistance is expected from resource-poorer villagers with restricted access to information. Moreover, the implementation of environmental regulations in these areas is often regarded as less strict. Facilities that have been obstructed by urban residents in particular are frequently moved to rural areas (such as the Liulitun, Panyu, and Songjiang projects).

Rural communities standing against anticipated pollution is a rather new development in Chinese environmental contention. In the countryside, pollution traditionally had to reach clearly discernible levels to spark action by the local population (Deng and Yang 2013; Jing 2000). Mobilization against anticipated environmental risks or health effects requires a certain level of awareness and information. However, in the case of waste incinerators the onset of contention is often triggered less by environmental or health concerns, but rather – similar to in urban communities – by property and monetary claims, mostly grievances about project-related land requisition or concerns about relocation measures and compensation payments. Regulatory and corruption issues similar to those expressed in the urban campaigns also lie at the core of many rural struggles. In several cases, the fear of anticipated environmental pollution or health effects from the

37 While the residents in Guangzhou's Huadu district (花都区, *Huadu qu*) staged contentious action nearly simultaneously to the Panyu residents, their case never attracted a similar level of (media) attention – much to the dismay of the residents (interviews HD1 2-7-13, HD2 7-7-13, HD3 8-7-13). Nonetheless, both communities eventually succeeded in achieving a relocation of the planned facilities within Guangzhou city.

planned project emerged after mobilization against the facility had already commenced. Moreover, as in urban areas mobilization against a planned MSWI plant is often facilitated by prior grievances related to previous environmental pollution in the area.

As in the urban cases, the modes of contention employed by rural communities are diverse and include everything from legal means to large-scale sit-ins or road blockades that have in some cases lead to serious scuffles with public security forces. Under the influence of the Chinese 'no burn' community, which urges local communities to avoid disruptive and potentially violent action, a shift towards legal means has been observed in recent years. In several cases, villagers have also closed ranks with the residents and owners of apartment complexes in the area, leading to collaborations across different social groups. In other cases, social cleavages have emerged between the different affected groups (cp. Chapters Three and Five).

Media reporting about rural anti-incinerator contention has mainly occurred in the context of environmental litigation cases where rural communities have used legal means to fight for their rights, such as in the case of Hebei Province's Panguanying village (cp. Chapter Four) or the series of environmental litigations by villager Xie Yong from Jiangsu Province's Hai'an county (海安县, *Haian xian*) under Nantong city (南通市, *Nantong shi*) jurisdiction. Some reporting has also occurred in light of the broader environmental and regulatory problems reflected in the individual cases. When rural cases feature in Chinese media, it is often a sign that the affected communities have received the support of the Chinese 'no burn' community, one of whose central strategies is to raise media attention towards the individual cases via their personal networks and outreach activities – similar to the policy entrepreneurs described by Mertha (2008). Most rural cases have only been mentioned in English-language or international media, or have not been reported at all. This is particularly applicable to cases that involve large-scale protests or clashes with security forces, such as that in Huangtutang village (黄土塘村, *Huangtutang cun*), Wuxi city (无锡市, *Wuxi shi*) in Jiangsu Province, where villagers blocked the construction site of a planned incinerator for weeks in March and April 2011 and got into a violent clash with security forces in May (interviews WX 6-5-13, ASW1 31-7-13, NGO NU4 8-11-12, NGO NU1 7-11-12, NGO NU3 25-7-13).

While the majority of local anti-incinerator conflicts are directed against planned or not yet completed plants, in recent years opposition against already operating plants has also increased. Apart from complaints about the smell and air pollution coming from these plants, such resistance often revolves around health effects – mostly high cancer or respiratory disease

Figure 2.1 Guangzhou Likeng incinerator, July 2013

rates – that are attributed to nearby incinerators.[38] While health effects are extremely hard to trace to individual sources such as MSWI plants – particularly since these are often located in areas with high pollution levels from other industry or waste facilities in the area – they have nonetheless become a major cause of local resistance. The first case that brought the issue of health effects related to incineration and particularly dioxin to the public attention and had a major effect on the ongoing debate about the pros and cons of incineration, was the struggle against the Likeng incineration plant in Guangzhou's Yongxing village (永兴村, *Yongxing cun*) (see Figure 2.1).

After the Likeng plant started operating as Guangzhou's first modern waste incinerator in 2005, cancer rates in the nearby village reportedly skyrocketed. While the villagers had been complaining about the plant for years, their contention reached new heights and national media attention in 2009 and 2010 after they received support from the urban Panyu district residents, who mounted their own campaign in late 2009 and referred to the Likeng incinerator as an example of the negative effects of incineration. The

38 The expansion of existing plants (such as the construction of a second incinerator) has also in several cases triggered contention that encompasses resistance against already operating plants.

media reporting about the case – mainly in the context of the well-reported Panyu case – sparked a nationwide discussion about the health risks of incineration, especially among experts (interviews LK 4-7-13, LK 9-7-13, NGO ECO1/PY1 1-7-13, NGO ECO1/PY1 10-7-13, GZ expert1 30-6-13, GZ expert2 2-7-13, Zhao 8-11-12; IPEN and Green Beagle 2015; Qiu 2009, 2010; Tam 2009; Wang and Li 2009; Zheng and Xuan 2010; Zuo 2013).

Another well-known case of health-related incinerator contention is the above-mentioned series of environmental litigations by the villager Xie Yong from Jiangsu Province. Since 2010, he has sued several parties, even up to the Ministry of Environmental Protection, after his son was born with cerebral palsy. Xie Yong and his wife attributed this disease to a nearby waste incinerator after a clinic doctor pointed to pollution as a potential cause. Xie received the support of several members of the Chinese 'no burn' community and the case attracted significant media attention. While the case has not been officially resolved, Xie and his family reached an out-of-court agreement with the company operating the plant in 2013 and received a monetary compensation in 2014 (interviews Xie 22-6-13, NGO CLAPV 18-10-12, NGO NU2 8-5-13, 26-3-14, 1-7-15; D_XYLC1; Balkan 2012). Health issues have also played a role in the Gaoantun case, where one of the main activists suffered from a severe respiratory disease which she attributes to the nearby (medical waste) incineration plant (interviews GAT 1-11-12, 15-6-13). Health problems also lie at the core of more recent contention by villagers in Hangzhou city's (杭州市, *Hangzhou shi*) Binjiang district (滨江区, *Binjiang qu*; since 2013, high cancer rates) and residents of Wuhan city's (武汉市, *Wuhan shi*) Yongfeng township (永丰乡, *Yongfeng xiang*) in Hanyang district (汉阳区, *Hanyang qu*; since 2013, significant occurrence of respiratory diseases). Both cases included major street protests (interviews HZ 23-6-13, WH 4-7-15, NGO NU2 1-7-15).

While several of the urban campaigns against planned or not yet completed waste incinerators (such as in Liulitun, Shenzhen, Songjiang, Huadu, and Panyu) could be regarded as 'successful' since the residents succeeded in obstructing the projects, the notion of 'success' is not so clear-cut. In several urban cases, the affected communities were more diverse and included not only homeowners but also villagers around the project sites that were largely located at the city's outskirts. While the homeowner communities are often rather homogeneous in their opposition against the projects, the interests of affected villagers and other social groups tend to be more diverse and can lead to strong cleavages that can render a campaign a success for one group but not for others (cp. Chapters Three and Five).

Moreover, many of the incinerator plants that were obstructed by urban residential communities were then moved to rural areas in the region where less opposition was expected – and, indeed, did not occur in most cases – from resource-poorer rural communities (e.g., in Liulitun, Songjiang, Panyu, and Huadu). In such cases, the success of one social group comes at the cost of another. This has sparked a fierce debate about social and environmental justice among China's 'no burn' community (interviews NGO WEC 23-10-12, NGO NU1 7-11-12, NGO NU2 8-5-13, NGO NU3 25-7-13, NGO FON1 19-10-12, Zhao 8-11-12, Xia 30-7-13, EnvJourn 23-4-13, ASW1 31-7-13).

Success is also hard to pin down in many peri-urban and rural cases and where pollution and health effects have already occurred. As scholars studying popular responses to pollution in China have pointed out, claims are often strategically adapted to what seems feasible in order to raise the contenders' chances of success (Deng and Yang 2013; Lora-Wainwright 2017). It is, for instance, questionable whether the local government's compliance with relocation or compensation demands can count as a thorough success if at least parts of the affected communities would have preferred to stay in the region but did not believe they had a chance for success or failed in their efforts to stop the project and hence settled for compensation claims (such as in Dagong or Likeng, cp. Chapter Five). As in other cases of environmental contention (cp. the contributions in Lora-Wainwright 2013c), success is difficult to pin-point and a complex matter in anti-incineration conflicts.

Overall, the developments in local anti-incineration contention during the research period point to a 'protest wave', as suggested by social movement literature. In August 2013, a *South China Morning Post* article warned that the country should count on growing opposition to newly announced incinerator plants due to spreading concerns about incineration-related health impacts by an increasingly well-informed public (Li 2013). In a similar vein, staff members of Green Beagle/Nature University, Friends of Nature, and Wuhu Ecology Center who systematically collect data on local anti-incinerator struggles noted that they were observing a rising number of local struggles in 2013 and 2014, suggesting a second peak after the 2009/2010 climax. Together with other members of the Chinese 'no burn' community, they also reported a significant increase in the number of local communities contacting them for assistance. Moreover, through numerous site visits to incineration plants across China the organizations found that many local cases of contention, particularly in rural areas, never reach the public's attention, hence suggesting a significant number of un-noted cases (interviews NGO NU2 1-7-15, NGO FON1 6-6-13, NGO WEC 23-4-13, Zhao 8-11-12, Xia 30-7-13).

It is important to point out, however, that when taking into account the massive number of planned and operating incinerator projects across the country, it is still the minority of affected communities that take action against the plants – even when assuming a substantial number of unknown cases. Moreover, since 2014 fierce public security reactions to more recent cases of contention – particularly those involving large-scale street protests like in Hangzhou's Binjiang district (2014), Guangdong Province's Luoyang town (罗阳镇, *Luoyang zhen*) in Buluo county (博罗县, *Buluo xian*) (2014), or Hubei Province's (湖北省, *Hubei sheng*) Xintao city (仙桃市, *Xintao shi*) (2016) – have been reported, and it is questionable how Chinese anti-incineration contention will continue to develop in light of the current political situation under Xi Jinping, who has served as the General Secretary of the Chinese Communist Party since 2012 and President of the People's Republic of China since 2013.

Overall, there are significant signs of diffusion processes and linkages between local anti-incinerator conflicts that had occurred by the time of research. An understanding of diffusion processes cannot be based on a macro-level investigation, however, since diffusion can easily be confused with 'the simultaneous or near-simultaneous emergence of contention in structurally similar situations' (Tarrow 2010: 204). While diffusion processes tend to generate waves or cycles of events, not all waves of events arise from diffusion processes. Rather, they might be independent responses to external events (Oliver and Myers 2003: 175). Hence, 'the outbreak of protests in multiple sites around a common set of issues or claims [...] does not suffice to demonstrate that diffusion has occurred [...]. Diffusion requires that movements across these multiple sites are linked together through activist networks, or at least informed and inspired by media-transmitted images or shared cultural understandings of popular struggles in other settings' (Givan, Roberts, and Soule 2010: 6).

The spatial and temporal shifts in local anti-incinerator contention that can be observed from a macro perspective could thus be caused by the changing number and location of announced projects – itself a response to the country's waste policies – or by changes in political opportunity structures, without being based on actual linkages or diffusion processes between the individual communities. A comprehensive understanding of the dynamics of diffusion within in the Chinese political context and of the role of the 'no burn' community at the local level therefore needs to be based on a more detailed investigation. The following three chapters present three case studies that draw out the nature and role of the horizontal and vertical linkages behind networked contention in more detail.

3 Learning between Homeowners

The Urban Case of Beijing Asuwei

The urban case of Beijing Asuwei shows how resource-rich urban communities can learn from each other, and how localized claims and grievances can transcend the local level and produce policy effects at higher political levels. This is the third in a line of resistance movements against operating or planned incinerator projects by urban residential communities in Beijing. Since these communities had close contacts with each other and various phases of their struggles overlapped, the Asuwei case has to be understood in the light of this broader development. The following section, which also provides the background for the Dagong village case outlined in Chapter Five, thus briefly introduces Beijing municipal waste policies and related contention before turning to the case of Asuwei. The Asuwei residents' linkages with other contentious communities and intermediaries are outlined in the context of the unfolding events, followed by an analysis of the role of these ties for the Asuwei struggle. I refer to the three cases of local contention against incinerator projects staged by urban communities in Beijing as 'homeowner' resistance, both because this is a common term used in the literature when referring to social contention by China's urban residential communities (Cai 2005; Cai and Sheng 2013; Chung 2015; Wang et al. 2013), and because the term *yezhu* (业主, 'homeowner') was a central term used for self-reference by the contentious community members under study.

Setting the Stage: Waste Incineration and 'Homeowner' Resistance in Beijing

With rapid urbanization and the expanding population, by the 1990s the treatment of household waste had become an urgent political issue for the Beijing municipal government. Garbage had for the last decades been dumped without treatment in hundreds of often illegal landfills around the city's outskirts. Referencing the city's six ring roads, these landfills were termed Beijing's 'seventh ring of garbage' by artist and filmmaker Wang Jiuliang in his oft-cited 2011 documentary film 'Beijing Besieged by Waste' – a term quickly picked up in the media (Feng 2011; J. Liu 2011). To improve the situation, the municipal government opened several large-scale sanitary landfills and hundreds of garbage collection centres across the city in the

1990s and early 2000s (Huo 2009; Watts 2010; Xie and Zhang 2011; *Xinhua* 2006; C. Yang 2011; Yu 2008; Zhang 2012).

By the end of the century, amassing piles of garbage, growing land problems, hiking real estate prices, and mounting complaints from neighbouring residents about the stench and pollution coming from the formerly remote landfills that were now enclosed by residential areas, forced the municipal government to look for alternative waste treatment strategies. Incineration, which significantly reduces the volume of waste while allowing the easy control of odours, seemed like a perfect solution to the garbage problem. Reflecting the national discourse and policies, municipal authorities and the state media soon praised incineration as a space-efficient and environmentally friendly way to deal with the city's waste problem (Wang 2001; *Xinhua* 2002).[39] In 2002, the Beijing government announced plans to promote incineration in order to improve the city's environment for the 2008 Olympic Games (Wang 2009; *Xinhua* 2006).

Legal Means and Lay Expertise: The Liulitun Campaign as 'Initiator-Movement'

It was not until 2006, however, that the planned location of one of these plants in Beijing's north-western Haidian district (海淀区, *Haidian qu*) became public after the project was included in the city's 11th Five Year Plan (2006-2010) along with three other incineration projects in the city's outskirts.[40] Construction for the Haidian district project was slated to begin in spring 2007 in the Liulitun neighbourhood in the vicinity of Beijing's largest landfill, which had been opened in 1999 (Beijing Municipal Development and Reform Commission 2006; Beijing Municipal People's Government 2006; Li et al. 2010; *Sohu News* 2007; *Xinhua* 2006).

Members of nearby residential communities, many of whom had petitioned about the stench coming from the landfill for years, fiercely

39 The Beijing government also started to try waste sorting in 1996 in an effort to reduce the amount of waste through reuse and recycling, with the environmental organizations Global Village of Beijing and Friends of Nature running sorting trials in several residential communities. These efforts remained largely fruitless, which officials, waste experts, and environmental organizations mostly attribute to a lack of processing facilities and effective municipal waste policies (interviews BMAC 7-11-12, Zhao 8-11-12, Nie 7-6-13, NGO GVB 18-10-11, NGO FON2 21-10-11, NGO NU2 28-9-11; Huo 2009; Xie and Zhang 2011). The first municipal garbage sorting policies were introduced in 2010 (Meng 2010a, 2010d; *Xinhua* 2010a).

40 Apart from the Liulitun incinerator project in the west, the plan included the Asuwei incinerator in the north, the Gaoantun incinerator in the east and the Nangong (南宫) incinerator in the south.

opposed the planned incinerator. They quickly began to organize both offline and via the residential communities' online fora (小区网络论坛, *xiaoqu wangluo luntan*). The resident community included many academics from the nearby Tsinghua and Beijing Universities, who had recently moved to the newly developed neighbourhood. They sent letters and petitions to various government departments and submitted an administrative redress to SEPA with the legal assistance of a lawyer named Xia.[41] After negative reactions from Beijing authorities, about one thousand residents staged a *jiti shangfang* (集体上访, 'mass petition') against the plant in front of the State Environmental Protection Administration on World Environment Day, 5 June, 2007. Two days later, the acting SEPA Vice-Minister Pan Yue issued a public notice to municipal authorities that called for a suspension of the project and a renewed environmental impact assessment with stronger public participation. This led to a suspension of the project plans (interviews Xia 6-11-12, Zhao 8-11-12; D_LLT1, D_NU1; Lang and Xu 2013: 836; Wang 2014; Z. Zhu 2007).[42]

The conflict rekindled in October 2008, when the *Beijing Evening Post* (北京晚报, *Beijing wanbao*) reported that the Liulitun EIA measures had been completed (D_LLT1). Liulitun residents immediately organized further actions. Most importantly, they submitted an extensive *yijianshu* (意见书, 'opinion booklet') about the project's hazards and EIA problems to government authorities in early 2009, which was also widely circulated on the Internet and quoted by the local media (D_LLT1, D_NU1; Guan 2010; Lang and Xu 2013: 837-838). Under mounting public pressure, the project was again called off by the MEP Pollution Prevention Division (污染防治司, *Wuran fangzhi si*) in 2009, demanding a further environmental impact assessment study and larger public scrutiny (*Xinhua* 2009a). In early 2011, the project site was finally moved to Dagong village in Sujiatuo town (苏家坨镇, *Sujiatuo zhen*), a remote mountain area at the border of the Haidian and Mentougou districts (门头沟区, *Mentougou qu*) about 20 kilometres to west

41 Xia became engaged in the case after being approached by Liulitun residents upon recommendation from the Center for the Legal Assistance to Pollution Victims (CLAPV). The residents had initially contacted the center for legal assistance. However, since the case did not fall into the parameters of CLAPV's activities, one of their staff members recommended lawyer Xia to the residents based on personal relations. While at first hesitant about the residents' chances of success and worried that the issue might turn into a mass incident, Xia eventually accepted the role of the residents' legal representative (interviews Xia 6-11-12, 30-7-13).

42 In the same notice, SEPA Vice-Minister Pan Yue also called for a halt and further environmental impact assessment of the disputed PX plant in China's coastal city of Xiamen, where large-scale demonstrations by residents against the proposed facility had occurred just a few days earlier on 1 and 2 June 2007 (Cody 2007; Z. Zhu 2007).

of the original location (Li et al. 2010; Global Times 2011a). The contestation at this new site is the subject of Chapter Five.

Despite being approached by residents, environmental organizations were not engaged in the Liulitun campaign for three major reasons: (1) the organizations regarding their engagement at the local level as too risky, both for themselves and for the local campaigners; (2) at this time, incineration had not yet emerged as a major public issue and was not in the focus of Chinese environmental organizations;[43] (3) several organizations, including Green Beagle, Nature University, and Wuhu Ecology Center, that played an important role in later cases had not yet been established. However, several future members of these organizations, particularly Green Beagle/Nature University staff, individually assisted the campaigners by providing information and reaching out to the media (interviews NGO FON1 19-10-12, 6-6-13, NGO NU1 7-11-12, NGO NU3 25-7-13, NGO WEC 23-10-12, Xia 6-11-12; cp. H. Liu 2011).

The Liulitun campaign was the first major case of local contention against a planned incinerator project in China[44] and had a significant impact on later cases, as demonstrated throughout the rest of this book. The Liulitun campaign can thus be regarded as somewhat of an 'initiator' or 'spin-off movement' from the perspective of social movement theory (Givan, Roberts, and Soule 2010; McAdam 1995; Soule 2004; Tarrow 2010) – an assessment that was also expressed by members of the Chinese 'no burn' community and local community members (e.g., interviews Zhao 8-11-12, NGO NU1 7-11-12, NGO NU2 18-10-11, NGO FON2 21-10-11, NGO WEC 23-10-12, ASW1 17-10-12, PGY1 4-11-12). In the words of Zhao Zhangyuan: 'Liulitun had a major impact. After Liulitun there was a chain reaction across the entire country. [...] When the news of Liuliutun's success spread, people everywhere reacted and mobilized their own "rights protection" (维权, *weiquan*) actions [against waste incinerators in their neighbourhood] as well' (Interview Zhao 8-11-12).

The notion of 'rights protection' or *weiquan* that professor Zhao refers to lies at the core of the Liulitun residents' self-perception. This notion, often used with the suffix 'activities' (活动, *huodong*) or 'campaign' (运动, *yundong*),

43 According to a Friends of Nature staff member, it was through being approached by Liulitun residents during the campaign that the organization first realized that incineration was an important issue that the organization should attend to (interview NGO FON1 6-6-13).

44 Due to the campaign's publicity, high-level political engagement, and status as one of the first major cases of urban local environmental contention against a planned facility – together with the Xiamen PX plant case, which occurred during the same time period – the Liulitun case features in various academic articles and media reports about China's urban environmental protests and so-called 'NIMBY' campaigns (Cui 2011; Johnson 2010, 2013a, 2013b; Lang and Xu 2013; J. Liu 2013; Otsuka 2009).

became the central identity frame for most of the contentious communities in this study (cp. the following case studies). It is also the core concept employed by the Chinese 'no burn' community to describe local anti-incineration contention (cp. interviews Xia 6-11-12, Zhao 8-11-12, NGO NU1 7-11-12, NGO NU3 25-7-13; Chen 2012; Mao 2012; Xia 2011; Zhao 2012a, 2012b). The concept has been used in the context of Chinese social contention across different issue fields for several years (Biao and Mosher 2012; Fu and Cullen 2008; Hung 2010).

Apart from their self-portrayal as *weiquan* defenders, several characteristic features of the Liulitun campaign were later adopted by other contentious communities standing against incinerator projects across the country. One was the comprehensive research conducted by community members, the results of which were compiled into the above-mentioned 'opinion booklet'.[45] While mainly targeted at demonstrating the 'irresponsibility' of siting an MSWI in Liulitun and the procedural flaws related to the project, such as in the EIA procedures, the booklet also detailed the plant's risks for the larger city area, the general hazards of incineration, particularly dioxin, and the international tendency to move away from incineration as a primary waste treatment strategy (D_LLT1; Guan 2010). Faced with the highly scientific and technical nature of the issue, the campaigners strongly relied on both domestic and international experts for information and certification, including Zhao Zhangyuan who had done research on waste issues earlier and was approached by the residents for his expertise (interview Zhao 8-11-12; D_LLT1).[46] Accordingly, the Liulitun residents employed an 'expert frame', portraying themselves as (lay) experts and, in their booklet, offering advice to the Beijing authorities about to how to deal with the city's waste problem (cp. Johnson 2013a, 2013b). Moreover, throughout their struggle, the residents relied mainly on 'peaceful' and legal means – assisted by the lawyer Xia who, like professor Zhao, also functioned as an important channel for communication between the residents and authorities (interviews Xia 6-11-12, 30-7-13, Zhao 8-11-12; D_LLT1).[47] In addition, the Liulitun residents mobilized powerful elite allies and effectively reached out to the media, which extensively reported about the case (Lang and Xu 2013: 837).[48]

45 The writing of an 'opinion booklet' is itself a common feature of the Chinese repertoire of contention employed by discontented local communities (cp. for instance Mertha 2008: 77).

46 *The Liulitun campaign was the first local anti-incineration campaign in which Zhao was actively involved as an outspoken expert to incineration.*

47 The 'mass petition' in front of the SEPA was regarded as on the boundaries of accepted legal behavior by the campaigners (Johnson 2010: 441).

48 Like Liulitun, in Xiamen it was academics, mainly a member of the Chinese Academy of Sciences and a professor at Xiamen University, who began to collect and study overseas

Learning from Liulitun: The Lost Gaoantun Struggle

The Liulitun campaign soon triggered a second campaign by urban residents living in Beijing's eastern Chaoyang district (朝阳区, *Chaoyang qu*). The city's first modern municipal solid waste incinerator (MSWI) had been under construction below the public's radar since 2005 as part of the Beijing Chaoyang Circular Economy Industrial Park (北京市朝阳循环经济产业园, *Beijing shi Chaoyang xunhuan jingji chanye yuan*), located in the vicinity of the Gaoantun Sanitary Landfill (高安屯卫生填埋场, *Gaoantun weisheng tianmai chang*), which had started operation in 2002 (see Figure 3.1). The Industrial Park also encompassed a medical waste incinerator that started operating in 2006 (interview GAT MSWI 9-11-12; D_GAT1, D_GAT2). While residents of the newly developed apartment complexes around the facility had been complaining about the stench from the landfill since 2005, it was not until after the Liulitun residents' 'mass petition' in June 2007 that they became aware of the nearly finished municipal solid waste incinerator in their neighbourhood (D_NU1; interviews GAT 1-11-12, 31-5-13; Johnson 2013a).[49]

After discovering the plant and learning about the hazards related to incineration from the Liulitun campaign and Internet searches, several residents from the nearby residential communities organized in opposition to the incinerator. According to one of the campaign leaders, it was

information about the dangers of PX plants after learning about the project proposal. They then summarized the details of their findings about the plants' environmental hazards, pollution risks, and related procedural flaws in a lengthy proposal submitted to the city authorities and the Chinese People's Political Consultative Conference in May 2007, demanding an EIA and stronger public engagement (Ansfield 2013; Cody 2007; Huang and Yip 2012; Johnson 2010; Zhao 2007; Z. Zhu 2007). This led to the organization of Xiamen residents via technology (in this case primarily text messaging), which mounted into large-scale peaceful 'strolls' on 1 and 2 June 2007. These demonstrations drew national publicity and led to a temporary project halt and its eventual relocation to the nearby city of Zhangzhou (漳州市, *Zhangzhou shi*). Information about the project and related contention was widely circulated on the Internet, including by local bloggers, and was thus easily accessible to other residential communities across the country. There are signs that the Liulitun residents had already learned from the Xiamen case (ibid.).

49 That the construction of the incinerator had not caused earlier opposition from neighboring residents can partly be attributed to the composition of the surrounding residential area. Since several of the nearby apartment complexes were still under construction when the plant's construction began, few people had moved into the newly developed residential communities at that time. While rural dwellers lived in the area, they did not have the capacity to oppose the facility (interviews GAT 1-11-12, 31-5-13, Zhao 8-11-12, Xia 6-11-12). While one resident had taken notice of the incinerator before, they had thought it was a factory based on the residential communities' lack of information about waste incineration at the time (interview GAT 1-11-12; cp. Johnson 2013a: 116).

Figure 3.1 Gaoantun incinerator, November 2012

particularly the media appearances, articles, and posts written by professor Zhao Zhangyuan that raised residents' awareness of the health impacts of dioxin emissions (interview GAT 1-11-12; Xu 2010). Emulating the Liulitun residents' actions, the Gaoantun campaigners primarily focused on legal issues related to the project's siting and environmental impact assessment, filing numerous complaints and petitions related to these issues (interview Xia 30-7-13; D_GAT3, D_GAT4; Johnson 2013a; *VOA* 2008; Wen 2009).

Due to lacking or negative responses from Chaoyang district and municipal authorities, the Gaoantun campaigners decided to step up their legal actions in fall 2007. Based on his engagement with the Liulitun case, they chose the lawyer Xia as their legal representative. In September 2007, Xia assisted the residents in submitting an administrative redress application with SEPA that demanded that the EIA approval for the project, given in 2004, be revoked due to flaws in the EIA process and the project's 'unsuitable' siting decision. The Gaoantun residents argued that their neighbourhood was even less suited for a waste incinerator than Liulitun had been, and demanded that SEPA terminate the project. In early 2008, however, SEPA announced its decision to uphold the project's EIA approval. Successive complaints about this decision submitted to the State Council also proved unsuccessful (interview Xia 30-7-13; D_GAT3, D_GAT4; Johnson 2013a; *VOA* 2008).

The Gaoantun residents' actions gained new momentum at the time of the Olympic Games, when the MSWI took up trial operations to cope with the skyrocketing amount of waste produced during the mass event. The acute stench during the MSWI's trial run at the end of July 2008 and from the large amounts of garbage reaching the landfill during the Olympic Games prompted the Gaoantun residents to resort to more disruptive means (interviews GAT 1-11-12, 31-5-13; D_GAT3, D_GAT4, D_NU1; Xu 2010; M. Yang 2009). When renewed complaints to the Chaoyang district authorities and the authorities' promises to stop the stench resulted in no major consequences, the Gaoantun residents used the residential communities' online fora to organize several 'Saturday strolls' (周六散步, *zhouliu sanbu*). On 30 August, several hundred participants blocked the road crossing that connected the residential area and the industrial park's access road, leading to scuffles with public security forces (interviews GAT 1-11-12, 31-5-13; *New York Times* 2008; Watts 2008).[50] While the Chaoyang district government officially apologized to Gaoantun residents for the stench and again promised to dissolve the issue a few days later – indeed, spending several million RMB to cover up the landfill and install a deodorizing mechanism to contain the smell (interview GAT MSWI 9-11-12) – residents felt that their more urgent concerns about dioxin emissions from the incinerator were still not taken seriously. This led to further 'strolls' on two Saturdays in September (6 and 20) (interviews GAT 1-11-12, 31-5-13).

After mounting pressure on the campaigners, the collective action eventually died down and some of the protestors moved to other residential areas. Campaign leaders were visited and warned off by public security staff, plainclothes police interrupted further organizational meetings, and the residential communities' bulletin board systems (BBS) and QQ-groups[51] were temporarily shut down (interviews GAT 1-11-12, 31-5-13, Xia 6-11-12; D_GAT3, D_GAT4). Several information disclosure requests and lawsuits against the operators of the different waste facilities that had been submitted by individual residents between 2008 and 2010 remained unsuccessful – including one by a young woman who suffered from a respiratory disease which she attributed to the waste facilities (interviews GAT 1-11-12, 31-5-13, Xia 30-7-13, NGO NU1 7-11-12, NGO NU3 25-7-13; D_GAT5 to D_GAT10; Huo 2009; S. Li 2010a; Meng 2010b; Wei 2011; Wen 2009; Xu 2010). In order to appease the neighboring residents, the incinerator's operating company did install a public real-time monitoring system in early 2010, however, which displays

50 According to a Gaoantun campaign leader, the residents deliberately held their 'stroll' after the end of the Olympic Games, both to avoid fiercer reactions from public security forces and to avoid impede the Games (interview GAT 1-11-12).

51 Tencent QQ is a Chinese instant messaging service.

real time figures of sulfur dioxide, nitrogen oxide and dust emissions on a large light-emitting diode (LED) screen at the industry park's main gate (interview GAT MSWI 9-11-12; Shuang Li 2010b).[52]

Both residents and members of the 'no burn' community attributed the Gaoantun struggle's lack of success and the authorities' fierce reactions to the protest, which eventually disheartened the campaigners, primarily to the fact that the incinerator had already neared completion by the time of contention, significantly raising the political and financial stakes for the municipal government. They also pointed out, however, that the Gaoantun residents never reached the same level of unity as the Liulitun and later also the Asuwei community did (interviews GAT 1-11-12, 31-5-13, NGO NU1 7-11-12, Xia 6-11-12, Zhao 8-11-12, BMAC 7-11-12; D_NU1).

Even though Gaoantun and Liulitun residents regarded collaborative action as beyond the boundaries of the politically permissible, they were in contact with each other and exchanged information throughout the course of the campaign (interviews GAT 1-11-12, 31-5-13; D_GAT11). As in the case of Liulitun, environmental organizations were not involved in the early stages of the Gaoantun struggle. However, individual organization members did assist the residents with information and media access (interviews GAT 31-5-13, NGO NU1 7-11-12; D_GAT11). After 2008, Green Beagle staff supported individual residents in their lawsuits, submitted comments and information disclosure requests on their behalf to the Beijing EPB, and organized lectures to provide Gaoantun residents with a public forum (interviews GAT 31-5-13, NGO NU3 25-7-13; D_GAT12, D_GAT13).

A Municipal 'Waste Crisis': Stepping Up Incineration

The suspension of the Liulitun project in 2007 was followed by a silence regarding further incinerator plans from the side of Beijing's municipal authorities (Meng 2009). This changed in early 2009, when Beijing officials publicly announced that there was a municipal 'waste crisis' following the 2008 report by the Ministry of Housing and Urban-Rural Development, which declared that more than one third of Chinese cities were facing such a crisis (cp. Chapter Two). This was supported by numbers provided by the Beijing Municipal Administration Commission (BMAC, 北京市

52 While regarded as a positive step by residents and members of the 'no burn' community, they nonetheless criticized the monitoring for ignoring the more hazardous substances of dioxin and heavy metals, for which no real-time monitoring system had been developed at the time (interview GAT MSWI 9-11-12; Shuang Li 2010b).

市政市容管理委员会, *Beijing shi shizheng shirong guanli weiyuanhui*), which is in charge of Beijing's waste treatment and policies. The BMAC announced that Beijing was producing 18,400 tons of waste per day, only 56 percent of which was covered by the city's disposal capacities. Moreover, according to the BMAC numbers, the amount of waste was increasing by 8 percent annually (Meng 2009). Without immediate measures to resolve the issue, Beijing's 13 landfills would be full by 2015, the Commission warned. Again, waste incineration was presented as optimal solution for this problem, and in mid-2009 the Beijing municipal government announced plans to increase the ratio of incinerated waste from 10 to 40 percent by 2015 by building seven new municipal solid waste incinerators across the capital's outskirts, bringing the total number of MSWIs in Beijing to nine (ibid.).[53]

Among the sites of the newly announced incinerator projects were, again, Liulitun – fueling the second phase of the Liulitun residents' resistance – and the vicinity of the Asuwei landfill (阿苏卫填埋场, *Asuwei tianmai chang*) close to Asuwei village (阿苏卫村, *Asuwei cun*) in Beijing's north-western Changping district (昌平区, *Changping qu*), which had been in use since 1994. After discovering the city's plans to build an incinerator in their neighbourhood in late July 2009, members of the residential communities surrounding Asuwei soon mobilized against the planned project and started the third 'homeowner' campaign against an incinerator project in Beijing, as detailed below.

The other planned facilities were sited in rather remote and sparsely populated mountain areas on the city's outskirts where the surrounding residents, most of them villagers, had little capacity to oppose the planned projects (interviews NGO NU1 7-11-12, NGO NU2 29-10-12, 8-5-13, 1-7-15,

53 An investigative article in *Southern Weekend* (Meng 2009) criticized that the announcement of a Beijing 'waste crisis' and plans for the construction of new incinerators were announced shortly after the '7th Solid Waste Advanced Salon' (第七届固废沙龙, *di qi jie gufei shalong*) organized by Tsinghua University's Department of Environmental Science in March 2009. The Salon brought together waste experts, Beijing officials, and business representatives. According to the article, the decision to increase the use of incineration was strongly influenced by vested economic interests on the side of experts and officials, who misled the public to believe this decision was based on an 'objective academic evaluation' of the situation. A similar critique was also raised by several members of the Beijing 'no burn' community (e.g., interviews NGO NU 1, 7-11-12, NGO NU2 29-10-12, NGO NU3 25-7-13, NGO FON1 19-10-12, NGO FON2 21-10-11, Zhao 8-11-12, ASW1 31-7-13). Moreover, the official numbers, particularly the amount of waste produced and the ratio of waste covered by the city's waste treatment facilities, were challenged by Nature University, which argued that the extent of the city's 'waste crisis' was exaggerated to lay the foundation for a rapid push toward incineration (D_NU1).

NGO NU3 25-7-13, NGO FON1 6-6-13, Zhao 8-11-12).[54] According to members of the 'no burn' community and a Beijing government waste expert, this siting decision was both to minimize the plants' impacts on surrounding residential communities and a deliberate strategy to reduce public opposition to the facilities, which had already placed major pressure on the Beijing government (ibid.; interview BMAC 7-11-12).

For the Broader Public Good: The Case of Beijing Asuwei

After Liulitun and Gaoantun, the Asuwei campaign was the third case of 'homeowner' resistance to a planned incinerator project staged by urban Beijing residents. Homeowner action against the plant mounted after a resident of a nearby residential community discovered the city's plans in July 2009. The homeowner campaign was preceded by several years of complaints and resettlement requests from the surrounding villagers, who had suffered from the stench and pollution of the Asuwei landfill since the 1990s.

At the time of its opening, the Asuwei landfill was located in a former agricultural area beyond the city's urban residential quarters at the border between Xiaotangshan town (小汤山镇, *Xiaotangshan zhen*) and Baishan town (百善镇, *Baishan zhen*). Today, the waste facility is still largely surrounded by fields, brush, and old villages. Starting a few hundred meters to the east of the landfill, however, the Xiaotangshan town area has in recent years developed into an affluent residential neighbourhood. Known for its hot springs and good environment, the area is now home to a large recreational area and several high-end residential communities (小区, *xiaoqu*),

54 The rather small Shunyi district (顺义区, *Shunyi qu*) incinerator, which had been built below the public's radar in a remote mountain area in the city's north-east, started operation as Beijing's second modern MSWI shortly after the Olympic Games in 2008 (Meng 2010c, 2010d). While there was opposition to the Nangong municipal solid waste incinerator (南宫生活垃圾焚烧厂, *Nangong shenghuo laji fenshao guang*) located in Beijing's southern Daxing district (大兴区, *Daxing qu*), it was mainly organized by environmental organizations rallying against the German Development Bank's (Kreditanstalt für Wiederaufbau, KfW) investment in the project (NEEC 2009). The organizations regarded the investment as highly irresponsible and not in line with the KfW's commitment to the 'sustainability of developing countries' (D_NG1, D_NG2). In August 2012, 18 Chinese environmental organizations sent a joint open letter to KfW and also approached the media (ibid.; Xu and Zhang 2012). While this led to a meeting with KfW staff in September 2012, the environmental organizations were not content with KfW's reply (interviews NGO NU2 8-5-13, NGO NU3 26-3-14; D_NG3, D_NG4). In September and October 2012, Nature University staff submitted information disclosure requests to the Beijing EPB, BMAC, and MEP, demanding the publication of the incinerator's EIA report. All of the requests were dismissed (interviews NGO NU2 8-5-13, NGO NU3 26-3-14; D_NG5 to D_NG7).

mostly commercial villa districts (别墅区, *bieshu qu*). Since it is located north of Beijing's Olympic Park, the larger residential area is called 'Aobei' (奥北, literally: North of Olympics) by its residents. The area encompasses about 5000 households, with a substantial number of residents that are members of the higher Beijing literary, art, media, academic, economic, and political circles.

It was the members of these upper-end residential communities who mobilized against the incinerator. Since the addition of the incinerator project raised the nearby villagers' hopes about receiving a timely resettlement – which was in their favour due to the landfill problems – strong cleavages emerged between the two social groups. From the homeowners' perspective, the struggle concluded after city authorities unofficially halted the incinerator project in 2012. The suspension of the project, which also caused the initial resettlement plans for the four villages surrounding the landfill to be set aside, led to strong resentment among the villagers, culminating in a week-long blockade of the landfill's access road in July and August 2013. The project suspension proved to be only temporary, however. Under pressure from the growing amounts of waste and the city's inability to handle it, renewed plans to resume the Asuwei incinerator project became public in 2014 and have since led to resumed contention by the homeowner community. Since the project's second phase and related resistance fall outside of the primary research period, which ended in 2013, this case study primarily focuses on the first phase of contention, although the second phase is briefly outlined at the end of this chapter.

The Prehistory: Landfill Complaints, Water Pollution, and the Villagers' Stance

In the 1990s, years before the announcement of the planned incinerator project, local farmers from the four villages surrounding the Asuwei landfill started complaining about its stench and about the garbage trucks passing through their living area (interviews Zhao 8-11-12, ASW1 31-7-13, ASW2 27-5-13, BMAC 7-11-12; D_ASW1 to D_ASW3; *China Speech* n.d.; Huo 2009; Sipan Li 2010; Xu 2010).[55] While their complaints were largely ignored for several years, they gained some backing in 2001. In that year, Zhao Zhangyuan, who had then just taken up the post of deputy director and general-secretary of

55 The four villages are Asuwei village under the jurisdiction of Xiaotangshan town, and Erdezhuang (二德庄), Niufangquan (牛房圈), and Baishan (百善) villages under the jurisdiction of Baishan town.

the Environment Committee of the Chinese Geophysical Society,[56] started to conduct research on urban underground water pollution. Through his research, Zhao started to become particularly interested in the problem of leachate[57] seeping from landfills and related groundwater contamination. He suggested to BMAC that a geophysical investigation of Beijing's large sanitary landfills should be conducted to find leakage points that should be mended to protect the environment (interview Zhao 8-11-12; D_ASW1 to D_ASW3; *China Speech* n.d.; Huo 2009; Wang 2009; Xu 2010).

Welcoming this idea, BMAC provided Zhao with an investigation fund and a group of experts who started their examinations at the Asuwei landfill. The experts soon discovered an underground hole in the landfill's outer walls that had caused significant pollution to the underground water in the Asuwei area. The monitoring of Beijing's other landfills produced similar results. Based on these findings, Zhao wrote a report to BMAC and SEPA and reached out to the media, including taking a *Xinhua* reporter on a site visit to Asuwei (interview Zhao 8-11-12; D_ASW1 to D_ASW3; *China Speech* n.d.; Huo 2009; Xu 2010). The leakage at the Asuwei landfill and the related water pollution in particular attracted some media attention, which, according to Zhao, caused the BMAC to halt the experts' examination work and conceal all related data. Disappointed by the government's reaction, Zhao conducted an investigation of the villagers' health on his own and discovered above-average rates of respiratory diseases and cancer in the villages around the Asuwei landfill, which he attributed to the pollution (interview Zhao 8-11-12; D_ASW2; *China Speech* n.d.; Xu 2010).

This provided the villagers with the grounds to stage several incidents of blocking waste cars from entering the facility. The incidents eventually prompted the Beijing government to provide the villagers with a minor compensation in 2001 (D_ASW2; Vanacore 2012; Xu 2010). In 2002, the Changping government further invested several million RMB for repair work on the landfill (D_ASW2). However, resettlement requests made by the villagers over the following years produced no results (interviews Zhao 8-11-12, ASW1 31-7-13, ASW2 27-5-13; D_ASW1 to D_ASW3; *China Speech* n.d.; Huo 2009; Xu 2010). For Zhao, who retired in 2003, these events lay the foundation of his growing interest in and research on the links between

56 For more information on Zhao Zhangyuan's background and overall role in China's anti-incineration contention, see Chapter Two.

57 'Leachate' is a term used in the environmental sciences, most commonly in the context of landfilling. It refers to liquids that, in the course of passing through matter, have dissolved or entrained environmentally harmful substances that may then enter the environment, for instance through leaks in the protective walls of sanitary landfills.

waste treatment strategies and harms to the environment and human health, which eventually led him to become one of the leading faces of China's 'no burn' community (cp. Chapter Two) (interview Zhao 8-11-12; *China Speech* n.d.; Xu 2010).

New hopes for a timely resettlement came to the villagers in June 2009, when the government announced plans to relocate the four villages surrounding the landfill by June 2012. There were two named reasons named for this relocation. First, the city planned to expand the protection zone around the landfill. With the existing landfill about to reach the limits of its waste handling capacities, the government planned to enlarge it. This necessitated a bigger protection zone. Second, the city planned to construct a waste incinerator in the vicinity. Excited by this news, the villagers started to prepare for their resettlement to multi-storey relocation homes, which, according to the relocation plan, were to be constructed nearby (interviews ASW1 31-7-13, ASW2 27-5-13; D_ASW2; ASW villagers microblog 2013a, 2013b, 2013c, 2013d; F. Liu 2013a, 2013b; Vanacore 2012; Xu 2010). Some of the villagers even started construction work on their houses to raise the compensation rate that would be provided for their homes (interview ASW 31-7-13; Sipan Li 2010; Xu 2010). However, the villagers' hopes for a timely resettlement were repeatedly disappointed over the course of the following few years (interviews ASW1 31-7-13, ASW2 27-5-13; D_ASW2; ASW villagers microblog 2013a, 2013b, 2013c, 2013d; F. Liu 2013a, 2013b; Xu 2010). In the eyes of the villagers, this was primarily due to the homeowners' opposition to the incinerator, which started at the end of July 2009 and which, in the eyes of the villagers, eroded the basis for their relocation (interviews ASW1 31-7-13, NGO NU2 8-5-13; *Jiang-Wai-Jiang* 2010c).

Beginning Homeowner Contention: 'NIMBY' Concerns, Information Haze, and the First Protest Actions

Homeowner contention started after a resident from the upper-end residential community Poly Ridge (保利垄山小区, *Baoli longshan xiaoqu*) went to the Xiaotangshan town government on 28 July 2009 to file a complaint about the noise from a highway to the south of her neighbourhood. By chance, she discovered a small notice on the government's public notice board that announced the planned construction of an electricity-generating municipal solid waste incinerator in nearby Baishan town. Construction by the state-owned company Beijing Huayuan Huizhong Environmental Technology Corporation (北京华源惠众环保科技有限公司, *Beijing huayuan huizhong huanbao keji youxian gongsi,* from now on referred to as Beijing Huayuan) was to begin at the end of 2009 (interviews ASW1 17-10-12, 31-7-13, ASW2 27-5-13; D_ASW4, D_NU1; Cui 2009, 2010c; Johnson 2013a; Sipan Li 2010; Xu 2010; Zhai 2010).[58] As part of the project's EIA procedures, the notice proclaimed a public announcement phase of ten days to solicit public opinion, which was already nearing its end at the time of discovery. The notice alerted the Poly Ridge resident, who, not allowed to photocopy the paper, took its picture with her cell phone. Back home, she immediately informed her neighbours and posted the photo and a comment on the residential community's online forum, asking whether anyone knew about the project and calling on her neighbours to oppose the planned facility (ibid.).

This news caused a major uproar among the Poly Ridge residents. Initial concerns centred mainly on the negative impact the facility would have on their property prices. As one campaigner put it: 'Beijing real estate prices are very high. Everyone had spent a couple of million RMB, at the highest tens of million, to buy the villas. So of course all the wealthy people living in the villa area didn't want that [the project]' (Interview ASW1 17-10-12). However, the residents were also angered by the government's means of communication. In their eyes, the officials had deliberately tried to circumvent the common people's (老百姓, *laobaixing*)[59] opinions (interviews ASW1 17-10-12, 31-7-13, ASW2 27-5-13; Cui 2010c; Liang 2011; Xu 2010; Zhai 2010). In the words of the same campaigner:

> The history of events was very coincidental. If she [the Poly Ridge resident] hadn't seen it [the notice], no one would have known. [...] They hadn't announced it online, it was just this tiny piece of paper in a dusty corner. [...] We never go to the town government. We don't even know where it

58 The Beijing Asuwei Municipal Waste Incineration Electricity Generating Plant (北京阿苏卫生活垃圾焚烧发电厂, *Beijing Asuwei shenghuo laji fenshao fadian guang gongcheng*) was a joint project of Beijing Huayuan as the operator and BMAC as the responsible unit (责任单位, zeren danwei). It was slated to have a handling capacity of 1200 tons of garbage per day. The plant was part of the planned Asuwei Circular Economy Park (阿苏卫循环经济园, Asuwei xunhuan jingji yuan), which would have a total handling capacity of 7000 tons of waste per day and treat waste from Changping and Chaoyang districts as well as the city's central Dongcheng (东城) and Xicheng (西城) districts, which do not have their own waste treatment facilities due to their dense population (interview BMAC 7-11-12; D_ASW2; Xie and Zhang 2011). Beijing Huayuan is a branch of the state-owned Beijing Environment Sanitation Engineering Group Co., Ltd. (北京环境卫生工程集团有限公司, *Beijing huanjing weisheng gongcheng jituan youxian gongsi*).

59 The traditional term *laobaixing* is the common self-reference used by community members in both urban and rural areas across this study. The term is particularly used to highlight the weakness and lack of efficacy of the 'common people', as opposed to the cadres and party-state officials in the context of an authoritative and unresponsive party-state.

> is. But they still think this counts as procedure (程序, *chengxu*). I just need to stick it up for ten days and then I'm legitimate (我就合法, *wo jiu hefa*). So at that time they just dodged the *laobaixing*. [...] You say it's for public welfare and go and deprive people of their rights. I don't think that's correct behavior. So at that time we were furious about the procedure. [...] They wanted to secretly build it and once it's built, it's reality. Then there is nothing to be done about it. (Interview ASW1 17-10-12)

This struck a nerve with the Poly Ridge residents, many of whom had learned about the nearby Asuwei landfill only after spending large sums on their houses. Feeling that they should have been informed about the nearby waste facility before purchasing their homes, the residents had repeatedly complained to the town government about the smell from the landfill and the growing numbers of waste cars passing by their homes. Hearing about the new incinerator plans, the residents were particularly angered about what they regarded as the government's attempt to furtively add another even more hazardous waste facility without soliciting their consent (interviews ASW1 17-10-12, 31-7-13, ASW2 27-5-13; Cui 2010c; Sipan Li 2010; Liang 2011; Meng 2010e; Xu 2010; Zhai 2010).

Within hours of the first post, the discussion on the community's online forum picked up speed and the number of informed residents grew continuously. While not clear on the hazards of waste incineration in this initial phase, some of the residents had heard about the earlier Liulitun and Gaoantun struggles and the negative connotation of the term 'incineration', which at that time was starting to become a widely discussed issue in the state media (interviews ASW1 17-10-12, 31-7-13, ASW2 27-5-13; Xu 2010). To find out more about the project's nature and implications, many of the residents conducted individual online searches. Since the majority of community residents had high educational backgrounds – including a substantial number of lawyers, academics, and entrepreneurs – the community possessed a high initial awareness of environmental problems and the capacity to immerse themselves in the complex and contested issue of incineration, including its environmental and health risks, technical details, and the legal questions related to the EIA procedures and their rights as residents (ibid.; D_ASW1 to D_ASW3; Cui 2009). Their first online searches produced a plethora of results, which they shared and discussed on the online platform. However, much of the information about the harms of incineration and particularly the dangers of dioxin that was publicly accessible on the Chinese Internet was highly conflicting. While the gathered information raised the residents' sense of alarm, they were confronted with a major information haze at this early stage (interviews ASW1 17-11-12, 31-7-13, ASW2 27-5-13; Xie 2010a; Zhai 2010).[60] As one of the residents recalled: 'We weren't very clear on what kinds of harms an incinerator could pose to the environment, so we all conducted online searches. We soon came across dioxin. But the government downplayed the harm of dioxin. According to the government there was a demonization of dioxin in society. One side said it was terrible, one side said it was not terrible. At that time this was a huge problem. We didn't know what to believe; which experts to believe' (Interview ASW1 31-7-13).

On the day after the notice's discovery, a small group of Poly Ridge residents went to the Xiaotangshan town government and demanded to see the mayor, to confront him about planning a large-scale and potentially hazardous waste incinerator in their neighbourhood without soliciting the people's opinions. Denied access by the town government officials, they demanded to meet the Changping district mayor and were referred to the district's petitioning department (信访部门, *xinfang bumen*). However, the residents' subsequent appeals to the department and phone calls and faxes to the service numbers provided on the EIA notice were left unanswered (interviews ASW1 17-11-12, ASW2 27-5-13; D_NU1; Sipan Li 2010; Xu 2010; Zhai 2010). Frustrated by the lack of effective communication channels with the government, the residents called the first 'homeowner meeting' (业主大会, *yezhu dahui*). At the meeting, the participants expressed their opposition to the planned construction project and – with the backing of the community committee (小区委员会, *xiaoqu weiyuanhui*), who had already prepared a written proposal declaring the residents' unanimous opposition[61] – decided to take up contentious action and unify the homeowners across the different residential communities in the larger Aobei area to increase the base of the opposition (interviews ASW 31-7-13, ASW2 27-5-13; Xu 2010).[62]

On 1 August 2009, several dozen Poly Ridge residents assembled at the community gate with their cars and drove around the area for several hours, displaying banners in their car windows that stated their opposition

60 Individual residents also took the initiative to visit the Gaoantun incinerator to gain a better understanding of incineration plants and were put off by the stench emitted by the facility, not realizing at the time that the major hazard of incineration plants was posed by odorless emissions of dioxin and heavy metals (interview ASW2 27-5-13).

61 This statement was then delivered to the town government.

62 Apart from Poly Ridge, which lies closest to the Asuwei landfill, there are about a dozen residential communities in the Aobei area to the east of the waste facility, most of them commercial villa districts. Together with Poly Ridge, the communities most actively engaged in the campaign were the Tang House Community (汤HOUSE小区), Napa Valley Community (纳帕溪谷小区, *Napa xigu xiaoqu*), Vancouver Forest Community (温哥华森林小区, *Wengehua senlin xiaoqu*), and Orange County Community (橘郡小区, *Jujun xiaoqu*).

to the Asuwei incinerator project and warned about the dangers of dioxin (interviews ASW1 17-10-12, ASW2 27-5-13; D_NU1; Sipan Li 2010; Xie 2010a; Xu 2010). Around that time, several Poly Ridge residents also started to distribute leaflets in the surrounding residential communities, including at a regular farmers market in the vicinity, both to mobilize opposition among the larger Aobei community and to reach out to the residents that did not frequent the community's online forum (interview ASW2 27-5-13). The residents further established two independent websites opposing the incinerator project (which security forces soon closed down), as well as an 'Aobei forum' (奥北论坛, *Aobei luntan*) to facilitate the online communication between the different Aobei communities (interviews ASW1 17-10-12, 31-7-13, ASW2 27-5-13, NGO NU1 11-10-11; *Jiang-Wai-Jiang* 2010a; Xu 2010).[63]

After being unresponsive on the previous days, the government was now fast to react to the residents' mounting contention. Only hours after the motorcade, in the evening of 1 August 2009, an official from the Xiaotangshan town government informed the residents that 20 homeowner representatives (业主代表, *yezhu daibiao*) were invited to meet with officials from the town, district, and city levels of government on the following day to discuss their concerns (interviews ASW1 17-10-12, 27-5-13; D_NU1; Sipan Li 2010; Xu 2010). Left with little time to prepare for the meeting, the homeowners selected representatives from the different residential communities and hastily prepared some points to raise with the government officials. At the meeting, the homeowners primarily challenged the secretive procedures, questioned the legality of the EIA process and the suitability of the site, and expressed their concerns about the environmental and health hazards emanating from the plant. However, not yet having delved deeper into the incineration issue, these latter concerns remained largely superficial, according to the residents' own assessment. The government in turn assured the residents that the plant would not pose any harm and defended the legality of the EIA procedures (interviews ASW1 17-10-12, 27-5-13; Johnson 2013a; Xu 2010;). BMAC officials nonetheless published a renewed announcement about the incinerator in the Chinese-language state-led newspaper *Beijing Daily* (北京日报, *Beijing ribao*) on 14 August, and announced a repeated ten-day period to solicit public opinions (interview ASW1 17-10-12; D_NU1; Sipan Li 2010; Xu 2010). After residents continued to challenge the legality of the EIA

63 The Aobei forum's original web address was: www.myaobei.com. However, during the course of events, most of the posts on the forum were deleted and the forum itself was temporarily shut down several times (interviews ASW1 17-10-12, 31-7-13, NGO NU1 11-10-11; *Jiang-Wai-Jiang* 2010a, 2010b, 2010c).

procedures through public petitioning channels, BMAC officials promised that the project would not begin construction before the EIA was officially approved (Johnson 2013a: 118; Xu 2010).

Coming to a Better Understanding of Incineration: The 'Aobei Forum' and Information Influx

By this time, the 'Aobei forum' had become a vibrant platform of communication for the larger resident community. While the residents of the individual *xiaoqu* ('communities') had started to hold frequent 'homeowner meetings' and established regular communication via personal meetings, phone calls, and text messaging (interviews ASW1 17-10-12, ASW2 27-5-13), communication across the different residential communities was primarily limited to the online forum at this stage.[64] Sensitive to the risks associated with openly criticizing a government plan and mobilizing contentious action, the majority of residents used online names to express their opinions (interviews ASW1 17-10-12, 31-7-13, ASW2 27-5-13). As one of the campaigners recalls about the importance of the forum for their communication and organization:

> At the time the website was established, I took on the web name [... XXX]. Everyone had an alias. Only after we had gotten familiar with each other, did we slowly find out who was who. [...] It's a Chinese specialty (中国特色, *Zhongguo tese*) that Chinese netizens are so many and the websites so popular, because we don't have a regular communication channel to express this kind of opinion. [...] The internet is a platform and if it hadn't been for the internet we couldn't have assembled. (Interview ASW1 31-7-13)

Throughout August 2009 and the following months, individual residents continued to delve deeper into the complicated and contested issue of incineration and shared and discussed their findings on the platform: 'After we came back [from the inspection trip to the Gaoantun incinerator], we started to search a large quantity of materials. [...] We wrote a lot of articles, a lot of things. We didn't publish them, we only posted them online. The thing is that all the homeowners wanted to understand, so we wrote a lot of things and sent them to the homeowners' (Interview ASW2 27-5-13).

64 Social media applications such as *weibo* or QQ, which played a crucial role in later contentious struggles and became an important tool of communication for the Aobei residents at later stages, did not play a significant role at the time (interview ASW1 31-7-13).

Through these inputs, the discussion on the forum reached continuously greater depths and levels of lay expertise. According to the campaigners, several developments and sources of information had an important impact on the community's awareness process and helped them to gradually pierce through the initial information haze. First, a wealthy lawyer and villa owner in the Napa Valley Community, surnamed Huang, soon emerged as eloquent opinion leader on the 'Aobei forum'. Under the online name 'Donkey Stan' (驴屎蛋, *lü shidan,* literally: donkey droppings) – a homophonic pun on his occupation as lawyer, in Chinese *lüshi* (律师) – Huang wrote numerous widely-discussed posts about the legal issues related to the EIA procedures, the results of his personal research on incineration, and appeals to the Aobei residents to stand against the planned facility. Due to his eloquence, well-rounded argumentation, and background as a lawyer, his posts helped to structure and clarify the overall discussion. While the campaigners repeatedly highlighted that the entire campaign evolved without a leader or head (头儿, *tou'er*) (interviews ASW1 17-10-12 31-7-13, ASW2 27-5-13), the homeowner community soon accepted Huang as their spokesperson (代言人, *daiyan ren*) (interviews ASW1 17-10-12, 31-7-13, ASW2 27-5-13; *CNature* 2011; Cui 2010c; Sipan Li 2010; Li and Xuyang 2010). As one of the residents put it:

> This lawyer Huang [...] is our spokesman. His proficiency is very high, also as a lawyer. He is very good. We all think that on a lot of issues his viewpoints are very reasonable (合理, *heli*). So if you say, who our head is, we didn't have a head. The "masses" (群众, *qunzhong*) were all spontaneous (自发, *zifa*). [...] We only had a spokesperson: that is Huang [...]. He made himself a name online, he wrote many articles [on the platform], very good, very reasonable. So everyone trusted him. (Interview ASW2 27-5-13)

Further impetus reached the homeowners at the end of August 2009, when the International Dioxin Symposium was held in Beijing and covered extensively in the state media. In particular, one half-hour long special feature about incineration and dioxin that was broadcast on the state-owned China Central Television left a deep impression on the lawyer Huang and other Aobei residents and helped them to better understand the environmental and health hazards related to incineration (interviews ASW1 31-7-13, ASW2 27-5-13; Xie 2010a; Xu 2010).[65] By that time, the growing number of local

65 The link to the program on the CCTV website (http://space.tv.cctv.com/video/VIDE1251815667747887; 4 February 2014) is no longer available. An extensive report about the program that reproduces long passages was published by, among others, *Sina News* (2009). On the

struggles and urban protests against waste incinerator projects, including in other Chinese cities such as Shanghai, Nanjing, and Shenzhen,[66] and the increasingly outspoken opposition by environmental organizations and experts had started to trigger a public and media debate about incineration as a primary waste treatment strategy for China.

The half-hour long special feature on the pros and cons of incineration featured a number of national and international experts, among them retired professor Zhao Zhangyuan, who had by now come to some media prominence as the frequently cited lead expert of the Chinese 'no burn' community. Apart from explicitly linking incineration to dioxin and cancer, the program also comprehensively reported on the Liulitun campaign. The rich issue-specific and critical information published in the program and other media reports during this period, and its certification by national and international experts, provided critical cognitive cues for the Aobei residents which helped them to create a more stringent collective understanding of the pending project and its related hazards and contributed to their feelings of entitlement and the formation of a 'contentious consciousness'. In particular, the dangers of dioxin shifted more towards the centre of the residents' concerns, which had previously revolved more around property prices and the feeling of being side-lined by the government. In the words of one campaigner:

> When they wanted to build this incinerator, we hadn't yet started opposing it, when we were still ambivalent (举棋不定, *juqi buding*), very coincidentally there was the 2009 World Summit on Dioxin (2009年

symposium, which was widely reported in Chinese media, see the website of the International Symposium on Halogenated Persistent Organic Pollutants at http://www.dioxin20xx.org/history.htm (4 February 2014). Reports about the symposium were published by the Research Center for Eco Environmental Sciences at the Chinese Academy of Sciences (2009) as the Chinese host institution and by news organizations such as Science Net (2009).

66 In April 2009, urban residents in Shanghai's Jiading district (嘉定区, *Shanghai Jiading qu*) protested against the expansion of the Jiangqiao incineration plant (上海江桥垃圾焚烧厂, *Shanghai Jiangqiao laji fenshao guang*), which was stalled later that year. In May 2009, residents from Nanjing city's Pukou district (浦口区, *Pukou qu*) successfully protested the planned construction of the Tianjingwa waste-to-energy plant (南京天井洼垃圾焚烧发电厂, *Nanjing Tianjingwa laji fenshao fadian guang*). Around the same time, Nanjing residents also opposed the construction of the nearby Jiangbei incinerator (南京江北生活垃圾焚烧发电厂, *Nanjing Jiangbei shenghuo laji fenshao fadian guang*). Also in May 2009, residents from Shenzhen city's Bao'an district (宝安区, *Baoan qu*) protested against the running Shenzhen Baigehu waste-to-energy plant (深圳宝安白鸽湖垃圾发电厂, *Shenzhen Baoan Baigehu laji fadian guang*). In August 2009, residents of Shenzhen's Longgang district (龙岗区, *Longgang qu*) further started to protest against an additional incinerator in their district (cp. Chapter Two and Appendix III).

> 世界二恶英大会, *2009 nian shijie ereying dahui*) in Beijing. So a lot of media reported about dioxin. So after that, there were three characters that popped up in our heads: 'dioxin' [二恶英, *ereying*]. [...] At that time experts hadn't yet emerged in large numbers, so it was the media that made us *laobaixing* know the term dioxin first. Second, via internet searches we knew that the harm of dioxin is very large. Third, we knew that very high levels of dioxin emerge from the burning of municipal waste. At that time we had an oversimplified reading of the issue. Once you said incineration it was dioxin, dioxin was cancer, and cancer was death. That was our logical reasoning. (Interview ASW1 31-7-13)

Based on his media presence and appearance in the CCTV program, the articles and blog posts written by Zhao Zhangyuan now became a central source for what the homeowners regarded as reliable information. His earlier research on the Asuwei landfill and the related water pollution in particular distinguished him as a source of independent expertise in the eyes of the Aobei residents. The campaigners expressed a general distrust of Chinese experts, whose interests they regarded as entangled with the industry in an 'expert-industry collusion' (Xie 2010a) and the public unattainability of reliable information (ibid.; interviews ASW1 31-7-13, ASW2 27-5-13; Zhai 2010). The homeowners soon found and downloaded Zhao's report on the Asuwei landfill problems, which provided them with concrete arguments regarding the unsuitability of the site for a further and potentially even more hazardous waste facility (ibid.; D_ASW1 to D_ASW3).

Apart from Zhao Zhangyuan as the stand-alone domestic expert openly opposing incineration, the work of international experts and organizations now started to play an important role in the Aobei residents' interpretation of the issue. Having spent a substantial period of time abroad, several of the residents were capable of assessing English-language information written by international experts, transnational organizations, and affected communities abroad (interviews ASW1 31-7-13, ASW2 27-5-13, NGO NU1 11-10-11, NGO NU3 25-7-13, Zhao 8-11-12; D_ASW1 to D_ASW3; Cui 2009; Xu 2010). This included international studies on the health impacts of MSWI emissions, particularly dioxin, such as the influential works by United States chemistry professor Paul Connett and reports on incineration and its harms by GAIA, Greenpeace, the World Bank, and the World Health Organization (ibid.; Allsopp, Costner, and Johnston 2001; Tangri 2003; World Bank 1999). These reports had already played a crucial role in anti-incineration movements in the United States, United Kingdom, and Taiwan (Rootes 2009b; Rootes and Leonard 2009; Walsh, Warland, and Smith 1997). The reports also fundamentally questioned the use of incineration as a waste treatment strategy and portrayed waste incineration as a 'dying technology' that was being gradually set aside or banned in many countries. Moreover, the materials included information about the emergence, background, and strategies of several anti-incineration struggles abroad (interviews ASW1 31-7-13, Zhao 8-11-12, NGO NU3 26-3-14; D_ASW1, D_ASW3; Allsopp, Costner, and Johnston 2001; Tangri 2003; World Bank 1999).

All this information was now shared and discussed on the online platform, with individual residents compiling and posting Chinese-language summaries of their findings to make the materials accessible to the larger homeowner community (interviews ASW1 31-7-13, ASW2 27-5-13). While becoming a core aspect of the Aobei residents' later framing and claim-making, at this stage the transnational diffusion of information primarily helped the community to come to a more nuanced understanding of the hazards of incineration and contributed to a heightened sense of danger and entitlement. Like the news about the growing numbers of domestic anti-incineration struggles, the knowledge about anti-incineration movements abroad also provided a further 'cognitive justification' for opposing the project (ibid.).

Learning between Homeowners: Liulitun as a Role Model

Another crucial source of information for the Aobei residents was found nearby. Not only was the Liulitun campaign extensively featured in the CCTV broadcast and mentioned in numerous media articles, but a plethora of information about the struggle was also available online. The 70-page 'opinion booklet' compiled by the Liulitun community earlier that year proved to be a particularly valuable source of information for the Aobei residents. The booklet provided not only information and critical cognitive frames on the problems of incineration in the Chinese context, but also outlined the problems with siting an incinerator in Liulitun and the flaws in the EIA procedures, which could partially be adapted for the Asuwei case. Moreover, the detailed descriptions of the Liulitun residents' actions (and the government's reactions) throughout the campaign – including the peaceful mass petition in front of SEPA and the writing of an 'opinion booklet' – provided a good starting point for the Aobei residents' own consideration of future strategies. In a similar vein, the (negative) experiences of the Gaoantun residents were also an important reference point for the Aobei community (interviews ASW1 17-10-12, 31-7-13, ASW2 27-5-13, NGO NU1 11-10-11).

The ties with Liulitun and Gaoantun residents did not remain limited to the nonrelational diffusion of media and online information. Soon, core members of the other two contentious communities started to become actively engaged on the 'Aobei forum' to jointly discuss the issues of incineration and waste treatment and their previous experiences (interviews ASW1 17-10-12, 31-7-13, ASW2 27-5-13, GAT 1-11-12). As one of the homeowners recalled about this horizontal communication and the Liulitun campaigns' relevance for the Asuwei struggle:

> At that time, the people from Liulitun emerged in large numbers. [...] In the beginning, most of the communication happened on the forum. At that time, they had all registered on the 'Aobei forum.' At that time, that forum was extremely popular in Beijing. [...] You wrote a post, asked a question and everyone would discuss, with a pretty high-level academic content. [...] They [the Liulitun residents] definitely were a very good and valuable reference for our resistance (抗争, *kangzheng*) in Asuwei. [...] Because at that time their research on incineration was deeper than ours. Because at that time they had already done resistance for over a year. They already had a theory and a very deep reflection [on the issue]. Moreover, they had some experts with immense logical reasoning. [...] They were very persistent and also very intelligent. These few families in Beijing had a very intelligent foundation [of taking action]: That was reasoning (讲理, *jiangli*) with the government rather than causing a mass incident (群体事件, *qunti shijian*) as their primary method to enlarge their influence. (Interview ASW1 31-7-13)

It was particularly this 'reasonable' (合理, *heli*) approach to expressing their claims – a central term repeatedly stressed by the Aobei campaigners in both (media) interviews and their writings – which was deemed successful and thus emulated by the Aobei residents. In their understanding, this approach was largely based on legal means and a 'rational' dialogue with the government, building on the residents' own lay expertise. A handful of homeowners thus started to write their own 'research report' (研究报告, *yanjiu baogao*) with a first (internal) preliminary version ready by the beginning of September 2009 (interviews ASW1 31-7-13, ASW2 27-5-13; D_ASW1; Xu 2010).

Taking to the Streets: Difficulties in Mobilizing Action and the '9·4 Incident'

Based on this deeper understanding of the risks of incineration, the Aobei residents became even more dissatisfied with the officials' claims at the 2 August meeting that the plant would be harmless. Against the backdrop of the Beijing media's continued promotion of the planned Asuwei Circular Economy Park, which was slated to break ground at the end of 2009, the Aobei residents decided to step up their contentious activities and take the Liulitun residents' peaceful 'mass petition' as an example. They heard about an exhibition that was to take place at the China National Agricultural Exhibition Center (中国农业展览馆, *Zhongguo nongye zhanlan guan*) on 4 September 2009, where the plans for the Asuwei Circular Economy Park would be on display. The Aobei residents decided to use this opportunity to stage their own peaceful protest against the facility (interviews ASW 1 17-10-12, 31-7-13, ASW2 27-5-13; D_NU1; *China Daily* 2009; Cui 2010a, 2010c; L. Li 2010a, 2010b; Liang 2011; Meng 2010e; Xu 2010; Xuyang 2009).

To ensure a large number of participants at this event, the actively engaged homeowners hoped to mobilize a larger crowd across the different affected residential communities via the online forum, further homeowner meetings, and the offline distribution of leaflets. To unify the larger Aobei community, facilitate personal face-to-face communication across the individual *xiaoqu*, and give more weight to their collective claims through an institutionalized structure, a group of residents started the process of establishing an Aobei residential community (社区, *shequ*) that would officially link together the different *xiaoqu* (interviews ASW1 17-10-12, 31-7-13, ASW2 27-5-13; Sipan Li 2010; Xu 2010).

The mobilization of a broader base of participants proved difficult, however. For one, a substantial number of villa owners only used them as a secondary residence, which turned out to be a major hindrance for contacting the homeowners or actively engaging them in the struggle. Second, many of the residents that belonged to the higher Beijing literary, art, media, economic, and political circles – some of them well-known public figures – refrained from becoming actively involved in the campaign, and especially from participating in contentious action, since they did not want to prejudice their position or had conflicts of interest (interviews ASW1 31-7-13, ASW2 27-5-13; Sipan Li 2010; Xu 2010). According to one of the campaigners, it was 'really only those without a name' who participated (interview ASW2 27-5-13). Even among the permanent and less prominent residents, many had busy lives and were not willing to or did not see an opportunity to dedicate much time to the issue. It soon became clear to the core group of Aobei residents that staging a large-scale action similar to the size of the Liulitun residents' 'mass petition' would prove difficult (ibid.; interview ASW1 17-10-12).

For these reasons, it was a relatively smaller group of about one hundred Aobei residents who gathered in the pouring rain in front of the Agricultural Exhibition Center on the morning of 4 September 2009. Many of the participants wore homemade t-shirts with slogans such as 'oppose waste incineration' (反对垃圾焚烧, *fandui laji fenshao*) or 'oppose construction in Asuwei, protect Beijing City' (反建阿苏卫，保卫北京城, *fan jian Asuwei, baowei Beijing cheng*). They also held banners displaying slogans such as 'in the names of wives and children, old and young, fiercely oppose the construction of the Asuwei waste incineration plant' (以妻儿老小的名义坚决反建阿苏卫垃圾焚烧厂, *yi qi er lao xiao de mingyi jianjue fan jian Asuwei laji fenshao chang*) – slogans that were similar to those used by the Liulitun campaigners (interviews ASW1 17-10-12, 31-7-13, 27-5-13; D_ASW1, D_ASW3, D_NU1; *China Daily* 2009; Cui 2010a, 2010c; L. Li 2010a, 2010b; Sipan Li 2010; Li and Xuyang 2010; Liang 2011; Meng 2010e; Wen 2010; Xie 2010a; Xu 2010; Xuyang 2009). Dedicated to their 'rational' and 'peaceful' approach, the residents deliberately refrained from loudly shouting out slogans and made sure not to obstruct visitors' entrance to the building. Instead, they stood silently to one side of the main gate. Building on the Liulitun residents' experience, the campaigners felt safe that such a 'peaceful demonstration' (和平示威, *heping shiwei*) would not evoke strong reactions from the public security forces (interview ASW1 17-10-12; D_NU1; Sipan Li 2010; Xu 2010).

However, the campaigners had neglected the fact that the 60th anniversary of the founding of the People's Republic of China, to be celebrated on 1 October 2009, was quickly approaching. This meant the anti-incinerator demonstration was happening at a sensitive time. Not long after the start of the demonstration, public security forces arrived at the scene and a scuffle with the demonstrators began. Six participants, among them the lawyer Huang and two elderly residents, were detained for 'disturbing the public order' (扰乱公共秩序, *raoluan gonggong zhixu*). Several of the other participants were questioned by the police and their names and phone numbers were noted down (interviews ASW1 17-10-12, 31-7-13, 27-5-13; D_ASW1, D_ASW3, D_NU1; *China Daily* 2009; Cui 2010a, 2010c; L. Li 2010a, 2010b; Sipan Li 2010; Li and Xuyang 2010; Liang 2011; Meng 2010e; Wen 2010; Xie 2010a; Xu 2010; Xuyang 2009).

The detained residents were released on the same evening and throughout the next day. Several Beijing media houses reported about the incident, therefore drawing public attention to the residents' claims. Nonetheless, the '9·4 incident' (9·4事件, 9·4 *shijian;* '9·4' refers to the date according to Chinese general practice) disheartened the Aobei community, and further activities such as a planned motorcade to the Bird's Nest Stadium were called off (interviews ASW1 17-10-12, 31-7-13; D_NU1; Sipan Li 2010; Xu 2010). Moreover, on 5 September, one day after the demonstration, the BMAC vice head Chen Ling announced at a visit to the exhibition that the Beijing government was firm on its plans to complete nine MSWIs by 2015, including the Asuwei incinerator – thus sending a clear message to the demonstrators. Chen promised that the government would adopt the most advanced technology and that the plants would meet EU standards, thus not posing any harm to the environment (Xu 2010; Xuyang 2009).

From Conflict to Communication: The 'Aobei Volunteers Research Group', Growing Lay Expertise, and Seeking Expert Advice

Despite its limited success, the 9·4 incident led to a change of strategies on both sides that eventually proved fruitful for the Aobei residents. At a waste treatment symposium at the end of October, Chen Ling and other Beijing government officials again repeated their determination to complete the Asuwei incinerator. They added, however, that the communication with the Asuwei residents would be improved to avoid a 'second Liulitun' (*China Daily* 2009; Xie 2010a; Zhai 2010). Shortly after the demonstration, the Beijing city government established an Office for the Reception of Visitors (接访办公室, *jiefang bangongshi*) at the Xiaotangshan town government, and Wei Panming, the BMAC official in charge of solid waste, ordered the town government to engage in closer communication with the residents. This measure was mainly attributed to mounting public and media pressure from the campaigners. The Aobei residents made use of this official line of communication and several meetings were held between the homeowners and government officials, which set in motion a process of more sincere communication and exchange of opinions over the course of the following months (interviews ASW1 17-10-12, ASW2 27-5-13; Xu 2010; Xuyang 2009). Communication between the residents and the government slowly improved and the homeowners had the feeling that their opinion was taken more seriously (interviews ASW1 17-10-12, 31-7-13, ASW2 27-5-13). As one of the representatives recalls:

> We raised some suggestions and views how the BMAC and the superintendents should do it [waste treatment]. They thought what we said all made sense. So the final outcome was that the government sent people to communicate with us a couple of times. Afterwards their feedback was that they thought us homeowners were very good homeowners. They are not unreasonable. They are very reasonable. They talk based on facts and

> evidence. This caused the city government to approve of us. So then the communication got better. (Interview ASW2 27-5-13)

The fierce police reaction at the Agricultural Exhibition Center and the government's continued determination to build the facility had nevertheless scared many residents away from continuing their participation in the struggle and destroyed their hopes of obstructing the plant. A core group of campaigners, among them the lawyer Huang, remained determined to fight against the incinerator and decided that a change of strategy on the part of the campaigners was necessary to produce a positive outcome (interviews ASW1 17-10-12, 31-7-13, ASW2 27-5-13; L. Li 2010a, 2010b; Sipan Li 2010; Li and Xuyang 2010; Meng 2010e; Wen 2010; Xu 2010). Realizing that continued conflict with the government would bring no results and that the already limited basis for contentious action had crumbled, they decided to fully embrace the 'rational' approach of persuading the government based on lay expertise, moving in a direction that was similar to the government's new communicative approach. In the words of the lawyer Huang:

> I was arrested during the demonstration. For me, it was very dramatic. [...] Suddenly you live in one house with those criminals. Actually, these few days, it's kind of funny, you can really say they were the turning point for my outlook on the waste issue. [...] The earlier opposition didn't have any benefits, did it? This made me rethink: If you want to express the standpoint of us masses (我们群众的思想, *women qunzhong de sixiang*), you must persuade the government (说服政府, *shuofu zhengfu*), make the government and the people (民众, *minzhong*) have a dialogue and a reasonable and fair platform, go consult and talk things over. Why fear disputes? So after this incident I really completely changed my mindset. (Interview ASW1 17-10-12)

Together with a core group of residents from different *xiaoqu*, Huang founded an 'Aobei Volunteers Research Group' (奥北志愿者研究小组, *Aobei zhiyuanzhe yanjiu xiaozu*) to delve deeper into the issue of incineration and revise and extend the initial research report. The research work was divided between the participants according to their individual abilities and areas of expertise and the results of their research and analyses were discussed in regular group meetings. In particular, the above-mentioned studies and research reports by international experts and organizations were more closely scrutinized through more careful reading and analysis (interviews ASW1 17-10-12, 31-7-13, ASW2 27-5-13; L. Li 2010a, 2010b; Li and

Xuyang 2010; Meng 2010e; Xu 2010). The engaged residents soon established close personal relations, becoming what Huang described as 'waste friends' (垃圾朋友，*laji pengyou*) (interview ASW1 31-7-13). As one of the other participants recalled about this new approach:

> After that [the incident] people from the ten surrounding *xiaoqu* all allied, held meetings, discussed, all very spontaneous (自发, *zifa*). [...] We understood that the most important thing is to be reasonable. To stress scientific knowledge. To practically and realistically go and talk based on the facts. To present proof taking a comprehensive policy approach at a technological and systemic level. [...] We all said doing this and that [contentious action] has no use. This [the construction] is decided by the National Development and Reform Commission (NDRC). The plans have already been issued. We can only convince people by reasoning, we can only handle this issue in a very rational manner. Use our maximum abilities to get this thing under control. (Interview ASW2 27-5-13)

The residents also held frequent symposiums and weekly meetings with the core campaigners of the Liulitun, and to a smaller extent Gaoantun, struggles. This helped the Aobei residents continuously expand their expertise on incineration, which was now considered their primary resource against the government plans. While the different contentious communities established an 'alliance of support' (interview ASW1 31-7-13), the Aobei residents refrained from asking them to directly participate in their activities. The restrictive political framework set clear limitations on further collaborative action:

> In 2009, they [the Liulitun residents] were like friends, we often had dinner together, invited them to Asuwei, eat and talk together, ten to twenty people. [...] At that time we were in negotiations with the government, we were very confident, because our knowledge was getting very deep. Our incineration knowledge was more sincere than that of the government. [...] But we didn't actually invite them [the Liulitun residents] to take part in our activities, because we know this thing. It's very risky, right? So we took action alone. [...] Of course from a morality and justice (道义, *daoyi*) perspective they definitely supported us. But we didn't notify them to take part. The risks are quite high. In ten percent of times, something bad happens. (Interview ASW1 31-7-13)

To further enhance their expertise on the issue, the campaigners also sought assistance from other parties. Based on his prior research and important role in the Liulitun campaign, the residents invited the retired professor Zhao Zhangyuan to give a presentation to the homeowners, which was well-attended by the resident community (interviews Zhao 8-11-12, ASW2 27-5-13). At the meeting, Zhao provided the residents with detailed information about incineration and its hazards, particularly in the Chinese context, further details of his earlier findings on the Asuwei landfill, and concrete geological and legal arguments against the siting of an incinerator project in the area, which were prominently incorporated into their argumentation and developing research report (interviews Zhao 8-11-12, ASW2 27-5-13; D_ASW1 to D_ASW3).

A group of homeowners also sought the advice of the Beijing Lawyers Group (北京律师团, *Beijing lüshi tuan*). With the help of a comprehensive presentation, which summarized the situation of waste incineration in Beijing, the details of the planned facility, and the environmental and geological situation of the Asuwei area – drawing heavily on the information provided by Zhao Zhangyuan – they presented their case to the lawyers and asked for legal advice (D_ASW2). While the residents decided not to pursue environmental litigation at that stage, the lawyers provided them with legal information that also found its way into their claim-making and research report. This included knowledge about Chinese laws and regulations relevant to their cause, international and particularly US environmental laws and regulations, and information about administrative and legal procedures related to the construction of waste facilities in other countries (interview ASW2 27-5-13; Xuyang 2009).

Not to be Trusted: The Limited Role of Environmental Organizations in the Asuwei Campaign

During this phase, individual homeowners established contact with the members of Beijing-based environmental organizations, among them Friends of Nature. As with the earlier Liulitun campaign, at the time environmental organizations were not yet actively engaged in the field of waste incineration. However, individual organization members did provide some minor assistance to the campaigners. The founder of Green Beagle, which was just being established at the time, published several articles about the campaign in the party-affiliated Chinese-language newspaper *Guangming Daily* (光明日报, *Guangming ribao*), where he was working as a reporter, to draw public attention to the issue and provide the residents with a platform for diffusing their claims. He also invited some of the campaigners to give a public presentation at his newly established organization (interviews ASW1 17-10-12, 31-5-13, 31-7-13, ASW2 27-5-13, NGO NU1 7-11-12, NGO NU2 14-10-12, 29-10-12, 26-3-14, NGO NU3 25-7-13, NGO FON1 6-6-13, Zhao 8-11-12). A member of the environmental organization Global Village of Beijing, who had a doctoral degree and specialized in waste, and who would later become a Green Beagle/Nature University staff member and an important intermediary in many of the later anti-incinerator struggles, was in close contact with some of the campaigners. Based in the United States at the time, he provided information to them via email. According to his own assessment, his input was of only minor relevance, however, since the group's capacity was already quite strong and the community had their own means of assessing information (interview NGO NU3 25-7-13).

Despite such individual assistance, many homeowners showed a strong reluctance to rely on environmental organizations. Several of the core campaigners expressed the opinion that the influence of environmental organizations was too small and that a better strategy was to turn themselves into experts and take ownership of the issue (interviews ASW1 17-10-12, ASW2 27-5-13; Liang 2011; H. Liu 2011). Moreover, individual campaigners expressed the belief that the approach of environmental organizations did not suit the campaign's 'reasonable' approach:[67]

> The difference between me and the environmental protection personalities (环保人士, *huanbao renshi*) is that I don't have such a strong smell of gunpowder. The environmental organizations only want to go and criticize the government [...]. But I don't think that has any use. If you oppose the government, then they won't listen to you, right? Moreover, sometimes their [the environmentalists'] views are quite childish [...]. So I thought that if you want to really do something, [...] you first have to become an expert in this field, then you will definitely have a voice. [...] Second, you need influence and that also depends on your expert knowledge. And third, you need the media. The media are a kind of resource. [...] But our NGOs [sic.] lack all that. (Interview ASW1 17-10-12)

Overall, the engagement of environmental organizations in the main phase of the Asuwei campaign was thus very limited and restricted to individuals. However, organization members refuted Huang's later claims, made in

67 That environmentalists are easily regarded as 'emotive', weakening their stance, is also reported by Mertha (2008: 146-147) in the context of opposition to the Nu River Project.

media interviews, that the environmental organizations 'did not show any sensitivity to our needs, [...] gave us no professional guidance, direction or moral support, and did not make any efforts to appeal to the public sense of responsibility on the issue' when needed most (H. Liu 2011; cp. also interview NGO NU3 25-7-13; Liang 2011).

For the Broader Public Good: Beyond 'NIMBY' Concerns and the Aobei Research Report

Throughout the fall and winter of 2009, the research group continuously deepened their expertise on incineration. With a better understanding of the issue, the campaigners came to the conclusion that the problem of incineration had to be tackled from a larger waste perspective if they were to find a thorough solution; they had to extend their claims beyond their own individual concerns. The group thus expanded its focus from merely developing a convincing argument against the construction of an incinerator in the Asuwei area to thinking about a broader solution for Beijing's waste problem, which they now regarded as inherent to the Asuwei issue. As one campaigner put it: 'Garbage is besieging the city. The government is not good, right. But it is not the government that has brought this about. It is every one of us, including me, who produce garbage every day. So shouldn't we all help the government, pull on the same string as the government, confront the issue together, discuss it together, solve it together? [...] So we seriously started to research the waste problem with our hearts and souls' (Interview ASW1 17-10-12).

With this broader outlook, the Asuwei residents started to deliberately set themselves apart from the Liulitun campaign, whose claims and framing were primarily focused on opposing the construction of an incinerator in their neighbourhood (D_LLT1; Johnson 2013a). Instead, the Aobei group broadened their claims beyond what they now criticized as mere 'NIMBY' concerns, with the explicit goal of using their lay expertise to assist the government in solving the garbage crisis for the broader public good – at both a local and a national level (interviews ASW1 17-10-12, 31-7-13, ASW2 27-5-13; D_ASW1, D_ASW3; Xu 2010). In (media) interviews and their research report, the campaigners portrayed themselves as fighting for the well-being of all of Beijing's residents and as (self-made) waste experts determined to advise the government with the ultimate goal of improving China's waste policies (interviews ASW1 17-10-12, 31-7-13, ASW2 27-5-13; H. Li 2010; Sipan Li 2010; Liang 2011; Meng 2010e; Xu 2010). In the words of one campaigner:

> We stated our reasoning very clearly: That the Beijing government would [...] not only sacrifice the interests of the Asuwei residents, but sacrifice the interests of all Beijing residents. So at that time our aim was very high, otherwise we would not have gotten anywhere. It's not just us wealthy people that have bought villas who are afraid of pollution at a 'NIMBY' [sic.] level, but we stood up for a much higher interest, for the whole people's entire interests (全体人民的整体利益, *quanti renmin de zhengti liyi*), for the entire interest of Beijing city. (Interview ASW1 31-7-13)

This broadening of the homeowners' framing and self-portrayal also had a strategic component. They were aware that, according to the perception of the broader public, the Asuwei case had a significant social justice component that was not working in their favour. In the campaigners' understanding, many Beijing residents blamed the wealthy people in the Aobei neighbourhood for using their ample economic and political resources to obstruct the construction of an essential incinerator in their posh neighbourhood, thus shifting the burden to a less privileged community with less capacity to oppose the facility. By broadening their claims to encompass the interests of all Beijing residents, they hoped to assemble a significantly broader base of support (interviews ASW1 17-10-12, 31-7-13, ASW2 27-5-13; Xu 2010).

At the beginning of 2010, after months of intensive research, rounds of revision, and discussion among the homeowners, which also continued on the 'Aobei forum', a final 77-page version of the research report entitled 'The Life and Death Decision of China's Urban Environment – Waste Incineration Policies and Public Will' (中国城市环境的生死抉择 – 垃圾焚烧政策与公众意愿, *Zhongguo chengshi huanjing de shengsi jueze – laji fenshao zhengce yu gongzhong yiyuan*) was completed and submitted to all relevant Beijing government institutions and disseminated to the media (D_ASW1). The report, which was the Aobei residents' main tool for expressing their claims to both the government and the public, shows an impressive level of depth and expertise. Framed as a collection of public opinions to be submitted to the government according to the EIA Law, the report's disclaimer summarizes several clear policy recommendations with the explicit aim of impacting both Beijing and national policy decision-making about waste.

The report starts by placing the Asuwei campaign in the context of the growing number of anti-incineration struggles in China, providing information on and photographs of incineration protests all over the country. This included the cases of Beijing Liulitun and Gaoantun, Shanghai Jiangqiao, Nanjing Jiangbei, and the then-most recent large-scale protests in Suzhou city's (苏州市, *Suzhou shi*) Wujiang (吴江区, *Wujiang qu*) and Guangzhou's

Panyu districts, where thousands of homeowners had protested against planned incinerators in October and December 2009, respectively (cp. Chapter Two and Appendix III). With this introduction, the campaigners framed themselves as part of a rising tide of contentious communities standing against incinerator projects, demonstrating that incineration was a hot-spot issue with the potential to impede China's social stability, and was thus deserving of high-level political attention. Incineration, the campaigners argued, had already produced major social unrest that threatened China's 'harmonious society' (和谐社会, *hexie shehui*) and was not in line with the Center's ideological concepts of 'people first' (以人为本, *yiren weiben*) and a 'scientific outlook on development' (科学发展观, *kexue fazhan guan*) – thus adhering to official discourse in their report.

This framing not only served to lend legitimacy to their claims and contentious activities – and made clear that the report was not directly opposing the government – but also displayed the homeowners' collective identity as part of a broader *weiquan* and anti-incineration community, which reflected in the interview narratives (interviews ASW1 17-10-12, 31-7-13, ASW2 27-5-13; D_ASW1, D_ASW3). The campaigners further set themselves in line with incinerator-siting conflicts and anti-incineration movements abroad – particularly invoking the experiences of and citing prominent figures from the United Kingdom, United States, Japanese, and Mozambique anti-incineration movements. By displaying the logo of the 'Global Movement against Waste Incineration', they depict themselves as part of this movement in their report (全球范围内抗议垃圾焚烧的示威活动, *quanqiu fanwei nei kangyi laji fenshao de shiwei huodong*) (D_ASW1, D_ASW3).

Like the Liulitun 'opinion booklet', the report then outlines arguments against the siting of the Asuwei incinerator based on both geological and legal arguments related to flaws in the EIA process. The plant, so the argument goes, would not only harm the Aobei area but all Beijing residents. Unlike in the Liulitun document, this is not the focus of the report. Instead, the core of the report is an extensive analysis of the harms of incineration for the environment and human health – adapted to the Chinese case – and a comprehensive outline of the policy recommendations summarized in the preface, which draw heavily on reports and studies by international organizations and experts and the experiences of other countries. Here, the pivotal role of the transnational diffusion of information and the campaigners' frame alignment with the international anti-incineration community becomes most visible. In a PowerPoint presentation complied by the campaigners to introduce their report to different audiences, the last slide is entirely dedicated to an acknowledgement of the central role of the works of United

States chemistry professor Paul Connett. The campaigners express their gratitude and respect for this 'forefather of anti-waste incineration' (反垃圾焚烧的前辈, *fan laji fenshao de qianbei*).

The central arguments against incineration given in the report are as follows. Drawing primarily on influential studies on incineration and human health by GAIA, Greenpeace, and the World Health Organization (Allsopp, Costner, and Johnston 2001, Tangri 2003, World Bank 1999), as well as the works of Paul Connett and other international experts, the campaigners describe waste incinerators as the main emitters of the carcinogen dioxin – the 'poison of the century' (世纪之毒, *shiji zhi du*). China, the report argues, is the main emitter of dioxin, with the emission rates of operating Chinese plants raising significantly above standards. Particularly in the Chinese context, with its lack of effective emission controls, high levels of corruption, and close entanglement of industry, politics, and the judiciary, the safety of the population cannot be guaranteed, they argue. This is enhanced by the low caloric value of Chinese waste and the lack of effective pre-sorting measures, which further hike emissions. To enhance the report's resonance with both Chinese policy-makers and the public, the campaigners refer to Chinese food safety problems such as the Sanlu milk powder scandal[68], arguing that the widespread construction of incinerators would significantly aggravate the issue of public risk. Overall, the campaigners conclude, continuing on the path of mixed waste incineration would be a 'severe criminal act against the people' (对人民犯下的严重罪行, *dui renmin fanxia de yanzhong zuixing*). Under these conditions, public security incidents (公共安全事件, *gonggong anquan shijian*) would continue to occur and the public would be forced to live under a 'constant security crisis' (安全陷入危机, *anquan xianru weiji*) (D_ASW1, D_ASW3).

Moreover, the campaigners argue, the economic side of incineration is a major impediment to China's sustainable development. The construction of waste incinerators is not only accompanied by enormous environmental and health costs and the waste of resources, but it also necessitates an immense financial investment. International studies and cases abroad have shown that this is an ineffective 'waste' (浪费, *langfei*) of taxpayer money and a major disincentive for more sustainable waste treatment strategies such as recycling and waste reduction. The high profit margins in the incineration

68 In 2008, government inspections revealed that the Sanlu Group (and other companies) had adulterated baby formula with melamine, a chemical that gives the appearance of higher protein content when added to milk. The food safety incident caused the death of several babies – with more than 50 000 hospitalized – and led to widespread concerns about food safety among the Chinese population (Branigan 2008).

industry also invite corruption and misconduct, hence going counter to the anti-corruption goals formulated by the party-state. This becomes obvious, the campaigners continue, in the vested interests of 'so-called' experts who distorted the truth and misguided policy-makers and the public into adopting the incineration strategy in the first place, thus going counter to the party-state's 'scientific outlook on development'.

Due to these problems, the campaigners conclude, many countries have already banned or significantly reduced incineration technology and rendered it a 'dying technology'[69] – and as a response international corporations subsequently pushed it into the Chinese market. The report claims that the push for mixed waste incinerators in China is a breach of several international laws and conventions that have been signed by China, including the Stockholm Convention dedicated to reducing hazardous emissions. As an alternate solution to the urgent waste problem, the report promotes RDF-technology,[70] front-end 'Zero Waste' policies – which encompass the 'three Rs' of waste reduction, recycling, and reuse – and the expansion of and official support for the non-incineration waste treatment industry in China. These are the central strategies also promoted by the international anti-incineration community. Over the course of several dozen pages, the report details the benefits of these strategies and the experiences of other cities such as San Francisco (United States), Canberra (Australia), and Edmonton (Canada) (D_ASW1, D_ASW3).

The report ends with a restatement of the policy recommendations summarized in the report's introduction. Apart from alternative waste treatment strategies, the campaigners also call for stronger public participation in environmental matters. Under the headline 'Viewpoint: Public participation is the only way for environmental protection and putting an end to corruption' (观点: 民众参与是保护环境, 杜绝腐败的唯一道路, *guandian: minzhong canyu shi baohu huanjing, dujue fubai de weiyi daolu*), they urge the government to better incorporate public opinion into the decision-making and policy formulation process. Citing then-Premier Wen Jiabao's urge to make policy decisions more scientific and democratic, they conclude:

69 This is a direct reference to the influential GAIA study 'Waste Incineration: A Dying Technology' (Tangri 2003).

70 The conversion of municipal solid waste into refuse-derived fuel (RDF) is a waste treatment strategy promoted in many Western countries in recent years. RDF is produced by shredding and dehydrating solid waste with a waste converter technology and consists largely of the waste's combustible components, such as plastics and biodegradable waste.

> The public's oppositional voice is not only an imitation of protests [amongst each other], it is also not merely the consideration of individual and partial interests, but it is the collection of the wisdom of the broader masses (更广泛的民众, *geng guangfan de minzhong*) that can perfect the policy system (政策系统, *zhengce xitong*). At the same time, the contemporary awakening of citizen awareness (公民意识, *gongmin yishi*) and the development of information technologies can be an effective way to obstruct the current occurrence of corruption. (D_ASW1)

From Outlaw to Media Star: The Phoenix TV Program, Huang's Japan Trip, and Cooperation with the Government

While communication with the government had improved, it was not until the submission of the research report that actual good relations developed between the Aobei residents and officials, which eventually led to the (temporary) shelving of the MSWI project (interview BMAC 7-11-12). While the research report deeply impressed the Beijing government officials, the unfolding cooperation and close relations between the Aobei residents and the government was primarily based on a rather coincidental meeting between Huang and a leading BMAC waste expert and National People's Congress representative surnamed Wang. This led to a stronger personalization of the campaign, and Huang soon became a well-known public figure.

Against the backdrop of the increasingly fierce public and media debate about the pros and cons of incineration and the growing number of local protests, Hong Kong-based Phoenix Television decided to produce a program on the topic 'Is the construction of waste-to-energy plants feasible in China?' (建垃圾焚烧发电厂在中国是否可行? *Jian laji fenshao fadian chang zai Zhongguo shifou kexing*?) as part of the well-known series 'Tiger Talk' (一虎·席谈, *yi hu yi xi tan*). The program was supposed to bring together experts, pro-incineration advocates, government officials, and affected community members. The recording for the program was set for 9 December 2009 in Beijing (D_ASW5). Rather unintentionally, the television program came to play an important brokerage function for the Asuwei campaigners and broader Chinese 'no burn' community by bringing together previously unconnected members of different contentious communities, experts, and officials.[71]

Among the participating local community members were Huang, as the representative of the Aobei residents, core members of the Liulitun and

71 A similar incident where a press conference unintentionally served as networking platform for activists is described by Zhu (2017).

Gaoantun campaigns, and members of the Likeng and Panyu communities from Guangzhou. The Likeng incinerator in Yongxing village on the outskirts of Guangzhou started operation in 2005 as the first incinerator in Guangzhou. In 2009, cancer rates were alarmingly high in Yongxing village, which the villagers – who had in previous years staged several partly violent and overall unsuccessful protest actions against the plant – attributed to pollution from the incinerator and the prior landfill in the vicinity. When residents in Guangzhou's Panyu and Huadu districts learned about the Guangzhou government's plans to build MSWIs in their neighbourhoods in September 2009, homeowners from both communities started mobilizing against the plants. To get a better understanding of incineration, Panyu homeowner representatives visited the Likeng plant and, upon seeing the villagers' health problems, staged a large-scale protest in November 2009, which was widely covered in state media across the country (cp. Chapter Two).

Like the Aobei residents, who were an important point of reference for the Panyu campaigners, the Panyu homeowners set up an online forum that became their main platform for discussion and organization. This 'Jiang-Wai-Jiang' forum (江外江论坛, *Jiang wai jiang luntan*), the residential community forum of the Lijiang Huayuan Residential Community (丽江花园社区, *Lijiang huayuan shequ*) in Panyu, became widely read by contentious communities and individuals concerned with incineration across the country in 2009 and 2010. It served as a major platform of exchange between Panyu and Aobei residents, with members of both communities active on and reposting threads from each other's forums (interviews NGO ECO1/PY1 1-7-13, NGO ECO1/PY1 10-7-13, ASW1 17-10-12, 31-7-13; *Jiang-Wai-Jiang* 2010a; *Global Voices* 2010). The Phoenix TV program event was the first in-person meeting of members of the Beijing and Guangzhou communities (interviews NGO ECO1/PY1 1-7-13, NGO ECO1/PY1 10-7-13, ASW1 17-10-12, 31-7-13). Since the Asuwei campaign was already at a more advanced stage, diffusion effects between the communities primarily went in the direction of the Guangzhou residents. For the Aobei campaigners, the case of Likeng – which was similar to the Gaoantun case – primarily demonstrated the inherent difficulty of proving causal linkages between health impacts and incineration or other waste facilities. This further encouraged their opposition to the plant (interviews ASW1 31-7-13, ASW2 27-5-13).

Of greater consequence for the Aobei residents' struggle than the meeting with the Guangzhou community members was the meeting between Huang and the BMAC waste expert Wang. The two men met casually in front of the studio rented for the Phoenix TV recording – as two smokers probably sharing a cigarette pause. They started up a conversation and realized that they got along quite well (interviews ASW1 17-10-12, 31-7-13, BMAC

7-11-12; D_NU1; Sipan Li 2010; L. Li 2010a, 2010b; Xu 2010). After the recording of the TV program – the broadcast was eventually cancelled due to political pressure (interviews ASW1 31-7-13, GAT 15-6-13, BMAC 7-11-12) – the two men established close personal relations and frequently met up for discussions and the exchange of opinions. The friendly relations between Huang and Wang also improved the broader relations between the Beijing government and the larger Aobei community. When the Aobei research report was finished in January 2010, Huang provided Wang with a copy. The BMAC official was deeply impressed by the level of expertise of the report and its recommendations for Beijing waste policies (interviews ASW1 17-10-12, ASW2 27-5-13, BMAC 7-11-12; D_NU1; Sipan Li 2010; L. Li 2010a, 2010b; Xu 2010).

After the recording of the TV program, Wang initiated a government-led inspection tour to visit incinerator plants in Japan and Macau, which were regarded as good examples of clean incineration technology. He hoped that the tour in February 2010 would increase the residents' support of incineration. Wang invited Huang to join the group as the city resident representative (市民代表, *shimin daibiao*). Further participants were BMAC and city government officials and two journalists from state-led media institutions (interviews BMAC 7-11-12, ASW1 17-10-12, ASW2 31-7-13; D_NU1; Cui 2010a, 2010c; L. Li 2010a, 2010b; Sipan Li 2010; Li and Xuyang 2010; Meng 2010e; *Xinhua* 2010c; Xu 2010). As Wang recalled about the significance of the TV program event and the government's rationale for the inspection trip:

> Just at this time, in 2009 and 2010, there were protests against the construction of incinerators in more than 30 cities. [...] The government was very agonized, very embarrassed. There were two alternatives. One was working with the *laobaixing*, get them to accept this [the projects]. One was not to build them in the same place anymore. There are these two alternatives, but I think there is only the alternative to communicate. Because we have turned from a nondemocratic society to a democratic and people's livelihood society (民主和民生的社会, *minzhu he minsheng de shehui*). The *laobaixings'* opinions should be respected, there should be communication. [...] So the government was also willing to communicate at the time. But in the beginning of the communication, the antagonism was very fierce. [...] So we did what you can call a national precedent. We did a TV program, a discussion with this Hong Kong TV station [...]. At that time, Guangzhou's [... representative], Beijing's Huang [...], Zhejiang's [... representative], we gathered all those famous opponents of incineration [...] to have a discussion with me. At that time the atmosphere was very tense. [...] The majority of people were all opposing incineration. [...] It felt like I was sitting in the

> defendant's seat. At that time I posed two questions: First, who doesn't produce waste? Please raise your hand. There was no one, everyone produces waste [...]. The second question I raised was: Who has been to a WtE plant? A real WtE plant in Germany or Japan or Singapore. No one had. I said I very much admire you being so persistent. But you essentially don't understand the incineration technology. [...] I suggest that we, the government, organize a group to go abroad and have a look at what a WtE plant actually is. After we've seen it and come back, don't hesitate to continue your resistance. [...] In the end, the discussion turned out pretty well. We all [...] established good contacts. Afterwards they came to me all the time, and I came to them all the time. [...] We have all become friends. (Interview BMAC, 7-11-12)

Huang took his position as citizen representative on the inspection trip very serious. In preparation for the journey, he opened a personal Sina blog for live reporting from the trip.[72] The first entry on his blog was a poem that he wrote and which has since become widely cited entitled 'Aobei does not believe in tears' (奥北不相信眼泪, *Aobei bu xiangxin yanlei*). He also collected questions from the Aobei community and other Beijing residents to take on the trip and pose to the operators of the WtE plants he visited and their neighbouring residents (interview ASW1 17-10-12; L. Li 2010a, 2010b; Li and Xuyang 2010). According to Huang, it was again the Liulitun campaigners who played an important role in helping him amend the questions (interview ASW1 31-7-13).

Accompanied by major state media reporting, the inspection group left for Japan and Macau on 22 February 2010 (e.g., L. Li 2010a, 2010b; Xinhua 2010c). By their return on 3 March 2010, Huang had turned into a media star. In the context of the spreading local opposition to waste incinerators, the inspection trip was hailed as a precedent for not only how the government had communicated with a contentious community and attentively listened to the concerns of the *laobaixing*, but also how this had even turned into cooperation between the government and the community members (ibid.; D_NU1; S. Li 2010a, 2010b; Sipan Li 2010; Xu 2010). As Huang recalled:

> I really became famous overnight. When we came back from the inspection trip, more than one hundred journalists were waiting for me at the airport. [...] China has never had such a thing before: An individual that first fights with the government [...], that goes out to oppose the government, is arrested by the government, and as a result is invited

72 The web address of Huang's Sina blog is: http://blog.sina.com.cn/lvshidaner (19 April 2019).

> by the government to go on an international inspection trip together. They [the media] thought this example was highly entertaining (好玩, *haowan*). (Interview ASW1 17-10-12)

While Huang's close personal relations with the government, his participation in the inspection trip, and his media prominence were met with initial criticism and distrust from the larger Aobei homeowner community, the immense media attention to that it brought their case was welcomed (interviews 17-10-12, 31-7-13, ASW2 27-5-13; Sipan Li 2010; Xu 2010). Other campaigners soon joined in the praise of the government's positive attitude and openness towards the *laobaixing* and presented the Asuwei case as a model for how siting disputes should be approached: 'This dialogue between the government and the people produced a series of effects and reactions. It's not like in many other places, where the public goes on the street to cause trouble, which the government firmly suppresses. And the result is that the popular resentment gets very serious and everyone's emotions are uncontrollable. Here, things went very well, that's maybe a special characteristic compared to other places. We can be a model for other places' (Interview ASW2 27-5-13). Huang's main personal conclusion from the trip was that incineration technology was a sophisticated technology. However, for it to be safely used in China, the country would have to introduce effective waste sorting measures (interviews ASW1 17-10-12, ASW2 27-5-13; Cui 2010c; Shuang Li 2010c; Meng 2010e; Xu 2010; J. Yang 2010).

From Lawyer to Environmentalist: Policy Impacts, the Foundation of Green Home, and a First Homeowner Success

The inspection trip and the improving relations between the Aobei homeowners and government officials brought about concrete effects not only for the Asuwei campaign but also by impacting city-level waste policies. This was enhanced by two national-level events around that time that had a major impact on the realm of waste. First, following the Panyu residents' large-scale protest in Guangzhou in November 2009, the Guangzhou government organized an expert symposium on waste treatment strategies and the suitability of the Panyu incinerator for Guangzhou at the end of February 2010. The leading 32 waste experts from across the country were invited to the forum. Among the participants were one of the main incineration advocates, professor Nie Yongfeng from Tsinghua University, and one of its fiercest opponents, retired professor Zhao Zhangyuan (interviews Zhao 8-11-12, Nie 7-6-13; *Guangzhou Daily* 2010; Qiu 2010; *Southern Metropolis*

Daily 2010; Xu 2010; Ye 2010). At the end of the symposium, the experts jointly signed a policy paper stating that all experts except one were in favour of waste incineration. The only expert openly opposing incineration was Zhao Zhangyuan. The event received major nation-wide media attention, with most headlines including the phrase '31:1' and explicitly naming Zhao as the sole expert opposing incineration technology. This laid the foundation for Zhao's national (media) prominence as the primary Chinese anti-incineration expert. The event raised public and media attention about waste incineration to new heights (cp. Chapter Two).

Shortly thereafter, in the first half of March 2010, the annual 'two meetings' (两会, *liang hui*) of the National People's Congress and the Chinese People's Political Consultative Conference (中国人民政治协商会议, *Zhongguo renmin zhengzhi xieshang huiyi*, CPPCC) took place in Beijing. Waste treatment and incineration were discussed as a major issue in the meetings. This was regarded by the Aobei campaigners as favourable circumstances for raising further public attention to the issue. As a people's congress representative, BMAC waste expert Wang presented the Aobei residents' research report to the NPC and Huang was interviewed by CCTV for a special feature on waste treatment strategies in the context of the 'two meetings' (interviews ASW2 27-5-13, BMAC 7-11-12; Xu 2010).

Shortly thereafter, at the end of March 2010, state media cited Beijing's vice-mayor Ji Lin with saying that the Beijing authorities were very concerned with the issue of waste treatment. While firm about continuing to use incineration technology, Ji announced that the Beijing government would introduce comprehensive waste reduction, classification, and recycling measures preceding the burning of garbage (Cui 2010b, 2010c; Shuang Li 2010b; Meng 2010c, 2010d; Xuyang 2010; Yang 2010; Zhai 2010). the state-led English-language newspapers *China Daily* and *Global Times* directly linked this policy change to the Japan inspection trip and the participants' recommendations (Cui 2010b; Zhai 2010). In virtually all subsequent media articles about the planned waste sorting system, lawyer Huang was quoted as expert on the issue (Cui 2010b, 2010c; Shuang Li 2010b; Meng 2010c, 2010d; Xuyang 2010; J. Yang 2010; Zhai 2010). A few days after the announcement, Aobei and Gaoantun residents came together to draft a proposal about garbage sorting, which was eventually submitted to the government in early 2011 (interviews ASW1 31-7-13, GAT 1-11-12; Shuang Li 2010b; X. Li 2011; Liang 2011). In April 2010 Huang flew to Guangzhou, where waste sorting measures were being introduced at the time, to discuss waste issues with Panyu campaigners and officials (interviews ASW1 31-7-13, NGO ECO1/PY1 1-7-13; Cui 2010c).

On 17 March 2010, BMAC published an official statement on the Asuwei incinerator project and for the first time officially replied to the Aobei residents' claims regarding the facility's EIA problems. The statement stressed that construction would not begin before the official ratification of the EIA report and that a future solicitation of public opinions would be conducted after an abridged EIA report was completed. Moreover, acute measures would be taken to combat the stench from the landfill – about which the Aobei residents had continued to issue formal complaints – by installing deodorant guns and additional plastic covers by May 2010, similar to the measures conducted at the Gaoantun landfill (H. Li 2010; Watts 2010; Xu 2010). Rumours that the Beijing government had shelved the Asuwei project also started circulating, but were repeatedly denied by government officials. In June 2010 the Beijing government announced on its website that the Asuwei construction would begin in 2010 despite the controversy, and reiterated its plans to complete nine MSWIs in Beijing by 2015. Huang immediately mobilized the Beijing media, who reported comprehensively about the news and gave Huang a platform to express his opinions as the representative of the Aobei community (interviews ASW1 17-10-12, ASW2 27-5-13; Cui 2010b; Meng 2010c, 2010d; J. Yang 2010; Zhai 2010).

Frustrated by the continued construction plans and concerned that the government's renewed push for incineration would set aside the introduction of effective waste sorting measures despite the government's earlier promises, Huang decided to take the issue in his own hands. He had come to the conclusion that effective waste sorting by individual households was not feasible in China and would need years of education. Further, he was convinced that the issue could not be left to the government alone. Huang thus decided to use his private funds to establish a recycling company that would sort and pre-treat unseparated household waste in Beijing, thus reducing the amount of incinerated waste and rendering the residue more suitable for burning. Over the course of the next months, he developed his ideas further and started the company registration process at the end of 2010. The official establishment of Huang's sorting station, named 'Green House' (绿房子, *Lü fangzi*), was completed in July 2011. A testing phase in which garbage from around 2000 neighbouring households was collected and treated commenced in August (interviews ASW1 17-10-12, 31-7-13, BMAC 7-11-12; D_NU1; *Beijing Evening News* 2011; Beijing News 2011a; Cheng 2011b; *CNature* 2011; Global Times 2011b; Meng 2010e; X. Li 2011; Liang 2011; L. Zhao 2012).

By then, Huang had gained the support of the Beijing government, which subsidized the project. The project was also applauded by the media. In state media and on the Internet, Huang was listed as one of the 2010 'Stars of the

Capital' (Wang and Balazovic 2010) and featured as 'Public Welfare Person 2011' (2011 公益人物, *2011 gongyi renwu*) (*Netease* 2011). TV features about his transformation from an arrested 'outlaw' to a government advisor and environmentalist were broadcast on CCTV 1 (2011), Phoenix TV (2014), Beijing TV (2013), and Sohu TV (2014). His plans were also welcomed by Beijing environmental organizations, particularly Green Beagle and Friends of Nature, who approved of his approach, recruited volunteers to assist Huang at 'Green House', and invited him for several public lectures (interviews ASW1 17-10-12, 31-7-13, NGO NU1 7-11-12, NGO NU2 14-10-12, 29-10-12, NGO NU3 25-7-13, NGO FON1 19-10-12; Beijing Evening News 2011; Beijing News 2011a; Cheng 2011b). In April 2011, Huang was nominated by Friends of Nature for the 2011 Green Person Award presented by the Beijing-based environmental organization Society of Entrepreneurs and Ecology (SEE) Foundation (Chinese name: 北京市企业家环保基金会, *Beijing shi qiye jia huanbao jijinhui*). Huang had successfully transformed himself from a lawyer and contentious homeowner into a government advisor and respected environmentalist, thereby lifting the Asuwei campaign well beyond the 'NIMBY' level.

After the construction of the Asuwei incinerator plant did not start in 2010 as had been officially announced earlier that year and when no further government news on the project was published throughout the years 2011 and 2012, the homeowners concluded that the project had been unofficially shelved by BMAC and the Beijing government. This was confirmed by BMAC waste expert Wang in late 2012 (interviews ASW1 17-10-12, 31-7-13, ASW2 27-5-13, BMAC 7-11-12). The established Aobei anti-incinerator community soon dissolved back into their individual busy daily lives and the residents were cautious not to pick at the fragile condition of the project's shelving. The Aobei homeowners had won a first victory, which in 2014 proved to be only temporary.

Becoming a Broker: The Role of the Asuwei Campaign for Other Contentious Communities

Due to its media prominence, strong presence on the internet, level of expertise, and (temporary) success in halting the project, the Asuwei campaign played an important role for other contentious struggles.[73] In the words of

73 Since an analysis of the specific impacts of the Asuwei campaign on other contentious struggles requires respective case studies, in this section the role of the Asuwei campaign is discussed with a focus on the Asuwei perspective. The section draws on the field visits to and interviews with the other affected communities conducted for this study, as well as my participation in several conferences organized by and interviews with members of the Chinese 'no burn' community.

Huang, 'You can say that Asuwei was the [incineration] dispute's blasting fuse in China. Asuwei is extremely well-known. [...] Asuwei has become a main battlefield, the Yan'an[74] for other Chinese cities' (Interview ASW1 2012-10-17).

The significant influence of the Asuwei campaign was confirmed by several members of the Chinese 'no burn' community and of affected local communities across the country. Apart from the myriad of (state) media articles and TV features on the campaign, information on the case also spread online via the widely-read Aobei forum and reposts on the just-as-prominent Panyu residents' *Jiang-Wai-Jiang* forum. On these internet platforms the members of other affected communities were able to personally approach the campaigners and ask for information and advice (interviews Zhao 8-11-12, NGO NU1 7-11-12, NGO NU3 27-5-13, NGO FON1 19-10-12, NGO WEC 23-10-12, NGO CLAPV 18-10-12, LK 4-7-13, 9-7-13, SJ4 17-5-13, FQ1 14-7-13)

Relations were particularly close between members of the Aobei and Panyu communities. These relations continued beyond the main phase of the Asuwei campaign; a frequent exchange of information and opinions – both online and at joint conference visits and workshops – persisted at the time of my field research (interviews ASW 1 31-7-13, NGO ECO1/PY1 1-7-13, NGO ECO1/PY1 10-7-13, NGO ECO2 10-7-13, BMAC 7-11-12, Zhao 8-11-12). A main reason for these sustained personal relations was that, like Huang, the Panyu campaign leader had also established his own environmental organization dedicated to sustainable waste treatment strategies in Guangzhou, which was named Eco Canton (宜居广州, *Yiju Guangzhou*). According to its founder, Huang 'was an example for us, also regarding waste sorting' (interview NGO ECO1/PY1 10-7-13). As such, the cases of Asuwei and Panyu show strong parallels. At the time of field research, they were the only two cases where Chinese local anti-incinerator siting disputes, which had started out from a 'NIMBY' perspective, had given birth to institutionalized organizations dealing with broader waste issues and the explicit aim of invoking policy change, thus shifting the local claims and concerns to higher levels.

Both campaigners also became important intermediaries for other cases. In the words of a Green Beagle staff member:

> [The Panyu campaign leader] and Huang [...] have both established their own environmental organization, they also got engaged in other cases. The best is probably Huang [...]. He didn't participate in any NGO [sic.].

74 Yan'an, which was the base of the Chinese Communists under Ma Zedong between 1936 and 1948, is celebrated by the Chinese Communist Party as the 'birthplace of the revolution'.

> He took things in his own hand and created something new and fresh as an individual. Everyone likes him. Everyone loves him. The media also really like paying attention to him. Whenever there is a case of resistance (反抗案例, *fankang anli*), he helps to develop it at any time. (Interview NGO NU1 7-11-12)

Since 2010, Huang also kept an active personal Sina microblog profile, which served as an important communication platform and means of information dissemination. In his role as a prominent (self-made) waste expert and spokesperson for the Asuwei campaign, Huang was contacted and asked for assistance by members of several affected local communities across the country:

> I am contacted by people almost every month. At the end of 2012, I got so many personal letters and emails and microblog posts of people I didn't know and who asked me to pay attention [to their case]. I very much invite other residents to contact me, to pack their bag and come to visit me in Beijing. I always try and meet them, also if they are in Beijing for other things. I will hold nothing back and tell them what to do, no problem. If they have no finances to come to Beijing, I will support them. I try to give whatever is asked, because I know that civil activities (*minjian huodong*) are not easy and in my heart I am anxious. [...] If they ask for me to come, I also try to visit them. Whenever there is an incident, I try to go there. Also people who want to do waste sorting contact me. [...] The most important thing I can do for them is to share my knowledge and experiences and provide them with media resources. Because I know the important media, also in the South, they often ask for my opinion. [...] Among all of us affected people, disadvantaged people, in Beijing, Shanghai, Jiangsu, Guangzhou, we have this tacit agreement that we support each other. (Interview ASW1 31-7-13)

Huang was contacted by and provided assistance to residents from several of the background cases in this study for which first-hand information was collected. Based on personal contact established at the Phoenix TV program event, for example, the villagers from Guangzhou's Likeng case invited Huang to pay a visit to their village and provide advice on how to continue their struggle in 2011. Huang visited the village twice and gave them advice on how to communicate with the government, sharing his experiences and offering suggestions. The villagers also asked Huang, in his position as a lawyer, to represent them in a lawsuit against the government and MSWI operator and to initiate a media campaign. Huang declined both requests for reasons of political sensitivity (interviews ASW1 31-7-13, LK 9-7-13).

Villagers from Huangtutang village, Wuxi city, in Jiangsu Province, who had blocked the construction site of a planned incinerator near their village for weeks in March and April 2011 and got into a violent clash with security forces in May, which left several villagers severely injured, sought advice from Huang and other intermediaries when visiting Beijing in 2011 (interviews WX 6-5-13, ASW1 31-7-13, NGO NU4 8-11-12, NGO NU1 7-11-12, NGO NU3 25-7-13). Based on his own experience, Huang told them to prioritize their own safety and engage in a dialogue with the government to settle the issue:

> I told them, first, since they had already caused such large trouble (闹得那么大, *nao de name da*), they have to strictly control things so that they don't develop in a vicious direction. You don't want a clash with armed forces (武力冲突, *wuli chongtu*). From my experience, if you disrupt the public order they take people into custody, sentence them, right? Second, I told them if you make such trouble, that doesn't fundamentally settle the problem, it's only a temporary compromise. Third, I said, you need to elect a group of people to function as your representatives during negotiations. Go talk with the government, ask them some important questions such as how much waste they want to burn, [...] whom they would resettle [...]. Disregarding of your trouble-making, return to the conference table! Only through reasonable discussions can you solve the problem, can you represent the interests of the *laobaixing*. (Interview ASW1 31-7-13)

Green Beagle staff – who were closely watching the Huantutang case and provided assistance to the villagers, since one staff member was a Huangtutang native – also asked Huang and other intermediaries to pay a visit to the village. While the lawyer Xia, professor Zhao Zhangyuan, and Green Beagle staff visited the village in April and early May 2011, the local situation had already become too sensitive by the time of Huang's planned visit to the locality (interviews WX 6-5-13, ASW1 31-7-13, NGO NU4 8-11-12, NGO NU1 7-11-12, NGO NU3 25-7-13, Zhao 8-11-12, Xia 30-7-13). Huang and Green Beagle staff also used their personal relationships with the media to initiate some reporting about the case. While several journalists visited the village and wrote reports about the unfolding events, the publication of the reports was blocked (interviews WX 6-5-13, ASW1 31-7-13, NGO NU4 8-11-12, NGO NU1 7-11-12, NGO NU3 25-7-13).

Residents from Shanghai's Songjiang district, where homeowners fought against the planned construction of an incinerator with several large-scale 'strolls' in May and June 2012, contacted Huang online and asked for his support (interviews ASW1 31-7-13, SJ1 18-5-13, SJ2 23-5-13, SJ4 17-5-13, 20-5-13). While he did not personally visit the site, he did help the Songjiang residents establish contact with several media houses to increase public attention to the case. Among others, Huang contacted a Phoenix TV journalist who had previously interviewed him. The journalist subsequently visited Songjiang and featured the case in a TV program about China's 'NIMBY movement' (Phoenix TV 2012). Huang also provided information and advice to the Songjiang residents via online communication (interviews ASW1 31-7-13, SJ1 18-5-13, SJ2 23-5-13, SJ4 17-5-13, 20-5-13). Residents from Huizhou's Huiyang district in Guangdong Province, and Shanghai's Fengqiao village in Fengxian district – where the local governments announced the construction of waste-to-energy plants in 2013 – also knew about and were in online contact with Huang, from whom they gathered information about incineration, the Aobei campaign, and suitable means of resistance (interviews HY1 11-7-13, HY2 23-6-13, FQ1 22-6-13, FQ1 14-7-13).

Apart from media reporting and the internet, information diffusion about the Asuwei campaign was facilitated by the inclusion of the Aobei residents' research report in a 'toolkit' of documents compiled by Green Beagle staff to assist affected communities. This collection of documents was distributed to all community members who sought the organization's assistance and was available for public download from an organization member's Sina microblog account (D_NU2). During my field research, I first heard about and obtained this 'toolkit' from Songjiang residents. Apart from the Aobei research report, the document collection included the central Chinese laws and regulations relevant to incineration and EIA procedures; Chinese-language versions of important international studies on the issue – including the influential GAIA, Greenpeace, and World Bank studies on incineration and health, as well as works by Paul Connett;[75] Chinese-language versions of documents and reports from anti-incineration movements and campaigns in other countries and regions, including Taiwan, the UK, the United States, and the 'Communities against Toxics' (CATS);[76] information booklets and handbooks for affected communities compiled by Chinese organizations,

75 Most of the Chinese translations of these documents were compiled by members of the anti-incineration campaigns in Taiwan, where the issue of incineration had been a hot topic and prompted widespread resistance since the 1990s (interview NGO NU3 26-3-14).

76 CATS is an organization opposing waste incinerators in the United Kingdom, Ireland, and Europe, with the aim of bringing together affected communities to share their experiences. CATS emerged in 1990 and is focused on providing assistance to affected local communities via information and contacts with experts and professionals.

including Green Beagle and the Wuhu Ecology Center; and a blueprint for an 'opinion booklet' that could be amended by the adopting community. As such, the Aobei research report and the broader 'toolkit' were not only a crucial resource for later contentious struggles, but also an important means of transnational diffusion. It ensured that international materials reached affected communities at the local level that had no other means of assessing such information.

The Aftermath: Social Cleavages, Villager Contention, and Renewed Construction Plans

The larger Aobei homeowner community considered the Asuwei campaign to be virtually concluded by 2012, after the project was unofficially shelved. The case was not settled for the Asuwei villagers, however. While the government had announced their resettlement in 2009, they had not yet been relocated by the official date in June 2012. Their repeated claims for compensation had not been met. The villagers' continued complaints and small-scale protest actions – there were some news that villagers had been arrested in June 2010 after blocking waste cars (*Jiang-Wai-Jiang* 2010a, 2010b, 2010c) – remained unsuccessful. The villagers' frustration culminated in a large-scale blockade of the waste facility's access road that lasted for several days from the end of July to the beginning of August 2013. The blockade was eventually dissolved by public security forces with the detaining of several villagers and did not produce a noteworthy outcome (interviews ASW1 31-7-13, NGO NU2 26-3-14, NGO NU2 1-7-15, NGO NU3 25-7-13; D_ASW6; ASW villagers' microblog 2013a, 2013b, 2013c, 2013d; F. Liu 2013a, 2013b).

The villagers blamed the homeowner community for selfishly opposing the incinerator without considering the well-being of the more disadvantaged group of villagers who had already had to cope with stench, pollution, and significant health impacts from the landfill for years (interviews ASW1 31-7-13, NGO NU2 1-7-15, NGO NU3 25-7-13; D_ASW6; ASW villagers microblog 2013a, 2013b, 2013c 2013d; F. Liu 2013a, 2013b). The Aobei homeowners refuted the claim that the delayed resettlement was linked to the shelved construction of the incinerator, instead arguing that the resettlement plans had always been based on the landfill's prior pollution problems and the government's plans to enlarge the scope of the Asuwei landfill. According to their assessment, the delay originated from a dispute between the different levels of government about the amount and origins of compensation funds (interviews ASW1 31-7-13, ASW2 27-5-13). Despite the cleavages between the two social groups, the villagers contacted Huang during their road blockade to seek his help for distributing their claims and photos of the incident via his microblog account and media relations. Reporting on the incident was not permitted, however, and the respective posts on Huang's microblog account were soon deleted (interview ASW1 31-7-13; live observation of Huang's Sina microblog account in July 2013).

The Aobei homeowners' hopes that the project had ultimately been set aside were disappointed in 2014, however. After the homeowner community had thought itself safe for almost two years, a renewed EIA announcement was made in December 2014. The two public announcement phases for the solicitation of public opinion as required by the EIA Law were set for December 2014 and February 2015. The construction of the Asuwei Circular Economy Park was to begin in mid-2015 and completed at the end of 2017 (interview NGO NU2 1-7-15; email communication with Nature University staff; D_ASW6 to ASW13; *Beijing Daily* 2015; *Global Times* 2014; Kong 2015). The renewed construction plans rekindled the opposition to the facility by some of the homeowners who had already engaged in the first stage of contention. Huang no longer participated in this second phase of resistance and there were rumours that he had changed sides and was now promoting the incinerator (interview NGO NU2 1-7-15; email communication with NU staff; *Global Times* 2014).

This time, the contentious homeowners were actively supported by Green Beagle staff who urged them to resort to legal means, drawing on their by-then-ample experiences with local anti-incinerator struggles (interview NGO NU2 1-7-15; email communication with NU staff; D_ASW6 to ASW13; *Global Times* 2014; Kong 2015). Organization staff and Zhao Zhangyuan helped the residents identify a lengthy list of problems in the EIA procedures and abridged EIA report (D_ASW11). Green Beagle staff subsequently represented the Aobei resident community by writing a letter to the waste incinerator's operating unit Beijing Huayuan and new EIA unit Sinoma Geological Engineering Exploration Academy Limited (Chinese name: 中材地质工程勘查研究院有限公司, *Zhongcai dizhi gongcheng kancha yanjiuyuan youxian gongsi*, in the following referred to as Sinoma Geological Engineering), to request a public hearing about the project (D_ASW7).

On 23 April 2015 a public hearing was held at the Beijing EPB with community members staging a small, peaceful protest outside the bureau (D_ASW8, D_ASW9; Kong 2015). Among the participants of the hearing were representatives of the facility's operating and EIA units, local government officials, Aobei homeowner and villager representatives, Green Beagle staff members, and Zhao Zhangyuan. The homeowner representatives, Green Beagle staff, and professor Zhao pointed out several flaws in the

EIA procedures, accusing the EIA unit of severe misconduct, and noted the unsuitability of the site for constructing a MSWI. Villager representatives stated their frustration that their grievances had been side-lined over the course of the last few years and repeated their claims of a timely resettlement and compensation payments (D_ASW6, D_ASW8, D_ASW9). The homeowner representatives were deeply dissatisfied with the responses of the operating and EIA units and government personnel, however, and the Beijing EPB approved the Asuwei incineration project five days after the hearing (D_ASW10, D_ASW12; *Beijing Daily* 2015).

At the end of June 2015, Green Beagle staff members assisted a homeowner representative in submitting an administrative redress application with the MEP to request that the Beijing EPB's project approval be revoked (interview NGO NU2 1-7-15; D_ASW12). This was declined by the MEP in mid-August 2015 (interview NGO NU2 1-7-15; D_ASW13). At the same time, Green Beagle together with seven other environmental organizations – including Friends of Nature – sent an open letter to the MEP, which requested that the EIA unit's EIA qualifications be revoked due to the flaws in the Asuwei EIA procedures (interview NGO NU2 1-7-15; D_ASW10, D_ASW11). Since 2009, these methods had become Green Beagle's primary strategy to counter planned incinerator projects. Flaws in the EIA procedures were a common issue that was also at the core of the Panguanying and Dagong village cases outlined in Chapters Four and Five.

Construction on the Asuwei incinerator broke ground in June 2015 without the prior relocation of the villagers. This led to a second large-scale blockade of the facility's access road by the villagers in early July. Again, the blockade was dissolved by public security forces after a few days, with several villagers taken into detention. After the incident, the Beijing government promised compensation payments to the villagers, which had not been distributed by the end of 2015. The core group of Aobei homeowners declared that they would continue their fight against the incinerator through legal means (interview NGO NU2 1-7-15; D_ASW6).

Analysis: The Role of Horizontal and Vertical Linkages in the Asuwei Campaign

The Asuwei case shows that both horizontal and vertical linkages played a significant role in Aobei homeowners' contention against the planned incinerator project. While strongly intertwined in terms of their impact on the development of the Asuwei campaign, in the following sections they are discussed successively for reasons of clarity. A noteworthy feature of the Asuwei case is the pivotal role of transnational diffusion processes, which is tackled in both sections. The concluding section discusses what the findings tell us about the prospects for scaling up local contention.

The Role of Horizontal Ties

Overall, the Aobei campaigners were well-connected with other affected communities in China through both close personal relations within Beijing and online communication and sporadic meetings across geographic distances. The case of Asuwei therefore demonstrates that a close domestic network of ties is emerging between different affected local communities – at least amongst urban residents – in which major diffusion processes are at play. This is greatly facilitated by the possibilities of digital communication across time and space. The Aobei community's ties with other local groups were not limited to the domestic level; they were also connected to affected communities and (trans-)national anti-incineration movements abroad. While these transnational linkages and diffusion processes were primarily limited to nonrelational online information flows, they show that Chinese anti-incineration contention is nonetheless embedded into and spurred by the global anti-incineration movement and incinerator siting disputes in different world regions.

In terms of the nature of the linkages, the Aobei homeowners' ties with other contentious communities were greatly facilitated by community members' good access to and accomplished use of the Internet. This permitted both the assessment of domestic and international online information and direct communication with members of other communities, including across geographical distances. While domestic information flows between the different communities were initially based on nonrelational channels via the Internet and state media, direct relational ties were soon established with members of the other Beijing homeowner communities (Liulitun and Gaoantun) both online, via the homeowners' 'Aobei forum', and offline, through regular meetings and symposiums. Ties with contentious communities in other parts of China (mainly Guangzhou Panyu) first developed online via social media. Through the brokerage role played by Phoenix TV, which brought together members of contentious communities from across China for their program, the Aobei campaigners also established direct face-to-face ties with community members from other regions (Panyu and Likeng in Guangzhou, Wujiang in Jiangsu Province). After Huang's transformation into a full-time environmentalist, the members of several other communities (e.g., Shanghai Songjiang, Huangtutang in Jiangsu Province, Huiyang in Guangdong Province) contacted Huang to seek his advice and assistance. They established online relational ties that were in some cases followed up by personal meetings in Beijing or through Huang's site visits.

In terms of these linkages' effects on the Asuwei campaign, the Aobei community's knowledge about other domestic and international anti-incinerator disputes provided important cognitive cues and facilitated the emergence of a 'contentious consciousness' for the Aobei residents. This fostered the formation of a collective identity as part of a broader anti-incineration and *weiquan* community and contributed to a heightened perception of opportunity. Moreover, the rapid increase in the number of local anti-incineration struggles across China produced favourable political opportunity structures for the campaign (such as during the time of the 'two meetings'). The Aobei campaigners referred to the growing number of domestic and international anti-incineration struggles to certify their claims both against the Asuwei incinerator and for broader policy change by presenting themselves as but one in a rising tide of local communities standing against a hazardous waste strategy pursued by misguided policymakers. This framing was also at the core of the community's self-portrayal as representatives of all Beijing residents and affected communities at large.

Despite the dense network of ties with other contentious communities across the country – many of which developed in later stages of the struggle – it was primarily the linkages with other Beijing homeowner communities, particularly the Liulitun campaign, that had a significant impact on the development of the Asuwei struggle. Online and media information about the Liulitun campaign, later complemented with personal ties with the Liulitun residents, helped the Aobei community pierce the initial information haze about incineration and the dangers of dioxin, providing crucial cognitive cues and critical frames that facilitated a collective interpretation of the situation and led to the onset of contentious action. The Liulitun campaign's positive outcome raised the Aobei residents' belief in the chance of success of a more contained strategy, while the negative outcome of the Gaoantun residents' more disruptive course of action dissuaded the Aobei campaigners from following a similar path. Regarded as a successful predecessor, the Aobei residents emulated the Liulitun residents' 'rational' and 'peaceful' approach based on predominantly legal means and the residents' own lay expertise. This included the writing of a research report as a central means of expressing their claims. China-specific information about the harms of incineration, EIA procedures, and the suitability of siting decisions provided by the Liulitun residents in both their 'opinion booklet' and personal discussions were valuable resources for the Aobei community. While the Aobei residents later explicitly distanced themselves from the 'NIMBY' concerns of the Liulitun campaign, it nonetheless had an important impact on the Aobei community's course of action. Collaborative action between the different Beijing homeowner communities beyond the exchange of information and a loose 'alliance of support' was, however, impeded by China's restrictive political framework and the associated risks of joint action.

Transnational diffusion processes played a similarly important role in the development of the Asuwei struggle. The online information coming to the Aobei residents from contentious communities and (trans-)national movements abroad played a similar role as the Liulitun campaign in providing the homeowners with reliable information about the environmental and health impacts of incineration, thus enhancing the formation of a 'contentious consciousness'. The information also certified critical domestic information that arrived at the residents via other pathways of communication, such as the information provided by Zhao Zhangyuan. Since the main role of transnational diffusion processes was played by international experts and environmental organizations and thus falls under the parameters of vertical linkages, it is discussed in the following section.

The Role of Vertical Ties

Closely interrelated with the horizontal linkages outlined above, vertical ties with domestic and international intermediaries played an important role in the Asuwei campaign. The primary domestic intermediary that impacted the Aobei community was Zhao Zhangyuan. Zhao first became known to the Aobei homeowners via nonrelational channels through both the media – particularly the CCTV report – and online information. The residents then established direct ties with Zhao by inviting him to the Aobei neighbourhood for a lecture. Based on his earlier research about the landfill problems in Asuwei and his role as an outspoken critic of incineration in the Liulitun campaign, the Aobei homeowners trusted Zhao to provide them with reliable information about the hazards of incineration and the Asuwei area's environmental situation, despite the overall information haze. This information not only fostered the community's sense of alarm and entitlement – facilitating a collective interpretation of the situation and 'contentious consciousness' – but also provided them with valuable resources for claim-making vis-à-vis the government and public. By certifying the homeowners' claims in media articles and on his personal blog, Zhao lent further legitimacy to the residents' claims.

A minor role was also played by the Beijing Lawyers Group, who provided the community with legal information after being personally approached by the residents. This information assisted the campaigners' claims-making in their research report and towards the government. The knowledge about both domestic and international environmental laws and regulations and the information about legal procedures related to the construction of waste facilities in other countries helped the residents base their argumentation on solid grounds.

Environmental organizations – which were not yet actively involved in the issue of waste incineration at the time of the campaign's onset – only played a minor role in the early stages of the Asuwei struggle. Their engagement at this time was primarily limited to the provision of information by individual organization members. Due to the homeowners' ability to gather relevant information, however, this did not have a noteworthy impact on the campaign.

During the second stage of contention in 2014 and 2015, Green Beagle staff in particular actively supported the contentious homeowners. By strongly advising the residents to resort to legal means, organization staff had an important impact on the residents' later path of action. Drawing on their prior experiences with other anti-incinerator siting disputes, Green Beagle staff lent the homeowners their issue-specific and environmental knowledge to identify major flaws in the EIA procedures. They subsequently represented the homeowner community in requesting a public hearing and stating their claims at the meeting – thereby certifying the homeowners' requests as environmental and waste experts. Organization staff also used their legal knowledge to assist the homeowner community with submitting a request for administrative redressal of the project approval to the MEP and – bringing together several environmental organizations – pressured the MEP to revoke the EIA unit's EIA qualifications due to flaws noted in the Asuwei EIA procedures. The organization's attempts to reach out to the media to raise public awareness about the case and increase pressure on the government were, however, of limited success. Together, the support provided by environmental organizations during the later stages of the Asuwei struggle constitute the organizations' main strategies for assisting affected local communities also in other localities.

A significant role was also played by transnational diffusion processes. While limited to online information flows, the studies and research work of international organizations, experts, and transnational anti-incineration networks – which had already had a major influence on anti-incinerator disputes and anti-incineration movements abroad – also played a crucial role in the Asuwei campaign. In particular, the studies on incineration and health by GAIA, Greenpeace, the World Health Organization, and the United States chemistry professor Paul Connett provided valuable information to the Aobei homeowners. This helped certify critical information about the hazards of incineration and fostered the community's awareness process. This international information played a central role in the homeowners' shift from localized 'NIMBY' to broader anti-incineration claims and provided the resources to make clear policy recommendations to the Beijing government. The experiences of other countries and cities abroad, as well as a knowledge of international environmental laws and regulations that were relevant for China's waste treatment strategy, provided the grounds for the campaigners' argumentation in the research report. The residents thereby aligned their framing, particularly their advocacy of 'zero waste' policies, with the international 'no burn' community.

Notes on Scale Shift

While the Asuwei case points to the slim prospects of the emergence of a Chinese issue-specific or broader environmental movement based on collaborative action between different affected communities, the Asuwei campaign was nonetheless able to produce claims and impact policy decision-making that went well beyond the local level. Collaborations and coalition-building with other contentious communities (such as the Liulitun residents or Likeng villagers), which could potentially serve as the basis for a broader issue-specific or environmental movement, were in the Asuwei case impeded by China's restrictive political framework and the related risks of joint action. Coalitions amongst urban Beijing homeowner communities were deliberately restricted to a loose 'alliance of support' that avoided any joint action beyond the exchange of information.

Nonetheless, the Aobei campaign came to play an important role as both an example and a diffusion node for other affected communities, facilitating the onset and development of contentious struggles against incinerator projects and operating plants in other localities. This was significantly enhanced by the possibilities of communication through digital technologies and social media. The inclusion of the Aobei research report in Green Beagle's 'toolkit' for affected communities also facilitated the Aobei campaign's role as a diffusion node for the transnational flow of information – thereby connecting resource-poorer Chinese communities with the transnational anti-incineration movement and global wave of anti-incinerator disputes.

While they started out with localized 'NIMBY' concerns against the siting of a hazardous waste incinerator in their neighbourhood, the Aobei residents soon broadened their claims beyond their individual community, presenting themselves as the representatives of all Beijing residents and attempting to impact local and national waste policies. This shift in framing and claims-making had a clear strategic component, aiming to widen the Aobei residents' base of support among the broader Beijing residents, but it still shifted the struggle's aims and implications beyond a merely localized level. While limited in its long-term effects, the Aobei homeowners succeeded in invoking some change to Beijing's waste policies, such as initiating more comprehensive waste sorting measures. They also contributed to the drawing of greater public attention to the issues of incineration and waste treatment.

With his transition to a full-time environmentalist and founding of 'Green Home', Huang institutionalized his supra-local engagement in the environmental and waste realm. He formed coalitions with pre-established environmental organizations that were already active in the issue field. In this regard, the case shows strong parallels with the Guangzhou Panyu case, where local resistance against a planned incinerator also led to the foundation of an environmental organization dedicated to city-level sustainable waste policies. Both organizations are also members of the Chinese 'Zero Waste Alliance', which was established in 2011 to promote national 'zero waste' policies (cp. Chapter Two).

The campaign's role in bringing the issue of waste incineration to the national political table and initiating waste sorting measures in Beijing demonstrates that local Chinese struggles can scale up to and produce results at higher political levels, as observed in Western countries. The establishment of an environmental organization dedicated to sustainable waste treatment strategies and its participation in the national 'Zero Waste Alliance' also shows how Chinese local campaigns can feed into broader issue-specific alliances with the aim of impacting national policy-making. However, as in Panyu, this transformation from 'NIMBY' claims to broader concerns about sustainable waste policies was limited to individual campaigners and did not produce a wider Not-In-Anybody's-Backyard (NIABY) attitude that could provide the foundation for a Chinese anti-incineration or environmental movement.

4 Making the Most of External Linkages

The Rural Case of Panguanying Village

The rural case of Hebei Province's Panguanying village shows the potential for horizontal and vertical linkages to foster Chinese environmental contention in contexts where diffusion effects can fully unfold.[77]

With a Little Help from their Friends: The Case of Panguanying Village

Panguanying village has a total population of about 1800 residents, most of them farmers, and is located in Liushouying town (留守营镇, *Liushouying zhen*) of Funing county (阜宁县, *Funing xian*) in the Qinhuangdao city area (秦皇岛市, *Qinhuangdao shi*) of China's north-eastern Hebei Province. The village lies about 30 kilometres to south-west of Qinhuangdao city and around 15 kilometres inland from the Nandaihe and Beidaihe tourist areas at the shore of the Bohai (渤海, *Bohai*) Sea. It is set amid large stretches of farmland. The region is an important agricultural production base for north-eastern China. Its produces grains (mainly corn), vegetables (such as cabbage and radish), and meat products (mainly pork). These agricultural products are largely sold in the major cities in the area, such as Qinhuangdao, Tianjin, and Beijing. Since the 1970s, various industrial plants and industry have settled in the region, including several paper mills, a fibre plant, a chemical fertilizer plant, an ore dressing plant, a pellet plant, and a large-scale slaughterhouse. These industries had already significantly contributed to the pollution of the local environment and the nearby Yang River (洋河, *Yanghe*), and Panguanying villagers reported that cancer rates in the village had been high for years.

First Awareness and the Onset of Action: Fighting for their Land

Around noon on 16 April 2009, a Panguanying villager doing farm work in his field noticed the village head and other local cadres measuring out and encircling collective village farmland that was tenured by several families

77 An abbreviated version of the Panguanying case is also presented in Bondes and Johnson (2017).

from Panguanying and a neighbouring village. Inquiring about the reason for their activities, the farmer learned about the local government's plans to build a waste incinerator on this stretch of farmland, a few hundred metres away from Panguanying village (see Figure 4.1).[78] Startled, the villager went to see one of the farmers that was directly affected by this land appropriation to ask whether the family's land had been sold, only to find him similarly surprised. Back at the field, the two villagers were informed that a total of about 70 *mu* of village land had been requisitioned by the government to build the waste facility and that the affected families were to be compensated with 34,300 RMB per *mu* (interviews PGY 1 4-11-12, 27-7-13; cp. D_PGY5 to D_PGY7).[79]

This news infuriated the affected farmer, surnamed Pan like many residents of Panguanying village, who started a heated debate with the local village head and refused to accept the sale of the collective land without prior consultation with the villagers. The village head, however, claimed that the local government was acting upon higher orders from the province, city, county, and town governments and that nothing could be done about it (interview PGY1 4-11-12). In the evening, a small crowd of villagers that had been summoned by the angered farmers confronted the village head about selling collective land without the prior knowledge of the village residents. Again, the village head referred to higher orders, cautioned the farmer Pan not to make trouble, and told the villagers to report to the responsible higher-level authorities if they wanted to voice their concerns (ibid.; interviews PGY2 4-11-12, 5-11-12).

Starting the following day, the villagers began to investigate the project details and the rightfulness of the land requisitioning. They visited and sent letters and petitions to higher-level authorities from the town- to the provincial-level governments via the official communication channels provided by the Chinese petitioning system and its 'letters and visits

78 According to official documents, the distance of the MSWI from Panguanying village is 750 metres (D_PGY1, D_PGY2, D_PGY32). Panguanying villagers as well as members of the environmental organization Green Beagle and professor Zhao Zhangyuan, who did an onsite investigation of the situation, argue that this number is embellished and that the actual distance amounts to only 516 meters (interview NGO NU2 29-10-12; D_PGY3, D_PGY4). This controversy, which extends to deviations in numbers regarding the MSWI's distance from other sites, including the Nandaihe tourist area, as well, is a central item in the later dispute, because it concerns the regulations for the siting of incineration projects.

79 *Mu* is the conventional unit for land area in China, translating to about 1/15 hectares or 667 square meters. The requisitioned 70 *mu* of land translate to about 4.7 hectares, 3.3 hectares of which belong to Panguanying village and the remaining 1.4 hectares to the neighboring village Xiaoying (D_PGY21, D_PGY32).

Figure 4.1 Construction site of Panguanying incinerator, July 2013

bureaus' (信访局, *xinfang ju*) (D_PGY6, D_PGY8, D_PGY9). Soon the farmer Pan had gathered a small group of villagers that were determined to lead the village in what they saw as a just fight for their rights. Among them was a knowledgeable elderly farmer from the same production team, also surnamed Pan but not a direct relative of the first farmer Pan, who soon became the second key figure in the villagers' struggle (interviews PGY1 27-7-13, PGY3 28-7-13). During this initial phase of events, the farmers were primarily angered by the local government's lack of communication with and 'cheating' of the villagers, what they believed to be an unlawful occupation of collective land, and concerns about the amount and proper distribution of compensation funds to the affected families; they did not initially pay much attention to the nature of the planned facility or the potential for related environmental or health risks. As the first farmer Pan recalled about this early phase: 'When we started we had no experience. They were taking away our land. We wanted to obstruct this from a land perspective, not let them build it. We have to eat from the land they wanted to claim. So we started to fight' (Interview PGY2 4-11-12).

Step by step, the core group of villagers surrounding the two Pans uncovered more details about the land requisitioning and the planned incinerator project, a Waste-to-Energy facility jointly planned in a build-operate-transfer mode by the Qinhuangdao city government and the private company Zhejiang Weiming Environmental Protection Co., Ltd. (浙江伟明环保股份有限公司, *Zhejiang weiming huanbao gufen youxian gongsi,* further referred to as Zhejiang Weiming).[80] In the course of their inquiries they found several flaws in the project's approval and decision-making processes. This particularly upset the villagers because of their prior experience with corruption by the local government and party cadres, including some misappropriation of village finances and earlier misconduct regarding the village's collective land (interviews PGY1 27-7-13, PGY2 5-11-12, 27-7-13, PGY3 28-7-13). Among other issues, the local, town, and county governments had apparently changed the original category of the slated construction land from 'basic farmland' (基本农田, *jiben nongtian*) to 'garden land' (园地, *yuandi*) to circumvent the national farmland protection policies – a change that the villagers regarded as absurd, since no (fruit) trees had ever been planted on the land which was mainly used by small groups of villagers (农民小组, *nongmin xiaozu*) to cultivate grains and vegetables for the regional agricultural market. In the villagers' understanding, this changing of category rendered the land requisitioning illegal.[81] Moreover, according to the villagers' research, the project had not been listed in the Qinhuangdao City General Plan (秦皇岛市城市总体规划, *Qinhuangdao shi chengshi zongti guihua*), Land Use Plan (土地利用总体规划, *tudi liyong zongti guihua*), or Environmental Sanitary Plan (环境卫生专业规划, *huanjing weisheng zhuanye guihua*) prior to the project's approval and beginning of construction, as required by state laws (D_PGY12 to D_PGY14).

These findings gave the villagers the grounds to demand the halting of the construction and the return of the land to its original state from the responsible authorities (D_PGY15). By mid-May 2009, however, various government departments from the local to the city level confirmed the lawfulness of and gave their consent to the incinerator project, thus leading to the project's approval by the Hebei Province Environmental Protection Bureau and subsequently the start of construction (D_PGY10, D_PGY16 to D_PGY19).[82] The continued petitioning by the villagers, drawing on whatever legal knowledge they could gather in the village, remained unanswered until

80 The plant was slated to handle a daily amount of about 650 tons of waste from across Qinhuangdao municipality. The total investment in the facility was about 220 million RMB (about 31.5 million euros) (D_PGY1, D_PGY10, D_PGY11, D_PGY32).

81 While the documents appear to support the villagers' view, it is hard to discern whether the land requisitioning was indeed illegal. The villagers' perception of its illegitimacy was, however, one of the main reasons for their actions at this stage.

82 A large number of the government documents cited in this case study were provided to the villagers by the court during one of their lawsuits.

September that year, when the Funing county government finally replied to the mounting public pressure by ordering a temporary construction halt based on the procedural flaws pointed out by the villagers, but in turn asking the farmers to refrain from any further petitioning (interview PGY1 27-7-13; D_PGY_21).

From Land to Pollution: The CCTV Broadcast and Learning from Beijing's Liulitun Campaign

In May 2010, after eight months of halted construction, the workers resumed construction on the plant (interview PGY2 28-7-13; D_PGY22; cp. *Phoenix Weekly* 2011; Shang 2013). This brought new momentum to the villagers' actions. Over the previous months, the spreading news about the planned incinerator had started to raise other concerns for the villagers. Apart from a local teacher who had heard about environmental harms associated with waste incineration, the former head of a local paper mill's environmental department – who had discarded incineration as a waste disposal strategy for the factory years earlier due to the related risks – started warning the villagers about the project's potential harm for the environment and human health (interviews PGY1 27-7-13, PGY2 5-11-12, PGY3 28-7-13; cp. Mao 2013; Shang 2013). With the help of an article about the dangers of dioxin in the *Farmers Daily* (农民日报, *Nongmin ribao*), he illustrated the risks related to the planned MSWI to the villagers:

> They [the villagers] didn't know about this issue then, so they couldn't do anything. No one understood this, how would the *laobaixing* understand this issue of environmental protection? [...] Originally, how would they know about dioxin, how would they know this can cause cancer? [...] I told them, because I understand this issue. And once they also saw this newspaper, they knew. It really helped our village. [...] Now everyone understands, the awareness on this issue has risen tremendously. It's an issue that concerns our vital interests, our future generations. (Interview PGY3 28-7-13)

Warnings about environmental pollution and health hazards, particularly cancer risks, found ample resonance with the villagers due to prior environmental and health problems in the village. Various industrial plants and local industry in the area, including the paper mill, a fertilizer plant, and a large-scale slaughterhouse, had already contributed significant pollution to the local environment and the nearby Yang River (洋河, *Yanghe*). Cancer rates in the village had been high for years; several family members of the core group of villagers had died of cancer (interviews PGY1 27-7-13, PGY2 5-11-12; D_PGY4; cp. Gao 2012; Mao 2012; Shang 2013).

The farmers' concerns reached new heights when they came across a China Central Television program broadcast in the context of the 2009 International Dioxin Symposium in Beijing in August that year, which had already played a major role in the contentious struggle in Asuwei (cp. Chapter Three). By that time, the growing number of local campaigns and urban protests against waste incinerator projects, and the increasingly outspoken opposition of environmental organizations and experts, had triggered a public and media debate about incineration in China. This debate now also reached Panguanying village. The half-hour long special feature on the pros and cons of incineration featured a number of national and international experts, among them the retired Beijing professor Zhao Zhangyuan. Apart from explicitly linking incineration to dioxin and cancer, the program also reported the case of the large-scale campaign against a planned MSWI project in Beijing's Liulitun neighbourhood, which the residents had successfully obstructed (cp. Chapter Three).

Deeply concerned after happening to see this broadcast on television, the first farmer Pan asked a younger, more tech-savvy relative to find the program and download it online (interview PGY1 4-11-12). Around the same time, the local teacher with prior knowledge about the harms of incineration also conducted an online search and found a plethora of materials about the Beijing Liulitun anti-incineration campaign (interviews PGY1 27-7-13, PGY2 5-11-12), most importantly the 40-page 'opinion booklet' compiled by the Liulitun residents to advocate their claims. The booklet contained the comprehensive results of their lay expertise regarding incineration harms, the residents' concerns regarding the planned project, and a detailed description of the course of events and applied modes of action during the campaign (D_PGY24).

After construction in Panguanying restarted, these findings led to a new wave of activities. Convinced that the source of such pollution had to be averted in their village and encouraged by the Liulitun campaign, the village leaders soon realized that not only were these materials invaluable resources for filling in their knowledge about both incineration harms and possible modes of action, but that they were also precious resources for mobilizing broader support among the village residents. The core group of villagers had by this time manifested around a solid kernel of three farmers. Another villager also surnamed Pan – and again no direct relative of the other two Pans – had been solicited to join the group by the elderly farmer Pan due to his

Figure 4.2 Villager signatures collected by the Panguanying farmers, November 2012

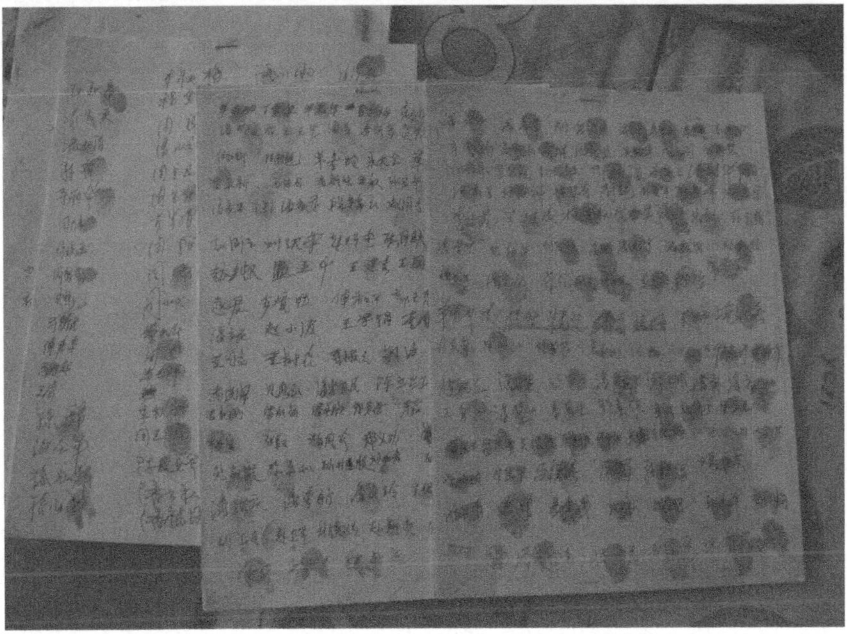

good reputation among the villagers.[83] Although not directly affected by the land requisitioning, this third Pan was deeply concerned by the CCTV report and decided to take a lead role in the fight against pollution in Panguanying (interviews PGY1 4-11-12, PGY2 4-11-12, PGY3 28-7-13; Shang 2013).

As a casual worker at the local railroads office at the time, he found some villagers to help him make dozens of paper copies of the Liulitun booklet at the railroads office (interview PGY2 5-11-12). The Pans also copied the CCTV broadcast to compact discs. Equipped with these materials, they started paying door-to-door visits to the Panguanying village residents in June 2010 and also successively reached out to the neighbouring villages within a 5-kilometre radius. Driving around the area in a minibus, they convinced not only the villagers but also many of the surrounding village committees of the threat of the pending pollution (interviews PGY1 4-11-12, 27-7-13, PGY2 5-11-12; Shang 2013). This strategy proved successful: the farmers collected a total of about 1500 villager signatures and the stamps and statements of 37 village committees in the area, all opposing the slated incinerator (D_PGY22) (see Figure 4.2).

83 While this third Pan was quite well respected among the community and managed to gain the villagers' active trust and support during later phases of the struggle, it is hard to pin down the exact source of his reputation.

As one of the Pans recalls about this phase of activities and the pivotal role played by the CCTV broadcast and Liulitun materials:

> Liulitun had a major influence on us. The Liulitun incinerator inspired us. [...] They reported about it on TV. Once I had seen that, oh dear, they absolutely mustn't build that waste plant here! [...] We copied the Liulitun materials and also copied the CCTV program on compact discs and distributed it to everyone in all the villages around. [...] If I hadn't had these things, the villagers wouldn't have believed me. They weren't clear on whether there would be pollution or not. They don't understand these kinds of things, right? But once they saw these things, ah, that waste plant really causes pollution, it was over, they didn't want to let them build it, they unanimously opposed it. (Interview PGY1 27-7-13)

Of course, the local, town, and city governments did not stand idly by as the villagers started to mobilize a collective opposition against the construction project they were determined to complete. Soon, cadres from the local and town governments started warning off the surrounding village committees, thus impeding the collection of further village stamps (interview PGY2 28-7-13; Gao 2012; *Phoenix Weekly* 2011; Shang 2013). To counter the spreading concerns about the safety of the incinerator, members of the town and city governments also visited the villages to promote the benefits of incineration, promising there would be no pollution and asking the villagers to 'sacrifice their small family to protect the large family' (舍小家保大家, *she xiao jia bao da jia*) (interview PGY1 27-7-13; *Phoenix Weekly* 2011; Shang 2013). However, the materials distributed by the Pans provided the villagers with critical information and an alternative cognitive framework that allowed them to critically assess the government's claims.

During this phase, the downloaded materials also proved to have another benefit. As one of the Pans remembers, cadres from the town government also started to exert personal pressure on the farmers, threatening to charge them with trouble-making and the illegal distribution of leaflets. In meetings with government officials, the Pans could, however, rebut these charges by arguing that distributing a program officially broadcast by the state-owned China Central Television could hardly be regarded as illegal and that they could also not be charged with the illegal distribution of leaflets, since they had not themselves written a single word (interview PGY2 27-7-13; Shang 2013). While the Pans report having been repeatedly visited during this time by local cadres – and at one point local mafia (黑社会, *hei shehui*) members – who tried to both threaten and persuade them to stop their activities, including issuing threats about having them arrested, it is noticeable that they were able to go about their campaigning activities relatively unimpeded. They were not actively stopped from completing their activities, and claimed in interviews that they were genuinely not worried by the pressure at this point, since they felt that they were doing nothing illegal and could not be touched (interviews PGY1 5-11-12, 27-7-13, PGY2 27-7-13).

Apart from the signature collection, the villagers also deliberately started imitating the Liulitun residents' strategies (interviews PGY1 4-11-12, 27-7-13, PGY2 27-7-13, 28-7-13). To give more weight to their claims, the farmers found a local university student to write their own version of an 'opinion booklet', which closely followed the Liulitun blueprint (D_PGY22). The villagers' claims now started to shift away from land and local corruption issues to centre more on environmental and health hazards. Mirroring the Liulitun booklet in both structure and content, the arguments in the villagers' booklet included the charge that the siting decision was unlawful not only due to the procedural flaws previously revealed by the villagers, but also on the grounds that it neglected major environmental and health threats for the approximately 30,000 residents living within a five-kilometre radius and for the close-by Nandaihe and Beidaihe tourist areas. The booklet also raised the problem of food safety that would arise from siting the incinerator in the midst of a large stretch of cultivated farmland (ibid.). Moreover, the social injustice of exposing the disadvantaged rural population to harms emanating from the incineration of city waste, which was mainly produced by urban residents, was explicitly criticized in the document:

> The government is encouraging the construction of a new socialist countryside and environmental protection. The *laobaixing* want to drink clean and unpolluted water, breathe fresh air and eat organic foods. These are probably also the goals the urban residents are pursuing, but they must not forget that at the same time of living a clean life themselves, this must not violate the rights and interests of peasants as 'disadvantaged group' (弱势群体农民的利益, *ruoshi qunti nongmin de liyi*).[84] We also want to exist, we also want a protected environment, we also want to live with

84 The notion of a 'disadvantaged group' was introduced in 2002 by former Premier Zhu Rongji (朱镕基), who admitted that Chinese society had produced a sizeable group of socially disadvantaged people, including peasants, whose rights and interests should be protected (Holbig 2002; Lin 2010; Yang 2003).

> dignity. [...] Society talks about being harmonious, not about constructing such projects at the cost of other people's harmony. (D_PGY22)

Moreover, the booklet also reflected the broader concerns and frames employed by incineration-opponents both in China and beyond. Based on the encompassing assertion of the general harms of incineration and the health threats posed by the 'unavoidable emission of dioxin', particularly within the Chinese regulatory and waste-specific context, the opinion booklet questioned incineration as a suitable waste treatment strategy, not only for China, but also on a global level, citing national and international scientific studies and invoking the experiences and anti-incineration movements of other countries (ibid.).[85]

In the document, the villagers portrayed themselves as but one in a growing number of affected local communities standing up for their legitimate rights (ibid.). As in the Asuwei case (cp. Chapter Three), the notion of *weiquan* – which was also at the core of the Liulitun residents' self-perception – became a central identity frame used by the farmers (interviews PGY1 27-7-13, PGY2 5-11-12, PGY3 28-7-13). Both in the Panguanying villagers' claims and their self-perception as part of a broader *weiquan* community, a process of frame alignment with the Liulitun residents, other contentious communities, and the domestic and international 'no burn' community had taken place. As one of the farmers now presented their cause: 'First of all, it's about public interests and rights protection (公益维权, *gongyi weiquan*), right? I have my personal rights: the right to health, the right to know. These rights have been given to the people by the Party and the people's government. But the local government has deprived us of these rights' (Interview PGY1 27-7-13).

Taken together, the newly compiled materials were the basis for a new round of petitioning. By mid-June 2010, the villagers started to personally deliver the materials to the town, county, city, and provincial governments – in one instance, with a group of more than 60 people. However, the villagers again received no or negative responses and the waste incinerator kept taking shape at the doorstep of their village (interviews PGY1 27-7-13, PGY2 5-11-12; Gao 2012).

85 The booklet here adopts the central arguments raised by the Chinese 'no burn' community and various international experts that: (1) Chinese waste has a lower caloric value than waste in most other countries, thus rendering Chinese waste highly unsuited for incineration; and (2) the regulatory framework in China does not permit the effective control and monitoring of toxic emissions or the enforcement of environmental regulations that would permit the effective management of such emissions.

Changing Strategies: Seeking External Help and the Turn to Environmental Litigation

Frustrated by these developments, the village leaders concluded that they had reached the limits of their own capabilities and that it was time to seek external help. In the Liulitun materials, the name of a Beijing-based lawyer, Xia, was frequently cited as the residents' legal representative who had played a crucial role in the successful outcome of the Liulitun campaign. Following the Liulitun example, the farmers decided to turn to Xia for assistance, convinced that 'if anyone could help us, it was him' (interview PGY2 5-11-12). Assisted again by the tech-savvy younger relative, the three Pans found the lawyer's contact information online and went to Beijing to personally present him with their collected materials and ask for help (ibid.; interviews PGY1 4-11-12, 27-7-13).

While at first hesitant to get involved since he saw few entry points for legal action and slim chances of success, the lawyer eventually decided to take up the villagers' cause free of charge after seeing the farmers' major previous efforts, their commitment to the public good of the village beyond individual claims for compensation, and the strong support among the village community (interviews Xia 6-11-12, PGY1 27-7-13, PGY2 4-11-12). Seeking advice on the issue from a renowned Beijing-based law professor specialized in the assistance of pollution victims, Xia hoped to turn the case into a precedent for environmental litigation (interview Xia 6-11-12).

As advised by Xia, the farmers now turned to more high-level legal means and – officially entrusting the lawyer with the responsibility of being their legal representative – launched a request for administrative redress (行政复议, *xingzheng fuyi*) to the Ministry of Environmental Protection and two other province-level institutions in August 2010, attaching copies of their compiled materials and requesting that the project approval given by the Hebei EPB in May 2009 be revoked (D_PGY12 to D_PGY14, D_PGY27 to D_PGY29).[86] Their claims centred on the procedural flaws discovered earlier – including the unlawful change of the nature of the land to 'garden land' and the siting of the incinerator in a densely populated area amid cultivated farmland – as well as another major flaw pointed out by the Beijing law professor: the project had not been listed in the Qinhuangdao City General Plan (秦皇岛市城市总体规划, *Qinhuangdao shi chengshi*

86 The other institutions are the Hebei Province People's Government and the Hebei Province Department of Land and Resources (河北省国土资源厅, *Hebei sheng guotu ziyuan ting*) (D_PGY25, D_PGY26).

zongti guihua), Land Use Plan (土地利用总体规划, *tudi liyong zongti guihua*), and Environmental Sanitary Plan (环境卫生专业规划, *huanjing weisheng zhuanye guihua*) prior to the project's approval and the beginning of construction, as required by state laws (ibid.; interviews Xia 6-11-12, 30-7-13; Xia 2011).

After initial difficulties, the MEP accepted the request for administrative redress in mid-September, requesting written statements from all involved institutions including the Hebei EPB and the construction unit Zhejiang Weiming, who disputed the charges raised by the villagers (D_PGY2, D_PGY30 to D_PGY32). On the scheduled day for the MEP's final decision in mid-December, the three Pans rented a bus at their own expense to bring about 50 villagers to Beijing (interviews PGY1 4-11-12, PGY2 4-11-12). Much to their dismay, however, the MEP decided to uphold the Hebei EPB's project approval, arguing that the villagers' claims were not sufficiently founded (ibid.; D_PGY10). Following this decision, the other two institutions also declined the farmers' requests for redress (D_PGY33 to D_PGY34).

Nonetheless, a new path of action had opened up to the villagers. During the procedures, the MEP had revealed the response statements of the engaged institutions to the farmers. In these statements, both the Hebei EPB and Zhejiang Weiming referred to an Environmental Impact Assessment (EIA) conducted in early 2009 by the Chinese Academy of Meteorological Sciences (中国气象科学研究院, *Zhongguo qixiang kexue yanjiuyuan,* in the following referred to as CAMS). This EIA, they claimed, included, as required by state laws: (1) the twofold public announcement of the project plans on the Panguanying village and Liushouying town committees' public notice boards in January and February/March 2009; (2) two inspection trips with villager representatives to the Wenzhou and Tianjin MSWIs in October 2008 and April 2009; (3) the soliciting of public opinions via the distribution of 100 questionnaires among the villagers, the vast majority of whom had allegedly agreed to the construction (D_PGY10, D_PGY31 to D_PGY32). This agitated the villagers since none of them had taken notice of the small paper bulletins, nor had the announcements been communicated to them by the village committee. Nor had any of them participated in the alleged opinion survey (interviews PGY1 4-11-12, 27-7-13, PGY 2 5-11-12).

The farmers' suspicion that there were serious flaws associated with the EIA was further strengthened by the assessments of other intermediaries from Beijing that were now starting to become involved in the case. Based on his appearance in the CCTV broadcast and his central role in the Liulitun campaign, the villagers were keen on obtaining the engagement of Beijing professor Zhao Zhangyuan. Upon the farmers' request and through the mediation of the Beijing-based environmental organization Green Beagle,[87] Xia introduced the Pans to the professor during one of their Beijing visits. Hearing about the widespread opposition among the villagers, Zhao agreed to come to Panguanying in November 2010 to conduct his own investigation of the situation (interviews Zhao 8-11-12, Xia 6-11-12, PGY1 27-7-13, PGY2 4-11-12, 27-7-13).

Based on his inspection of the area and the project site, communication with the villagers, and a close examination of the farmers' compiled materials and the abridged EIA report – by law publicly accessible on the Hebei EPB website, but previously unknown to the farmers – Zhao concluded that not only was the location unsuited for the construction of an incinerator, but that the entire EIA was severely flawed. Enabled by his prior experience as an expert in other EIA procedures, he identified several striking mistakes in the EIA report and other government documents, which led him to the conclusion that the involved institutions, above all the EIA unit CAMS in collusion with the Liushouying town and Panguanying village governments, had practiced serious forgery in the EIA process (interviews Zhao 8-11-12, PGY1 27-7-13, PGY2 5-11-12, 27-7-13, NGO NU3 25-7-13; Gao 2012; *Phoenix Weekly* 2011).

His main points of criticism, which he outlined in an expert commentary provided as a supplement to the farmers' administrative redress with the MEP and which he also published in an extensive report about the case on his personal Sohu blog, encompassed the charges that: (1) the EIA unit used erroneous evaluation methods and purposely misrepresented the strained local environmental conditions and the extent of emissions to be expected from the plant, thus coming to a false conclusion about the project's feasibility; (2) the project approval was based on procedural flaws, including the land and plan issues pointed out before, as well as an incorrect measure of the sanitary protection belt zone required around the incinerator; (3) the report lacked the mandatory discussion of the necessity of an incinerator in this area and the general pros and cons of incineration, as well as a convincing description of how the emissions would be managed and controlled; (4) since none of the villagers knew about the public participation questionnaires and univocally opposed the project, this part of the EIA must be faked, as had

87 The main Green Beagle staff members engaged in the Panguanying case changed to Nature University after its establishment in 2011 to lead the new organization's 'School of Waste' (cp. Chapter Two). Since Nature University was not yet established throughout the early stages of the Panguanying struggle, the staff members are here referred to according to their initial affiliation throughout the case study.

previously been the case in other localities (interview Zhao 8-11-12; D_PGY4; Zhao 2012). Although it was conveyed to the MEP in a supplementary letter by the lawyer Xia a few days before the MEP's final decision in December (D_PGY13), Zhao's expert assessment did not change the MEP's decision to uphold the project approval.

During his visit to the village, professor Zhao also briefed the farmers on more general environmental issues and their health impacts; incineration and its harms, particularly the cancer risks emanating from dioxin; and local anti-incineration struggles in other localities, such as in Beijing Gaoantun, Beijing Asuwei, Guangzhou Panyu, and Guangzhou Likeng (interviews Zhao 8-11-12, PGY1 27-7-13, PGY2 5-11-12, 27-7-13). According to one of the farmers, it was through Zhao's explanations that they first realized not only the scope of problems related to the EIA, but also the extent of pollution in the area and its relation to the high cancer rates in the village (interview PGY1 27-7-13).

While not yet based on a personal inspection trip to Panguanying, similar information also reached the villagers from the members of Green Beagle, whom Xia had introduced the villagers to during one of their Beijing visits (interviews Xia 6-11-12, NGO NU2 18-10-11, 29-10-12, 26-3-14, NGO NU3 25-7-13, PGY1 4-11-12, 27-7-13, PGY2 5-11-12). The organization members' assistance to the villagers during this stage of engagement mainly consisted of two parts. First, organization staff provided the villagers with further information about incineration and its harms, about the experiences of other affected local communities, and about possible courses of action and their legal rights as stipulated in environmental laws and regulations, such as the 2002 Environmental Impact Assessment Law and the 2008 Regulations on Open Government Information, which entitle citizens to apply for the disclosure of (environmental) information.[88] In their communications with the villagers, the Green Beagle members strongly urged the farmers to pursue a legal course of action to ensure their personal safety and avoid violent clashes with state security forces, as had occurred in other localities such as Guangzhou Likeng (ibid.; interview NGO NU1 7-11-12).

Second, the Green Beagle staff assisted the farmers with disseminating their cause to the public and drawing media attention to the case, both to exert pressure on the government institutions and to increase the villagers

88 The compilation of documents provided to the villagers by Green Beagle staff encompass articles and blog posts on incineration and its harms by different experts including Zhao Zhangyuan and Green Beagle staff member Mao Da; media articles and blog posts about other cases including Beijing Gaoantun, Beijing Asuwei, Guangzhou Likeng, Guangzhou Panyu, Nanjing Jiangbei, and Shenzhen Longgang; and relevant environmental laws and regulations (D_PGY36).

protection by focusing outside attention on them. During a visit of one of the Pans to Beijing, the organization asked the farmers to present their case at one of their regular public lectures and also invited media representatives to participate (ibid.). In early 2011, one of the journalists present at this lecture, a journalist from the Hong Kong-based *Phoenix Weekly*, was one of the first media representatives to visit Panguanying and report about the events in the village (interview NGO NU2 1-7-15; *Phoenix Weekly* 2011). When they became concerned about their personal safety during later events, the farmers relied on their relations with the media established during this meeting and via other mediation by Green Beagle to ask for protective media attention (interviews PGY1 4-11-12 27-7-13, PGY2 4-11-12). Beyond organizing the lecture in Beijing, Green Beagle staff also disseminated the villagers' claims through the organization website and via their personal blogs and microblog accounts (interviews NGO NU1 7-11-12, NGO NU2 8-5-13, 1-7-15, NGO NU3 25-7-13). Still, media attention to the case remained limited during this phase.

A Temporary Victory: The Flawed EIA Report and the Villagers' First Success in Court

The new information and the intermediaries' support and advice encouraged the villagers to step up their legal actions. After the MEP decision to uphold the Hebei EPB's project approval, the three Pans decided to launch an administrative lawsuit in the Shijiazhuang City Qiaoxi district People's Court (秦皇岛市桥西区人民法院, *Shijiazhuang shi Qiaoxi qu renmin fayuan*) at the beginning of January 2011, again keeping the lawyer Xia as their legal representative (D_PGY37, D_PGY38). Following the advice of Xia and the Beijing-based law professor, the three farmers centred their allegations on the newly discovered EIA flaws pointed out by professor Zhao (D_PGY41, D_PGY42).[89] As Xia recalls about this strategic decision:

> The farmers cared most about the land. They love it dearly and were very distressed. [...] So they wanted to continue with the land issue. The land issue was basically a muddled war, they [the responsible government departments] had all cheated. [...] In China this is very common, no one

89 From a legal angle, three entry points had emerged: (1) the EIA fraud; (2) the land issue, focusing on the unlawful rededication of farmland; and (3) the reversed plan issue, i.e., the project's approval without prior listing in the relevant city plans (interviews PGY1 4-11-12, Xia 6-11-12, 30-7-13; Xia 2011).

> investigates this, and the *laobaixing* also have no way of investigating. So land, I usually don't get into that. So I gave the villagers the advice: If you continue with land then don't get your hopes up. Do it from an environmental side. That has chances of winning. (Interview Xia 6-11-12)

Due to an unexpected development, this strategy proved successful. During the evidence collection procedures for the lawsuit, an exchange of documents between the MEP, the Hebei EPB, the villagers, and the court took place. To the great surprise of Xia and the villagers, the Hebei EPB released not only an encompassing collection of internal government documents related to the project, but also the full EIA report in early February 2011 (interviews PGY1 4-11-12, PGY2 5-11-12, 27-7-13, Xia 6-11-12, 30-7-13; Gao 2012; *Phoenix Weekly* 2011; Shang 2013).[90] This included the 100 public participation questionnaires allegedly distributed among the villagers to solicit their opinions on the project, as well as the protocol of a villager representative meeting convened by the village committee in March 2009, where the villager representatives had allegedly given their signatures in consent to the construction of the incinerator (D_PGY43, D_PGY44).

These documents provided the three Pans with the basis for collecting conclusive evidence of forgery practiced during the EIA's public participation process. As their first step, the farmers visited the about 30 villager representatives and party members that had allegedly signed to signal their agreement with the plan at the 2009 meeting. While the signatures proved to be real, the participants in the meeting testified that they had unanimously opposed the project. Their signatures were given as confirmation that they had received a participation fee of 10 RMB, and later attached to a false meeting protocol by the village head to claim that the participants had given consent (D_PGY45).

In the second step, the farmers also paid door-to-door visits to all of the villagers from Panguanying and surrounding villages whose names and signatures appeared on the questionnaires. All of the visited villagers testified in a written statement on the original forms that they had 'never seen this questionnaire, do not know who signed it, and do not agree with the construction of the incinerator' (D_PGY46 to D_PGY47). Moreover, the

90 In interviews, both Xia and Green Beagle staff repeatedly highlighted that such a release of a full EIA report is very rare in China and was not expected. For all of them, it was the first time they were able to see a full EIA report. They regard this release as a sign that the Hebei EPB was likely unaware of the flaws in the EIA report (interviews Xia 6-11-12, 30-7-13, NGO NU2 8-5-13, 26-3-14, NGO NU3 25-7-13; Mao 2012, 2013).

Pans discovered that the survey contained several other mistakes: various alleged respondents had left the area many years before; one was a convicted criminal on the run; others had died several years ago; and some names were altogether fictional. The questionnaires that did name actual inhabitants of the villages contained mistakes such as the wrong gender, date of birth, or level of education of the respondant (D_PGY46; Gao 2012; Shang 2013).

At the beginning of March 2011, the Pans presented this newly collected evidence to the village and town governments to confront them with their misconduct, as well as to the Hebei EPB, the MEP, and the Shijiazhuang court as evidence for the ongoing lawsuit. After the release of the documents the farmers had also notified Green Beagle staff, who now helped them write a lengthy letter to the MEP. The letter outlined in detail not only the mistakes in the EIA public participation measures, but all charges raised by the villagers including the procedural flaws, their environmental and health concerns, and all points of criticism regarding the EIA report as raised by professor Zhao. The letter invoked the expertise of both the professor and the Green Beagle organization as certification for the villagers' claims and concluded that, should the MEP not investigate this kind of forgery, it would be a 'betrayal of law-based governance' (对依法治国的背叛, *dui yifa zhiguo de beipan*) and a 'destruction of the environmental law' (对环境法治的摧残, *dui huanjing fazhi de cuican*) (D_PGY3). The farmers also contacted the *Phoenix Weekly* journalist, who now came to Panguanying and wrote a first lengthy article about the case (*Phoenix Weekly* 2011).

Upon receiving this new set of evidence, the Hebei EPB and Zhejiang Weiming ordered an immediate construction halt to the project (D_PGY48, D_PGY49; Mao 2013). According to Xia, who was in communication with both parties at the time, both the Hebei EPB and Zhejiang Weiming were caught by surprise and angered by these disclosures, since they had been assured by the EIA unit CAMS and the lower-level governments that all procedures had been conducted according to the requirements (interviews Xia 6-11-12, 30-7-13). In mid-May, the court notified the three Pans that a hearing would take place on 30 May (D_PGY50).

During this phase, and particularly in the days leading up to the hearing, cadres from the village, town, county, and city governments again tried to ensure the completion of the project by both pressuring and attempting to bribe the farmers into dropping their lawsuit. Cadres from the local and county governments visited the farmers' relatives and urged them to convince the Pans to back out. They also offered to relocate the village should the villagers agree to the construction. A high-ranking city government official offered the Pans monetary and other benefits such as jobs and free health treatment for some of their relatives, and paid a visit to the lawyer Xia to convince him to withdraw the lawsuit. After these attempts were unsuccessful, the cadres directly threatened the farmers (interviews PGY1 4-11-12, 27-7-13, PGY2 5-11-12, PGY3 28-7-13, Xia 30-7-13; Shang 2013). However, with the backing of their families and the larger villager community and the support of the intermediaries, the villagers persisted in their litigation.

Three days before the court hearing, on 27 May, the Hebei EPB revoked their official approval of the project EIA by their own initiative after it had become clear that the court would rule in favour of the villagers based on the conclusive evidence of the EIA flaws. The EPB ordered a project halt until a new EIA process could be conducted, and in the meantime suspended all EIA applications for Qinhuangdao city construction projects (interview Xia 30-7-13; D_PGY42; Mao 2013).[91] Having achieved the aims of their litigation, the Pans withdrew their lawsuit in early June (interviews PGY1 4-11-12, PGY2 4-11-12; D_PGY42).

Among the engaged intermediaries and environmental lawyers, this outcome was regarded as a major victory, not only for the people of Panguanying, but also as a precedent for successful national-level environmental litigation by an affected community, which they hoped would have wider-reaching impacts (interview Xia 30-7-13; Chen 2012; Gao 2012; Mao 2013; *Phoenix Weekly* 2011; Xia 2011; Zhao 2012). In a lengthy post on his personal Sohu blog, Zhao Zhangyuan called the Panguanying case a 'new model of environmental protection based on public participation in Chinese modern history' (中国近代史上公众参与环保的新典型, *Zhongguo jindaishi shang gongzhong canyu huanbao de xin dianxing*):

> They won! That this 'disadvantaged group' organized such a fierce campaign for justice is definitely a sign of progress for Chinese society! The Qinhuangdao incinerator case signifies the shift from urban residents participating in Chinese environmental protection to the rural masses (农村民众, *nongcun minzhong*). With the spread of environmental pollution [...], rights protection activities by the masses (民众维权活动, *minzhong weiquan huodong*) are flaring up all over the country and the rights protection level is continuously rising. [...]. At the same time it [the Panguanying case] also shows the successful work of lawyers, NGO

91 According to the assessment of the lawyer Xia, this move was facilitated by an unrelated change of personnel in the EPB's leadership. Since the new leadership was not personally responsible for the earlier decision to approve the project, there was less internal pressure to uphold the decision (interview Xia 30-7-13).

> organizations (NGO 组织, *NGO zuzhi*) and numerous experts and scholars fighting for justice. This is an epitome (缩影, *suoying*) of the frequent resistance against incinerator projects in China in recent years. (Zhao 2012)

Particularly for the staff of Green Beagle and the lawyer Xia, who hoped to persuade other affected communities to resort to legal means rather than disruptive and potentially violent actions, the successful lawsuit in Panguanying provided a valuable example (interviews Xia 6-11-12, 30-7-13, NGO NU1 7-11-12, NGO NU2 29-10-12, 26-3-14, NGO NU3 25-7-13).

By this time, the Pans had already spent a significant amount of money and time on the campaign. While they felt that the money was well spent, they had nonetheless significantly strained their financial resources. One of the Pans, who had held an outside job at the time of the struggle's onset, reported that he had instead started cultivating land at home to be more flexible with his time. Another Pan, who produced corn, radish, and cabbage to export to Korea and Japan on a contract, had to leave some of his land lying fallow due to the large demands of the activities. His family reportedly sold three pigs around the time of the lawsuit to cover some of the expenses. The villagers repeatedly offered to contribute money to the campaign, but according to the Pans, they declined taking the offered support (against their wives' wills) out of fear of losing face should the lawsuit not be successful. While Green Beagle offered some financial support for the Pans' trips to Beijing during the later phases of the struggle, including covering their train tickets and lodging, the frequent trips to the capital posed significant inconveniences for the villagers, who said they took the slowest trains, stayed in shabby rooms, and only ate one warm meal per day – and even then mostly at the invitation of their Beijing allies – to reduce costs (interviews PGY1 27-7-13, PGY2 27-7-13; D_PGY40; Shang 2013). While the Pans were proud that they had managed these expenses largely on their own, they hoped they could now return to their normal lives.

New Battlefronts: Stepping up NGO Engagement and Shifting to the National Level

It was only shortly after their court win that new problems mounted on various fronts, problems that attracted wider engagement from environmental organizations, since they were related to the broader activities of the Chinese 'no burn' community. On the website of the Clean Development Mechanism (CDM), coordinated by the United Nations Framework Convention on Climate Change (UNFCCC), Green Beagle staff discovered that Zhejiang

Weiming and its partner in the United Kingdom, Eco-Frontier Carbon Partners Limited, had applied for the Panguanying project to be granted CDM status in mid-June (D_PGY11). Worried that this would facilitate a renewed EIA process for the project and as part of Green Beagle's broader campaign against the classification of incinerators as a 'clean' energy source, the organization mobilized other organizations active in the China Waste Information Network (cp. Chapter Two).

At the end of July and again in December 2011, eight Chinese environmental organizations under the lead of Friends of Nature and Wuhu Ecology Center, including Green Beagle, submitted a critical comment on the project to UNFCCC – the organizations' standard procedure for opposing Chinese incinerator projects applying for CDM status. The submission was supported by the transnational organizations Global Alliance of Incinerator Alternatives and CDM Watch. The comment pointed out the flaws in the Panguanying EIA process, questioned the general adequacy of incinerators to receive CDM status, and called on the UNFCCC not to support a project that was in open violation of the Chinese Environmental Impact Assessment Law (D_PGY51, D_PGY52; GAIA n.d.).

Also in June 2011 and as part of a broader campaign targeting violations of the EIA law, five of the above organizations including Green Beagle and Friends of Nature, together with professor Zhao, petitioned the MEP through an open letter to withdraw the grade-A qualification of CAMS as an EIA unit based on the fraud practiced in Panguanying and another locality (FON 2011a; Mao 2013; L. Liu 2013b). In mid-May, the abridged EIA report for a planned incinerator in Beijing's Dagong village, also conducted by CAMS as the appointed EIA unit, had been published on the website of the Haidian district EPB. Based on a visit to Dagong village and an examination of the report, members of Green Beagle and Friends of Nature as well as professor Zhao, all of whom were engaged in the case early on, concluded that this EIA contained flaws similar to those in the Panguanying report (cp. Chapter Five). Armed with these parallels, the organizations decided to target CAMS in a joint fashion and turn both cases into the basis for a broader campaign against EIA-related fraud, since they had also come across similar problems in other localities (interviews NGO NU1 7-11-12, NGO NU3 25-7-13, NGO FON1 19-10-12). In July, the organizations submitted a detailed 'list of incidents of violations against regulations by Chinese incinerators' (中国垃圾焚烧厂违规事件列表, *Zhongguo laji fenshao chang weigui shijian liebiao*) to the MEP, which included Dagong, Panguanying, and several other incinerator projects (D_DGC1). As part of this campaign, professor Zhao and the environmental organizations reached out to the media to attract public attention to the issue (D_DGC2). A wave of reporting about the organizations' charges against CAMS followed in a number of Chinese media outlets.[92] While this campaign did not produce major responses from the MEP or other institutions at the time, it gained new momentum in late 2012.

A third line of activities, this time directly targeting the construction unit Zhejiang Weiming, started in early September 2011. At that time, the MEP announced on its website that it had accepted Zhejiang Weiming's application for an environmental examination of the company so it could enter the stock market through an initial public offering (IPO) (D_PGY53; Gao 2012; Mao 2013).[93] Afraid that a stock market entry by Zhejiang Weiming would result in renewed construction, the villagers and environmental organizations petitioned the MEP to reject the company's application, pointing out that it had been engaged in a case of 'severe illegal conduct related to the environment' (严重环境违法行为, *yanzhong huanjing weifa xingwei*) (interviews PGY2 5-11-12, NGO NU3 25-7-13, Xia 30-7-13; D_PGY48, D_PGY49; *Caijing* 2012; Mao 2013; Gao 2012). With the help of Green Beagle staff, the Pans sent a lengthy letter to the MEP Pollution Prevention Division on 14 September 2011, explicating in detail the prior events in the village and their successful lawsuit (D_PGY48). Two days later, the same five environmental organizations sent a similar letter to the MEP to certify the villagers' claims and increase pressure on the Ministry (interview NGO NU3 25-7-13; D_PGY49; Mao 2013). However, the MEP still announced in mid-December that the company had passed the environmental examination, arguing that it was not Zhejiang Weiming that had conducted the flawed EIA and breached environmental laws (D_PGY49).

In February 2012, it was gradually revealed that Zhejiang Weiming was indeed planning to restart the construction of the incinerator, as anticipated by both villagers and intermediaries (interviews PGY1 4-11-12, PGY2 4-11-12; Gao 2012). In May and June 2012, high-ranking Zhejiang Weiming personnel contacted Green Beagle staff and the lawyer Xia to discuss the company's

92 Articles about the issue were, among others, published in the *National Business Daily* (每日经济新闻, *Meiri jingji xinwen*) (Z. Li 2011; *National Business Daily* 2011), *Legal Daily* (法制日报, *Fazhi ribao*) (Qie 2011a, 2011b, 2011c), *The Beijing News* (新京报, *Xin jing bao*) (Guo 2011; Jin 2011b), *China Business News* (第一经济日报, *Diyi jingji ribao*) (Zhang 2011a, 2011b), and the *Southern Metropolis Daily* (南方都市报, *Nanfang dushi bao*) (X. Yang 2011), which was alert to the issue because of the incineration cases in Guangzhou.

93 Before launching an IPO and entering the stock market, companies have to be evaluated by the MEP, which has to certify that the respective company abides with the environmental laws and regulations in order to reduce the risks of investors.

further plans for the incinerator. The company invited organization staff, the lawyer Xia, and professor Zhao to visit one of their incinerators near Shanghai (interviews Xia 6-11-12, NGO NU1 7-11-12, NGO NU2 25-7-13; Mao 2012). For the environmental organization, this was a sign that their efforts to support the villagers and their broader campaigns against incineration and lacking EIA implementation were being taken seriously. Organization staff and the lawyer Xia used the opportunity to certify the villagers' claims and concerns and appeal to the company to engage in direct communication with the villagers (ibid.). The company subsequently sent personnel to Panguanying to directly consult with the Pans and explore the chances that the villagers would agree to a renewed construction of the project under the condition of lawful procedures (interview PGY2 5-11-12). By that time very familiar with their rights, the farmers insisted that a new EIA process, particularly a new round of public participation measures, would have to be conducted and that the villagers would under no circumstances agree to the construction (interviews PGY1 4-11-12, PGY2 5-11-12; cp. Mao 2012, Shang 2013).

Nevertheless anxious that the project would be implemented against their will, the Pans launched another request administrative redress and, after its failure, two successive lawsuits against the MEP's decision to ratify Zhejiang Weiming's market entry – again supported by lawyer Xia, the Beijing-based law professor, and this time the China Lawyers Association (中华律师协会, *Zhonghua lüshi xiehui*) as well. Green Beagle staff assisted the villagers with attracting media attention to the developments (e.g., Y. Li 2012; Z. Li 2012). While reaching the Beijing Municipality Higher People's Court (北京市高级人民法院, *Beijing shi gaoji renmin fayuan*) in the appeal – another precedent, according to the lawyer Xia – the case remained unsuccessful and was conclusively dismissed by the Beijing Higher Court in September 2012 (interviews PGY1 4-11-12, PGY2 5-11-12, Xia 6-11-12; D_PGY49, D_PGY57, D_PGY59 to D_PGY61).

Back to Local: Running for Village Elections and Mounting Pressure

Faced with these setbacks and in light of the looming plans to resume the incinerator project, the farmers turned to another course of action which they hoped would finally succeed in obstructing the project. According to the villagers' logic, one way to thoroughly halt the project and return the land to its original state for farming was to officially assume the position of village head. This authority could then be used to obstruct any renewed construction, which had to be approved by the village committee (interviews PGY1 4-11-12, 27-7-13, PGY2 5-11-12, 27-7-13, PGY3 28-7-13; Chen 2012; Mao 2012, 2013; Shang 2013).

By the end of 2011, the village head Qiao, a fierce project proponent and central figure in the related fraud, had been forced to resign from office by the infuriated community, leaving the position open, with interim authority transferred to the village's party branch secretary. The periodic village election in Panguanying was scheduled for February 2012 and the Pans decided that one of them should sign up to run in the election campaign. Since the elderly Pan had suffered a stroke in 2011 that had left him bedridden, attributed by the villagers to the major pressure he experienced during the EIA lawsuit, it was the well-reputed former railroad-worker Pan who signed up as the nominee for the position of village head (ibid.; Liu 2012a, 2012b, 2012c).[94] According to the candidate Pan, the local governments up to the county level were anxious about his nomination because of his broad-based support among the villagers, but did not dare to prohibit it (interviews PGY2 5-11-12, 27-7-13). Instead, the scheduled election was repeatedly delayed during the following months until public pressure from the villagers and growing outside attention eventually forced the responsible government departments to fix the election date for 29 November 2012 (interviews PGY1 4-11-12, PGY2 5-11-12, 27-7-13; Chen 2012; Mao 2012, 2013; Shang 2013). In the months and weeks leading up to the election, not only the Pans but also the broader villager community were exposed to growing pressure from the former village head and his collaborators at the village and town levels. According to the villagers, many of them were pressured by local and town cadres to vote for one of the other candidates put up by the former village committee, which severely intimidated some of the villagers (interviews PGY1 5-11-12, PGY2 5-11-12, PGY3 28-7-13; Chen 2012; Shang 2013).[95] Pressure against the Pans peaked when local bullies, apparently hired by the former village head and his supporters, threw rocks at the house of one of the Pans and physically threatened some of his family members in the days before the election (interviews PGY1 27-7-13, PGY2 27-7-13).[96]

In the eyes of the villagers, the intensifying struggle in the village was no longer only about the incinerator but instead centred on broader local political entanglements. According to their assessment, the village

94 While he, like the other Pans, was not a party member, laws and regulations also permit independent candidates (Alpermann 2010a, 2013).

95 This was reported by several villagers during the author's field visit to Panguanying at the beginning of November 2012, three weeks before the scheduled election.

96 After this incident, Pan installed a safety camera directed at the front gate of his house so that he would have videotaped evidence of any further physical attacks. The camera was still in use at the time of the author's second field visit in July 2013.

committee under the leadership of the former village head Qiao, as well as government cadres up to the county level, were afraid that the mandatory examination of the village finances after the inauguration of a new (independent) village head would bring to light severe financial misconduct involving various levels of the government (interviews PGY1 4-11-12, 27-7-13, PGY2 5-11-12, 27-7-13, 28-7-13, PGY3 28-7-13). During the summer and fall of 2012, the villagers became increasingly outspoken in demanding their right to vote (选举权, *xuanju quan*), including displaying banners in front of the Funing county government building (interviews PGY1 4-11-12, PGY2 5-11-12, 27-7-13; Chen 2012, Mao 2012, 2013; Shang 2013). In interviews, the core villager group reported that the election issues further politicized their thinking and increased their distrust in the party-state. While their anger was mainly directed at the local government, they also criticized the broader political system for creating an environment where corruption could not be efficiently ferreted out and where the effective implementation of the country's environmental laws was virtually impossible (interviews PGY1 4-11-12, PGY2 5-11-12, 27-7-13, PGY3 28-7-13).

During this phase, Green Beagle staff played an important protective role for the Pans. Against the backdrop of the mounting pressure, and anxious about Zhejiang Weiming's pending plans to resume the project, the organization decided to launch another dissemination campaign to raise outside attention to the events in the village, both to exert pressure on the local government departments and to increase the safety of the Pans:

> They [Zhejiang Weiming] wanted to resume construction, they hadn't given up hope. [...] So against this background we thought the situation was quite critical. [...] So at that time we thought that we as public environmental organization definitely had to get involved, help them, intervene; otherwise these two people [the Pans] could experience a major threat (威胁, *weixie*). So we further publicized this issue, made the outside world pay more attention, in order to make things a bit safer for the two. (Interview NGO NU3 25-7-13)

In September 2012, Green Beagle staff paid their first personal visit to Panguanying to gain a better understanding of the local situation and demonstrate to the local government that the situation was being closely followed from the outside. At the same time, organization staff started to solicit public attention to the issue and called on the media to report on the upcoming election. Their main efforts focused on distributing background information on the case and the evidence collected by the villagers regarding the EIA fraud through their personal relationships with media representatives, their networks in the environmentalist and 'no burn' communities, and social media – including the social media accounts of both the organization and its staff members (interviews NGO NU1 7-11-12, NGO NU2 29-10-12, NGO NU3 25-7-13; Mao 2012).

According to organization staff, a window of opportunity in the broader political environment helped their media campaign take off (interviews NGO NU1 7-11-12, NGO NU3 25-7-13). Around that same time, the MEP was conducting a national examination of several hundred EIA units in an attempt to strengthen the effective implementation of the EIA law, leading the Ministry to urge the media to report on EIA-related malpractice at the local level. In this context, the case of Panguanying was a welcome story picked up by numerous news media outlets, including the Chinese-language party-led newspapers *People's Daily* (L. Yang 2013) and *China Youth Daily* (中国青年报, *Zhongguo qingnian bao*) (*China Youth Daily* 2013) and the widely-read commercial financial magazine *Caixin Magazine* (Cui 2013a, 2013b). The organization's efforts also initiated a steady stream of external visitors – including journalists, other intermediaries, and researchers – to Panguanying in the second half of 2012 and early 2013. The organization also used these favourable circumstances to again petition the MEP regarding the EIA unit CAMS in January 2013 via an open letter jointly signed by eleven environmental organizations and an online signature collection conducted through the Nature University website (interview NGO NU2 26-3-14; L. Liu 2013b). While the MEP claimed to have lowered CAMS' EIA qualification to grade-B at the end of 2012 in the context of their national examination, CAMS was still listed as grade-A on the MEP website in 2014 (interview NGO NU2 26-3-14; Liu 2013b).

During the election, Green Beagle staff took on the role of election observers. Faced with mounting pressure, the candidate Pan made a phone call to Green Beagle staff in the days before the election to ask for assistance and external attention (interviews PGY1 27-7-13, PGY2 27-7-13, NGO NU2 26-3-14; Chen 2012). One day before the election an organization member arrived in Panguanying, bringing with her a journalist from the party-led *People's Daily*-affiliated newspaper *Global Times*. On election day the procedures started as scheduled in the morning at the local elementary school. The former village head Qiao and cadres from the town and county governments held a speech asking the several hundred assembled villagers to trust the government's capacities to guarantee a fair election. Several dozen public security forces from the Funing county and Liushouying town police stations were also present to maintain order (ibid.; Liu 2012a, 2012b, 2013a).

However, the villagers' concerns proved to be well-founded. About halfway through the election, when it became clear that Pan would receive the majority of votes, three armed men burst into the polling station and destroyed the ballots and voting boxes. According to the villagers, the thugs were hired by the former village head Qiao and his collaborators from the town government (interviews PGY1 27-7-13, PGY2 27-7-13, PGY3 28-7-13, NGO NU3 25-7-13; Chen 2012; Liu 2012a, 2012b, 2013a). The villagers and external observers reported that the security forces did not intervene and let the disrupters escape unchallenged. Infuriated by their passivity, several hundred villagers blocked the school's main gate, refusing to let the security forces depart, and demanded an explanation from the head of the Liushouying town police station, who promised to investigate the issue. After several hours, Pan asked the villagers to relent at the urging of the lawyer Xia, who had told the Pans in a phone call to ensure that none of the villagers violated any laws, so as to ensure their safety. Due to public pressure, a new election date was set for one month later, on 29 December 2012 (ibid.).

To pressure the responsible government departments and increase the safety of the villagers through public attention, Green Beagle staff reported on these events throughout the day via their social media accounts, including the microblogging platform Sina Weibo. These posts were shared by the other engaged intermediaries such as the lawyer Xia and professor Zhao, as well as the broader 'no burn' community. On the day after the first election, one of the organization members wrote a lengthy report about the event and its background on her personal Sina blog, including photographs and documentary evidence (Chen 2012). The *Global Times* journalist who was present also reported on the issue in the newspaper's English-language edition (Liu 2012a, 2012b).

On 29 December, the second election took place as scheduled, in the presence of three Green Beagle staff members and the *Global Times* journalist. This election was also interrupted, and no new village head elected. This time more than 100 partially armed town and county public security forces were ordered to the location. The election proceeded without disturbances, with some of the villagers holding up banners reading 'please return the right to vote to the villagers of Panguanying' (Liu 2012c, 2013a). At the time of the vote counting, however, there was a scuffle between villagers, public security forces, and the town government cadres who were responsible for conducting the election. To count the votes, government-installed election staff had withdrawn into one of the classrooms and drawn the curtains, prohibiting any villagers from entering to observe the procedure – a clear breach of the laws and regulations, which require a public vote count

(Alpermann 2010a, 2013). Since the villagers suspected forgery, some of them, including the candidate Pan, forced their way into the classroom by breaking the window and discovered that the election staff had hidden numerous ballot slips in their clothes to distort the results (ibid.; interviews PGY1 27-7-13, PGY2 27-7-13, PGY3 28-7-13, NGO NU3 25-7-13; Mao 2013). In the afternoon and again on the following day, several dozen villagers went to the Funing county government to demand that the election personnel be held responsible for the forgery and that a new election date be set (ibid.). As during the first election, Green Beagle staff disseminated information about the situation throughout the day and the *Global Times* journalist also reported on the events (Liu 2012c).

Coming to an Impasse: Quieting Down and Setting an Example

In the days after the election, the villagers collected what they regarded as conclusive evidence of the sabotage and forgery during the elections. These materials, sent to the Funing county government on 3 January 2013, encompassed videos, photographs, and written testimony, including an eye-witness report by Green Beagle staff (interview PGY1 27-7-13; L. Liu 2013a). Five days later, on 8 January, three villagers including one of the Pans were summoned to the Funing county public security bureau to discuss potential charges against them for 'disrupting public order' (ibid.). Pan, who claimed that he had deliberately stood apart from the scuffle to avoid violating any laws, immediately contacted the Green Beagle staff who had returned to Beijing (interview PGY1 27-7-13). Since the EIA lawsuit, Green Beagle had become the villagers' major communication channel with the outside world. The Pans themselves had very limited Internet access and had gained only a little proficiency in using social media despite the organization's recommendations (interviews PGY1 27-7-13, NGO NU1 7-11-12, NGO NU2 26-3-14). Organization members again disseminated the news via their networks and social media channels. The *Global Times* journalist who had reported about the elections earlier also wrote another article about the developments (L. Liu 2013a). During the hearing at the public security bureau, the farmer Pan referred to Green Beagle staff and the *Global Times* journalist as eye-witnesses. According to Pan, it was their protective role – together with several dozen villagers who surrounded the public security bureau to testify that Pan had not been involved in the scuffle – that prompted the head of the bureau to drop the charges (interview PGY1 27-7-13).

In mid-January, the Pans submitted a petition regarding the election problems to the Funing county and Liushouying town governments (interviews

PGY1 27-7-13, PGY2 27-7-13; D_PGY62). The town government replied in mid-March, stating that it had conducted the election procedures according to state law. In the notice the government claimed that the villagers had violently disrupted the election, attacked the election personnel, destroyed the ballots, and impeded the vote counting. According to the letter, the election was thus considered invalid by the county government. Since no new village head had been elected, the notice further announced that a county government cadre would be transferred to Panguanying to take the position of first village secretary and that two town cadres would be dispatched to the village to establish a working group to be stationed in the village and preserve order. Moreover, the county and town governments would find an opportune time to hold another election (D_PGY63; Shang 2013).

The struggle in Panguanying had come to an impasse. After the instalment of the interim village government, local political life virtually came to a halt and the village community, which was now under strict surveillance, did not take any further actions (interviews PGY1 27-7-13, PGY2 27-7-13, PGY2 28-7-13, PGY3 28-7-13). In the summer of 2013, renewed plans for the continued construction of the incinerator by the town government briefly surfaced and the Pans were placed under intensified observation. However, in a meeting with town cadres, the candidate Pan again insisted that the villagers would continue to oppose the construction and would seek the assistance of their own experts, namely Zhao Zhangyuan, should new EIA procedures be conducted.

Since several government departments and the construction unit Zhejiang Weiming were no longer supporting the project, the plans were finally set aside (interviews PGY1 27-7-13, PGY2 27-7-13, PGY2 28-7-13, PGY3 28-7-13, NGO NU3 25-7-13, NGO NU2 26-3-14, Xia 30-7-13). In mid-August 2013, the Liushouying town government informed the villagers that they were considering dismantling the construction site and that a date for doing so would be decided. By 2015, however, no official timeline for dismantling the construction site had been announced, and action related to the incinerator had quieted down on all sides. A new election date had not been announced (interview NGO NU2 1-7-15).

While the incinerator issue had not been comprehensively resolved and the farmers had not obtained their ultimate goal of returning the project site to its original state for farming, they nonetheless felt victorious for having obstructed the completion of the incinerator (interviews PGY1 27-7-13, PGY2 27-7-13, PGY3 28-7-13). The widespread media and public attention to the case, the frequent visits of journalists and researchers to the formerly rather isolated village, and the great value attached to the case by the engaged intermediaries and other environmental lawyers had boosted the Pans' self-esteem. While their claims as presented in 2012 centred mainly on the villagers' rightful struggle for their rights and a sense of belonging to a broader '*weiquan*' community (interviews PGY1 4-11-12, PGY2 5-11-12; D_PGY22), in 2013 their self-perception had broadened beyond their individual struggle. In their narratives they started to portray themselves as fighting for social justice and environmental protection and their case as an exemplary example of local environmental contention, which they hoped would have a national impact and encourage other communities to stand up for their rights and the environment (PGY1 27-7-13, PGY2 27-8-13, 28-7-13).

This perception was strengthened by the Pans' participation in a 'symposium on NIMBY movement cases' (邻避运动案例研讨会, *linbi yundong anli yantaohui*) organized by Nature University in May 2013 (Nature University n.d.). The symposium brought together the members of several contentious communities, journalists, lawyers, environmental activists, and scholars working on the issue to discuss the fact that 'China has already completely entered an age of NIMBY campaigns' (中国已经全面进入邻避运动时代, *Zhongguo yijing quanmian jinru linbi yundong shidai*) (Mao 2013). At this symposium, Green Beagle staff, the lawyer Xia, and the Pans presented the case of Panguanying as a central topic on the agenda (ibid.; China Solid Waste Net 2013; Fu 2014). Panguanying had turned into a nationally known 'model case' for local environmental contention. To manifest the exemplary character of their campaign, in 2013 the farmer Pan expressed plans to erect a monument for all the external helpers – experts, lawyers, media, and organization staff – that had assisted their struggle and as a symbol that 'three uneducated farmers can also achieve such things' (interview PGY1 27-7-13).

Analysis: The Role of Horizontal and Vertical Linkages for the Panguanying Villagers

The case of Panguanying shows that both horizontal and vertical linkages played a significant role in the villagers' contention against the planned incinerator. While strongly intertwined in their impacts, in the following sections they are discussed separately for reasons of clarity.

The Role of Horizontal Ties

While the villagers' ties with other contentious communities remained limited and were mainly restricted to nonrelational ones, these linkages still played a pivotal role in the development of the villagers' resistance. This case demonstrates that horizontal diffusion processes between different localities do indeed impact local environmental contention in China and contribute to the geographic spread of local resistance, even in the countryside where diffusion channels are more limited than in urban settings.

In particular, linkages with the Beijing Liulitun campaign played an important role in the Panguanying case. As summarized by one of the village leaders: 'We drew a lesson from them [Liulitun], we learned from them. We used their strong points to mend our shortcomings. Because they understand everything. So we also drew a lesson from their methods to do this. [...] If it hadn't been for the Liulitun plant issue, we really wouldn't have known how to find this [information], from where to start' (Interview PGY1 4-11-12). Here, the mass media and the Internet proved to be the main channels of diffusion during the initial stages of the contention. While the majority of villagers in Panguanying had no access to or ability to engage with the Internet, the technological knowledge of individual community members such as the local teacher and tech-savvy younger villagers sufficed for using the Internet as a valuable information source. These nonrelational linkages were later complemented with ties mediated by the supra-local intermediaries in the case, who served as important channels for the transmission of information about anti-incineration struggles in other geographic localities.

The impact of the Liulitun campaign on the Panguanying villagers' struggle was primarily based on the information transmitted via the CCTV broadcast and the online materials found and distributed by individual community members, most importantly the comprehensive Liulitun 'opinion booklet'. The information contained in these materials regarding both the details of the Liulitun campaign and about waste incineration and its hazards played a crucial role for the villagers' awareness process. They contributed to the shifting of claims and framing from a focus on land and local corruption issues to concerns about environmental pollution and health hazards. The information certified the village leaders' initially diffuse concerns regarding environmental and health risks, which found ample resonance in the prior environmental pollution issues and high cancer rates in the village. Both the CCTV broadcast and the opinion booklet provided alternative interpretive frames and cognitive cues questioning waste incineration as a panacea for China's waste problem. This permitted a critical assessment of government assertions, such as the repeated promises that no pollution would emanate from the plant. The materials thus provided the initial cognitive cues and justification for the villagers' environment-related claims and contentious actions.

The materials also fostered an alignment of the Panguanying villagers' interpretive frames and claims with those employed by the Liulitun residents and the broader 'no burn' community – as demonstrated in the Panguanying villagers' own opinion booklet, which was modelled on the Liulitun blueprint. The CCTV program and Liulitun materials further contributed to a shift of the villagers' contentious identity from centring on fighting for the legal rights to their land and transparent information (cp. the notion of 'rightful resistance' as introduced by O'Brien and Li (2006)) to being part of a broader Chinese '*weiquan*'-community and the Chinese 'no burn' community.

With regards to the mobilization and organization of action, both the CCTV program and the Liulitun materials were indispensable resources for the village leaders in mobilizing the support of the villagers and the surrounding village committees. By certifying the village leaders' interpretations and assertions vis-a-vis the larger villager community, the materials enabled large-scale signature collection as an important petitioning resource. They also had a protective function, shielding the farmers from local government pressure and charges of trouble-making and the illegal distribution of leaflets. Moreover, the technical and issue-specific information on waste incineration provided in the Liulitun materials served as a first source of knowledge and expertise that was critical for the villagers' environment-focused action, thus filling in essential resource-gaps.

The Panguanying villagers directly emulated various aspects of the Liulitun activities. The choice to write their own opinion booklet to promote their claims and advocate for them via-a-vis the responsible government institutions and construction company was directly based on the Liulitun blueprint. The villagers' turn to environmental litigation and the engagement of the lawyer Xia was also based on the information about his successful engagement in the Liulitun campaign. One of the village leaders later regarded this as the most important decision of the whole struggle (interview PGY2 28-7-13). The (nonrelational) ties with the Liulitun campaign thus greatly facilitated the Panguanying villagers' access to justice. The success of the Liulitun residents further strengthened the villagers' belief in their own chances of success, particularly through legal action, thus changing the farmers' perceptions of threat and opportunity. In a similar vein, the Liulitun materials also encouraged the village leaders to establish contact with Zhao Zhangyuan. His field visit had important implications for the further course of their actions. While the villagers knew of his role as an anti-incineration expert via the CCTV broadcast, his engagement in the Liulitun campaign fostered their choice to actively engage him in their struggle.

The horizontal linkages formed by the Panguanying villagers with other contentious communities mostly remained limited to nonrelational ties. Despite the important role played by the Liulitun campaign, the village leaders showed no interest in establishing closer relational ties with Liulitun residents. In their eyes, they had already learned everything relevant from the online materials and the advice provided by the intermediaries, thus rendering any further personal exchange unnecessary (interviews PGY1 4-11-12, PGY2 5-11-12, 27-7-13). While collaborative action with Liulitun residents was unlikely since the Liulitun campaign had already come to an end, no active exchange, collective claims, or collaborative action emerged with simultaneous ongoing struggles, such as in Dagong where the villagers faced similar issues of EIA fraud practiced by CAMS. While joint claims regarding CAMS' EIA fraud in both Panguanying and Dagong later emerged, they were only produced by the environmental organizations to foster their broader EIA campaign and did not lead to direct relational ties between the two communities.

The lack of relational ties was not caused by a lack of opportunity, at least in the case of the Beijing struggles. During the village leaders' frequent visits to the capital, Green Beagle took on a brokerage role by introducing the farmers to members of the other contentious communities – in particular members of the urban anti-incineration campaigns in Beijing's Liulitun, Asuwei, and Gaoantun neighbourhoods. These initial relational ties were not followed up or intensified by the Panguanying villagers and did not play a significant role in their contentious actions. In the case of the urban Beijing homeowner campaigns, where large-scale street protests were at the core of activities, the lack of relational ties and collaborative action can partly be attributed to the different opportunity structures and diverging path of action employed by the Panguanying villagers, who relied mainly on legal means and dismissed large-scale 'strolls' due to a perceived lack of the required people 'mass' and because the related risks were regarded as too high (interviews PGY1 4-11-12, 27-7-13, PGY2 28-7-13). Moreover, the village leaders regarded the Gaoantun and Dagong cases in particular as negative examples of unsuccessful local contention, pointing to the perceived lack of persistence and unity among their residents. This discouraged any interest in a more active exchange (interviews PGY1 5-11-12, PGY2 28-7-13; D_PGY22).

The Role of Vertical Ties

Closely interrelated with the villagers' ties to other contentious communities, linkages with members of the Chinese 'no burn' community played a pivotal role in the development of environmental contention in Panguanying. The case demonstrates that collaborations between the 'two facets' of Chinese environmental activism can generate powerful effects in environmental contention in China.

The main intermediaries actively engaged in the Panguanying struggle were the Beijing-based environmental professor Zhao Zhangyuan, the Beijing-based lawyer Xia, a second law professor who specialized in assisting pollution victims, and members of the Beijing environmental organization Green Beagle, all of whom established direct relational ties with the village leaders and made visits to the site. The important role of these supra-local actors was repeatedly highlighted by the village leaders. As summarized by one leader: 'All the ones that donated themselves to the issue, the experts,[97] the professors, the lawyers, the media: Their help was immense. That waste plant, if it hadn't been for these people, we absolutely couldn't have done this, we wouldn't have known what to do. [...] Really, it was only with their help that we could do this. They gave us so much help I can't even name it' (Interview PGY1 27-7-13).

The first intermediary to impact the Panguanying villagers' struggle was Zhao Zhangyuan. Like the Liulitun campaign, he played an important role in the initial stages of the villagers' awareness process via nonrelational channels. In the CCTV broadcast and Liulitun materials, he provided critical information and interpretive frames about incineration and its risks. His status as a renowned professor and incineration expert helped to certify this information both in the eyes of the village leaders and the broader villager community. This fostered the farmers' cognitive justification for taking contentious action centring on environmental concerns. Zhao also played an important role in the external certification of the villagers' struggle. Based on his field visit, he wrote an extensive report about the happenings in Panguanying on his personal blog, thus disseminating and certifying the villagers' cause. He was also cited as an expert in the majority of the Chinese media articles published about the case, thus lending credibility to the villagers' claims and actions. Moreover, Zhao proved to be a valuable resource in the villagers' negotiations with the government: in unofficial meetings with local and town cadres, the farmers 'threatened' to engage professor Zhao as an official expert should the government decide to restart the construction process (interviews PGY1 27-7-13, PGY2 27-7-13). Zhao also provided the villagers with important resources that enabled the mobilization of contention. During his visit in early winter 2010, he

97 Members of environmental organizations, including Green Beagle, were generally referred to as 'experts' by the villagers.

offered issue-specific and technical information needed for environment-related action. Based on his own investigation of the local environmental situation and his critical reading of the abridged EIA report, he pointed out various flaws in the EIA procedures that became the basis for the villagers' environmental litigation, hence facilitating their access to justice.

No less important, albeit more focused on legal assistance, was the role played by the lawyer Xia. He was contacted by the villagers due to his successful role in the Liulitun campaign. Based on his experiences with the Liulitun case, he advised the villagers to shift the focus of their legal action from land to environmental litigation based on the EIA flaws. This significantly impacted the villagers' course of action and contributed to their success. Apart from providing legal advice, legal representation, and assistance in the collection of evidence, Xia also took on an important brokerage role by personally introducing the village leaders to both professor Zhao and members of Green Beagle. Moreover, like Zhao Zhangyuan he published articles and held lectures on the case, thus disseminating and certifying the villagers' cause for the public and the media. During later phases of the struggle, Xia also represented the villagers in their communication with the construction unit Zhejiang Weiming.

The third major supra-local party was Green Beagle, introduced to the villagers by the lawyer Xia during one of their Beijing visits. Organization staff played a multifaceted role during the struggle despite entering at a later stage. In the initial phase of their engagement, the organization mainly focused on disseminating information about the villagers' struggle via their online networks, at issue-related meetings in Beijing, and by reaching out to the media and urging them to report about the case. The organization brokered ties between the villagers and media representatives that proved helpful throughout the struggle: whenever the villagers came under pressure, such as during the local elections, these contacts could be activated to ask for media attention to pressure the government and provide protection to the village leaders. The steady stream of visitors initiated by Green Beagle staff in 2012, including other intermediaries and researchers, also played an important protective function for the village leaders and helped to publicize their cause.

Green Beagle's focus on the dissemination of information and media outreach was complemented with further functions during the later stages of events. During their first field visit to Panguanying in September 2012, members of the organization played a similar role as Zhao Zhangyuan in providing issue-specific knowledge, collecting evidence related to EIA flaws, and advising the villagers about how to mobilize and organize their actions. At the time of village elections, they took on the role of election observers and played a protective function for the village leaders. During the farmers' visits to Beijing, organization staff provided them with both financial and practical support and invited the villagers to issue-specific meetings. Apart from offering a public forum to advocate for their claims, these meetings also had a brokerage function by bringing the campaigners together with members of other contentious communities – albeit not followed up on by the villagers, as outlined above – and providing access to other members of the Chinese 'no burn' community. Green Beagle also advocated for the villagers' claims in petitions and (open) letters to the MEP and other national-level institutions. Like the lawyer Xia, the organization further took on a representative function in the villagers' communication with Zhejiang Weiming.

A number of other national and international environmental organizations were also loosely engaged in this case – namely the Beijing-based environmental organization Friends of Nature, the Anhui-based organization Wuhu Ecology Center, and the transnational organizations GAIA and CDM Watch. All of these organizations are part of the 'China Waste Information Network' established in 2009 and coordinated by Wuhu Ecology Center and GAIA, and their engagement remained restricted to participating in advocacy activities without establishing direct relational ties with the villagers.

Notes on Scale Shift

The case of Panguanying shows that the horizontal and vertical ties that have emerged in the field of anti-incineration contention play a closely interrelated role in the spread and development of local contention. As also demonstrated by the Asuwei case, the linkages between different contentious communities remain severely limited and have little potential for fostering collective claims or collaborative action beyond individual localities. Other than in Asuwei, this is not primarily due to the restrictive political framework, but instead due to a lack of interest on the part of the villagers. While they did to some extent align their interpretive frames and claims with the broader 'no burn' community and developed a contentious identity as part of this larger social group and the Chinese '*weiquan*' community, this identity was mainly used to further their localized claims and did not produce broader collective claims or actions. Based on these findings, the chances of scaling up local contention by building on horizontal alliances among different communities seem to be rather slim.

More promising for fostering Chinese environmental contention at higher political levels are the vertical linkages established between the Panguanying villagers and members of the national 'no burn' community. These ties helped increase pressure on the responsible government institutions all the way up to the national level and contributed to a broader public and media awareness of the problems associated with waste incineration and broader regulatory failures. The Panguanying struggle, which reflected broader environmental and regulatory problems that were typical in the siting of large-scale construction projects – including local corruption, flaws in the EIA process, and lacking transparency and communication – significantly furthered the intermediaries' cause. The supra-local actors' engagement in the local campaign fed into their broader work advocating for a greater awareness of the risks of incineration, a more critical assessment of the government's waste strategy, better implementation of environmental laws and regulations at the local level, the promotion of environmental litigation, and greater transparency and communication between local governments and their constituencies. Moreover, the case offered the intermediaries a publicly presentable example of the growing environmental and '*weiquan*' awareness in China and a successful case of a rural community standing up for their rights and demanding public participation in environmental policy – a case, which could also be presented to other affected communities in order to raise their believes in the chances of success of contentious action (such as in Dagong, see Chapter Five).

While the Panguanying case thus points to the limitations of scaling up local contention based on collaborative efforts or alliances among different affected communities, it shows that the vertical linkages between the diverse actors in the environmental sphere from the local up to the (trans-)national level can strengthen a national-level issue network that tackles not only the risks related to waste incineration but also broader environmental issues under the conditions of a restrictive political setting.

5 The Limitations of Linkages

The Peri-Urban Case of Beijing's Dagong Village

The peri-urban case of Dagong village outlined in this chapter shows the limitations of linkages. While environmental organizations and various Beijing intermediaries engaged with the Dagong case early on and tried to mobilize the communities around the project site, no unified or sustained local contention emerged. This case shows that even when strong linkages exist, diffusion effects are strongly dependent on the local context and fail to unfold if not anchored in sustained local contention. Moreover, the case shows the 'uneasy alliance' (McAdam and Boudet 2012: 135) between environmentalists and local communities where the preferred means of action and priorities of the two groups diverge.

'Resistance is Futile': The Case of Dagong Village

Dagong village in Sujiatuo town was selected in 2010 as the new location for the obstructed Liulitun incinerator (cp. Chapter Three; Haidian District People's Government 2010a, 2010b; Beijing News 2010b, 2010c, 2010d; Shuang Li 2010a). Located about 20 kilometres west of the original project site, the small village with about 100 households and 400 residents lies in a picturesque mountain area of Beijing's Western Hills (北京西山, *Beijing xishan*) at the border of the Haidian and Mentougou districts.

Nestled against the slope of the Yangtai Mountain range (阳台山, *Yangtai shan*), with traditional houses lining each side of a steep main road, the village towers a few hundred meters above the incinerator site with an open view over the Beijing city basin. The official distance between the village and the project site is listed as 460 metres. The road leading to Dagong curves its way around the incinerator before ascending to the village. In the hills along the mountain ridge around Dagong are several large orchards where cherries, apples, pears, plums, and walnuts are cultivated for the regional fruit market. The locals also practice beekeeping and sell apicultural products, as is common for fruit-growing regions.

Figure 5.1 Main road of Dagong village, June 2013

A Tale of Urbanization: The Dagong Village Area and its Diverse Social Groups

The social forces in the area and their claims and interests expressed during the incinerator episode are quite diverse. While the villagers of Dagong grow fruit, walnuts, and vegetables for their own consumption, they no longer cultivate the village's collective land. As reported by villagers and others from the area, the collective land was sold to the state-owned construction mogul Beijing Jinyu Group Co. Ltd. (北京金隅集团有限责任公司, *Beijing Jinyu youxian zeren gongsi*) in the late 1980s, which opened a quarry in the vicinity and employed many of the villagers as workers. At that time, the village land was declared as state-owned and subsequently industrial land and the Dagong villagers' residency permits were, in the words of the villagers, changed from those of farmers (农民户口, *nongmin hukou*) to those of urban residents (城镇居民户口, *chengzhen jumin hukou*)[98] (interviews DGC1 22-7-13, DGC2 23-7-13, DGC3 23-7-13, DGC4 23-7-13, DGC5 29-5-13, DGC7 1-6-13, NGO NU2 4-6-13; Beijing Municipality First Intermediate People's

98 These residency permits are officially termed agricultural *hukou* (农业户口, *nongye hukou*) and nonagricultural *hukou* (非农业户口, *fei nongye hukou*).

Court 2014). However, several of the villagers complained during my field visits that the money from the Jinyu land deal had never been distributed among the households, thus feeding into an atmosphere of distrust of the (local) authorities (interviews DGC2 23-7-13, DGC4 23-7-13). Apart from its remote location, the availability of industrial land was one of the main reasons given by Beijing authorities for the selection of Dagong as the new site for the Haidian district incinerator, since it rendered the often-contested state occupation of farmland unnecessary (interview BMAC 7-11-12; Cui 2010d; Wen 2011).

While some of the Dagong villagers opposed the incinerator project during its early phase, the larger villager community soon settled for seeking compensation because they regarded any resistance against the incinerator as futile. Similar to the road blockade by the Asuwei villagers (cp. Chapter Three), the Dagong villagers' protest actions were mainly directed at pushing for a timely relocation (interviews DGC2 23-7-13, DGC3 23-7-13, DGC4 23-7-13). Again like in Asuwei, this soon produced cleavages – in this case, with the environmentalists engaged in the case (interviews NGO NU1 7-11-12, NGO NU2 4-6-13, NGO NU3 25-7-13, NGO FON1 19-10-12, Zhao 8-11-12).

Dagong village is the closest of five villages within two kilometres of the incinerator that were all slated for relocation between 2013 and 2015.[99] While Dagong village falls directly within the 300- to 500-metre protection zone around the incinerator and is hence eligible for environmental relocation, the planned resettlement of the other four villages was officially attributed to the area's restructuring process under Haidian district's 'new village construction' (新农村建设, *xin nongcun jianshe* or 新农村改造, *xin nongcun gaizao*) urbanization scheme (interviews DGC5 29-5-13, 1-6-13, DGC6 1-6-13, DGC71-6-13; D_DGC3, DGC_4; Haidian District People's Government 2012; Yi 2011a; Zhang 2011a).[100] Nan'anhe village lies a few hundred metres east of Dagong. Separated from Dagong village by a small valley and the Sixth Ring Road, Nan'anhe has a direct view of the construction site and produced some individualized contention against the incinerator project. Similar to the Dagong villagers, however, the larger Nan'anhe villager community regarded opposition to the project as futile, and instead focused mainly on making

99 Apart from Dagong, the other villages are Nan'anhe (南安河村, *Nan'anhe cun*), Beianhe (北安河乡, *Beianhe xiang*), Zhoujiaxiang (周家巷村, *Zhoujiaxiang cun*), and Xugezhuang (徐各庄, *Xugezhuang*), all in the jurisdiction of Sujiatuo town in Haidian district.

100 This local policy is part of the country's 'Building a New Socialist Countryside' (社会主义新农村建设, *shehui zhuyi xin nongcun jianshe*) program. For further information on the program and China's urbanization policies and processes, see Ahlers 2014; Ahlers and Schubert 2010, 2013; Meyer-Clement 2016; Trappel 2015.

relocation claims (interviews DGC5 29-5-13, 1-6-13, DGC6 1-6-13, DGC7 1-6-13). At the time of my field research there were no signs of contention from the other three villages. According to the other social groups and engaged intermediaries, this could be attributed to the general lack of awareness about the project and acquiescence to the relocation plans (interviews DGC1 23-7-13, DGC5 29-5-13, DGC6 1-6-13, DGC7 1-6-13, DGC8 1-6-13, DGC9 22-7-13, NGO NU2 4-6-13, 1-7-15, NGO FON 19-10-12, Xia 6-11-12, Zhao 8-11-12).

The villagers' overall lack of resistance has to be understood against the backdrop of the broader socio-economic development of the region. Although it had not reached the Dagong and Nan'anhe area prior to the project start, the region was already in the midst of a major urbanization process before news of the planned incinerator surfaced. As part of Haidian district's 'new village construction' plans, many of the villages to the immediate east of the Dagong and Nan'anhe area were being relocated at the time of field research in June and July 2013. While Dagong, Nan'anhe, and the other villages around the incinerator site had not yet been included this plan, the villagers regarded their eventual relocation as inevitable within the broader restructuring process (DGC1 22-7-13, 23-7-13, DGC2 23-7-13, DGC3 23-7-13, DGC4 23-7-13, DGC6 1-6-13, DGC7 1-6-13). Regardless of their personal preferences, the villagers' claims thus mainly centred on obtaining proper compensation payments rather than on obstructing the incinerator.

Another residential unit included in the project's relocation plans, albeit on a voluntary basis, was the Western Hills Fengjingyuan Jinyu Relocation Buildings (西山枫景苑金隅回迁楼, *Xishan feng jingyuan Jinyu huiqian lou*, hereafter referred to as Jinyu Relocation Buildings). The buildings are located a few hundred metres south of Dagong village in the jurisdiction of Mentougou district. The buildings were constructed by the Jinyu Group in 2009 and 2010 to relocate residents from a nearby area as part of the area's industrialization process. At the time of the incinerator's first announcement in late 2010, the residents had only recently moved into the apartments and quickly became one of the main social groups opposing the project (interviews NGO NU2 4-6-13, 1-7-15, NGO FON1 19-10-12, DGC5 29-5-13, 1-6-13; D_DGC5 to D_DGC10; Beijing News 2010d; Haidian District People's Government 2012; Jin 2011b; Yan 2011).

The incinerator's local opponents also included two units under state jurisdiction. Among the more outspoken opponents were the management and several dozen permanent residents of Dajue Temple (大觉寺, *Dajue si*), located a few hundred metres northeast of the construction site. Like Dagong village, the well-known Buddhist temple lies at the foot of the Yangtai Mountains, concealed from from Dagong village's line of sight by one of the mountains' foothills. Founded in the eleventh century and home to several cultural and artistic relics, the temple is an official state-protected historical site under the jurisdiction of the Beijing Municipal Administration of Culture Heritage (北京市文物局, *Beijing shi wenwu ju*) and subordinate to the Ministry of Culture (中华人民共和国文化部, *Zhonghua renmin gongheguo wenhua bu*). With an official distance of 780 meters from the project site, the temple is the second-closest unit to the incinerator after Dagong village. Concerned about the cultural relics and the temple's permanent residents, the temple staff openly opposed the project in its early stages, but were soon muzzled by the project's proponents (interviews NGO NU2 6-4-13, DGC5 29-5-13, DGC9 22-7-13; D_DGC3, D_DGC5 to D_DCG10; Jin 2011a; Qie 2011a, 2011d, 2011e; Yan 2011).

The second social group under state jurisdiction that spoke against the incinerator project, although not publicly, were the residents of a military base located about two kilometres from the project site. With about 1000 permanent residents, mainly military staff and their families, the unit was the largest community affected by the project. However, like the contention by the Dajue temple staff, the military staff's opposition was also not sustained due to political pressure (interviews NGO NU2 6-4-13, 1-7-15, NGO FON1 19-10-12, Zhao 8-11-12, DGC5 29-5-13, 1-6-13, DGC9 22-7-13; D_DGC8 to D_DGC11; Qie 2011e).

The construction site is also surrounded by several nature parks and reserves, with Jiufeng National Forest Park (鹫峰国家森林公园, *Jiufeng guojia senlin gongyuan*) only a few hundred metres above Dagong village and three other parks within five kilometres of the site.[101] While the parks' management and staff were included in the EIA process and initially expressed concerns about the planned incinerator's environmental impact, there were no signs of contention from them at later stages (interviews DGC5 29-5-13, DGC9 22-7-13, NGO NU2 6-4-13, 1-7-15, NGO FON1 19-10-12, Zhao 8-11-12; D_DGC6, D_DGC9, D_DGC12; Beijing News 2010d; Yi 2011a; Zhang 2011a).

Overall, the Dagong area is characterized by a diversity of social groups with diverging claims and interests, which rendered the formation of unified or sustained local contention difficult in the rather restrictive local political setting.

101 Nearby nature parks and reserves include the Miaofeng Mountains Forest Park (妙峰山森林公园, *Miaofeng shan senlin gongyuan*), Miaofeng Mountains Scenic Area (妙峰山名胜风景区, *Miaofeng shan mingsheng fengjing qu*), Yangtai Mountains Nature Area (阳台山自然风景区, *Yangtai shan ziran feangjing qu*), and Beijing Fenghuangling Nature Park (北京凤凰岭自然风景公园, *Beijing fenghuang ling ziran fengjing gongyuan*).

From Liulitun to Dagong: First Project Announcements and Early Environmentalist Engagement

The first news that the Haidian district government was looking for a new site for the Liulitun incinerator project surfaced in February 2010. According to news reports in two state-owned newspapers, the Chinese-language *Beijing Times* (京华时报, *Jinghua shibao*) and the English-language *China Daily*, BMAC was considering moving the facility to an exhausted mine in a remote area of Northwestern Beijing at the border of the Haidian and Mentougou districts – indicating that it was already targeting the Dagong area (Cui 2010; Wen 2010). A few months later, in June 2010, the incinerator project was introduced on the Haidian district government website – now explicitly naming Dagong village as the project site (Haidian District People's Government 2010a). The waste-to-energy plant, given the euphemistic name 'renewable energy power plant' (再生能源发电厂, *zaisheng nengyuan fadian chang*) was planned as part of the Haidian District Circular Economy Industrial Park (海淀区循环经济产业园, *Haidian qu xunhuan jingji chanye yuan*) on a build-operate-transfer (BOT) basis by the state-owned company Beijing Lühai Environmental Protection Co. Ltd. (北京绿海能环保有限责任公司, *Beijing lühai neng huanbao youxian zeren gongsi*, in the following referred to as Beijing Lühai) – the same company that was responsible for the earlier Liulitun project.[102] This web entry remained largely unnoticed by both the public and environmental organizations.

The first broader public announcement of the incinerator plans in Dagong village was made on the Haidian district government website and in several Beijing state media outlets on 16 November 2010 as part of the project's EIA procedures (Haidian District People's Government 2010b; Beijing News 2010b, 2010c; Shuang Li 2010a). The announcement now included the mandatory ten-day period for the solicitation of public opinion on the project, which would end on 29 November. The announcement further named CAMS as the EIA unit – the same unit engaged in the EIA fraud practiced in Panguanying village, albeit not yet discovered in detail by the Panguanying villagers and engaged intermediaries at the time (cp. Chapter Four).

The first critical voices about the planned incinerator came from the state-led newspaper *Beijing News*, which had just functioned as the primary media outlet for the government's EIA announcement. Two days after the

102 The incinerator was slated to treat a total amount of 1800 tons of garbage per day. The total investment amounted to 1.26 billion RMB, and was provided by the Haidian district government (Beijing News 2010b, 2010c, 2010d; Haidian District People's Government 2010b; Shuang Li 2010a).

first announcement, the newspaper reported that none of the interviewed Dagong villagers and new residents of the nearby Jinyu Relocation Buildings, or a spokesperson from the nearby Miaofeng Mountains Scenic Area, had yet been notified of or heard about the incinerator plans and that they had expressed opposition to the project during the journalist's site visit – with one of the residents referring to the prior opposition to incinerator projects in other parts of the city (Beijing News 2010d). While some of the interviewed Dagong villagers were already expressing hope that they would be relocated should the project materialize, the Miaofeng Mountains spokesperson said the park staff would consider submitting their opinions on the project (ibid.).

Stronger opposition to the project began a few days later, from environmental organizations and the Beijing 'no burn' community. A Green Beagle staff member had discovered the project announcement on the Haidian district government website on 19 November, during a routine check for new developments on the Beijing incinerator front (interviews NGO NU2 4-6-13, NGO NU3 25-7-13, NGO FON1 19-10-12, Zhao 8-11-12; D_DGC6, D_DGC13).[103] Fast to react to the news, Green Beagle immediately informed staff members from the other two main Beijing-based environmental organizations that were actively engaged in waste issues at the time, Friends of Nature and the Beijing Global Village Environmental Education Center. Together, staff members from the three environmental organizations paid a site visit to the Dagong area on 22 November 2010 to get a better understanding of the local situation. Similar to the *Beijing News* journalist, the organization members found that neither the residents nor the village committees of several nearby villages, including Dagong, nor the residents of the Jinyu Relocation Buildings, had been informed about the construction plans or been asked for their opinions about the project (ibid.; D_DGC8, D_DGC14). Besides conducting this site visit, Green Beagle also reached out to the media and asked professor Zhao Zhangyuan to write an expert statement about the project's expected environmental impacts based on his initial understanding of the situation (interview Zhao 8-11-12; D_DGC15).

One day after their inspection trip, on 23 November, the organizations held an urgent public symposium about the planned incinerator project in Beijing, bringing together journalists, several members of the Beijing

103 The main Green Beagle and Global Village of Beijing staff members engaged in the Dagong case transfered to Nature University after its establishment in 2011 to lead the new organization's 'School of Waste' (cp. Chapter Two). They are the same staff members who engaged in the Panguanying case (cp. Chapter Four). Since Nature University was not yet established during the early stages of the Dagong struggle, the staff members are here referred to according to their initial affiliation.

'no burn' community, and campaigners from the earlier anti-incinerator struggles in Liulitun and Asuwei. Since no close ties had been developed with members of the social groups in the Dagong area yet, none of the affected residents participated in this first meeting (interviews NGO NU2 4-6-13, NGO NU3 25-7-13, NGO FON1 19-10-12, ASW1 31-7-13; D_DGC13; Liu 2010; Wu 2010). At the symposium, the organization members reported about their site visit to the Dagong area and criticized the lack of transparency and public communication with the affected communities. They raised doubts about whether the EIA's public participation procedures were carried out as required and called for a public hearing on the project (D_DGC13, D_DGC14, D_DGC16).

At the meeting, Zhao Zhangyuan further criticized the site selection based on his inspection of satellite maps of the area. According to Zhao, the Dagong site was no more suited for an incinerator than the Liulitun area, as both were upriver and upwind of Beijing city, with Beijing's major water systems – the Yongding River (永定河, *Yongding he*) and Jingmi Diversion Canal (京密引水渠, *Jingmi yinshuiqu*) – nearby. Zhao presented his first preliminary calculations of the expected environmental impact on several Beijing city districts should the incinerator be built in Dagong village. Moreover, presenting national and international research, both Zhao and the organizations' staff warned of the health and environmental impacts of incineration, particularly dioxin, and fundamentally criticized China's push for incineration technology rather than promoting recycling, reduction, and reuse measures as had been promised by the Beijing authorities earlier that year (D_DGC13 to D_DGC17). In a similar vein, campaigners from the prior Liulitun and Asuwei struggles, including the lawyer Huang (cp. Chapter Three), challenged the necessity of Beijing's construction of more incinerators and called for more comprehensive waste sorting measures instead (D_DGC13).

The online edition of the independent commercial magazine *Caixin* and the state-led English-language newspaper *China Daily* both reported on the symposium, critically assessing the planned incinerator project (Liu 2010; Wu 2010). The broader public and media reactions about the proposed project remained very limited at this stage, however. The organizations' subsequent submission of public opinion statements to the Haidian district government and the project's construction and EIA units, as invited by the EIA announcement, also received no response (interviews NGO NU2 4-6-13, NGO NU2 1-7-15, NGO NU3 25-7-13, FON1 6-6-13, Zhao 8-11-2; D_DGC14, D_DGC15, D_DGC18, D_DGC19). In its statement, Friends of Nature named several consulted experts – including Zhao Zhangyuan, the lawyer Xia, and a public health and hygiene professor from Peking University's medical department – and requested that the government also disclose the identities of its experts and allow an open debate about the project (D_DGC14).

Apparently unimpressed by the Beijing 'no burn' community's opposition efforts, Haidian district Party secretary Zhao Fengtong confirmed on 16 January 2011 that the Dagong project would be the official substitute for the halted Liulitun incinerator, with construction beginning in July 2011 and finishing at the end of 2012 should there be a positive EIA report. As the first official announcement that the publicly debated Liulitun incinerator had been shelved once and for all, this announcement attracted some reporting from state-led Beijing media outlets (*Global Times* 2011a; Liu et al. 2011; Wen 2011; Yi 2011a). According to the reports, Party secretary Zhao and BMAC officials had declared the Liulitun site as 'unsuitable' (不合适, *bu heshi*) and selected the more remote and unproblematic Dagong location. This change of sites was portrayed as a thoughtful government decision dedicated to social stability, listening to the (Liulitun) residents' opinions and minimizing the environmental impacts (ibid.). Due to the earlier publicity of the Liulitun campaign, this was a good framing to minimize public opposition to the new project among the broader Beijing resident community.

Haidian district Party secretary Zhao further declared that the project had thus far met little local opposition and that any resistance to the incinerator was based on a lack of understanding about the real situation or guided by selfish concerns. Haidian district waste could 'wait no longer', he argued, with the Liulitun landfill predicted to reach the limits of its capacity at the end of 2012. Should opponents still step forth, the Party secretary promised to take them on an inspection trip of the Liulitun landfill and the Gaoantun incinerator (Wen 2011; Yi 2011a). A large LED screen would also be installed at the facility's gate to display the monitored emission data, like at the Gaoantun incinerator, thus rendering public concern about hazardous discharges from the plant unnecessary (ibid.; *Beijing Morning Post* 2011; *Global Times* 2011a). The Beijing authorities had clearly learned from the resistance to the city's prior incinerator projects and were intent on not letting public opposition get in the way of another facility. This was confirmed by a BMAC official, who argued that if the city could not build the Dagong plant, it would have detrimental implications for all of Beijing's future incinerator plans (interview BMAC 7-11-12). After the (at that point in time) successful opposition in Liulitun and Asuwei, the political stakes of the Dagong incinerator struggle had significantly increased.

A Second Round of Environmentalist Action: Further EIA Announcements and the Abridged EIA Report

Things were rather quiet until 16 May 2011, when the Haidian district government published the second mandatory EIA announcement via its website and Beijing state media – this time drawing on the *Beijing Morning Post* (北京晨报, *Beijing chenbao*) as primary media outlet (*Beijing Morning Post* 2011; FON 2011b; Han 2011; Yi 2011b). The government informed the public that the EIA report for the project had been completed by CAMS, certifying that it would have no major environmental or health impacts on the project's surroundings or Beijing city at large. Construction on the project would begin in the next weeks and four villages, including Dagong, would be relocated starting at the end of the year with the government providing housing. A second public opinion solicitation phase was set until 27 May (ibid.).

Without having more than the rather rudimentary project announcement at hand in November, the 'no burn' community's range of activities had been limited. Now that they had access to the abridged EIA report, the environmental organizations and other members of the Beijing 'no burn' community could start taking further action. While the abridged EIA report was not published online as required by the EIA Law and had to be obtained in paper at the construction unit's office in the Dagong area, the environmental organizations were able to get hold of it on 24 May (interviews NGO NU2 4-6-13, NGO FON1 19-10-12; D_DGC14; FON 2011a, 2011b). Here, the organization's parallel engagement in the Panguanying case (cp. Chapter Four) proved helpful. Through this case and with the assistance of professor Zhao Zhangyuan, who was well versed in EIA procedures due to his own prior work as an EIA expert, Green Beagle had recognized the value of subjecting (abridged) EIA reports to close scrutiny and using potential flaws as valuable grounds for contention (interviews NGO NU2 4-6-13, NGO NU3 25-7-13, Zhao 8-11-12). A closer reading of the Dagong report indeed produced similar flaws as found in the Panguanying report – not too surprising, since CAMS was the EIA unit in both cases.

Under time pressure due to the nearing end of the public opinion solicitation phase, Friends of Nature again gathered several members of the Beijing 'no burn' community for a meeting in Beijing to discuss the case on 27 May 2011. They again brought together members of the engaged environmental organizations, professor Zhao Zhangyuan and the lawyer Xia – the main intermediaries in the ongoing Panguanying case – as well as a public hygiene professor from Peking University's medical department and a befriended journalist from the state-led *China Inspection and Quarantine Times* (中

国国门时报, *Zhongguo guomen shibao*). The points raised at this meeting were similar to those made at the November meeting, now adding, among others, a more detailed critique of technical aspects such as the (in their eyes) insufficient treatment of fly ash and slag residue outlined in the report. The participants also criticized the fact that the EIA report could not be obtained online as required by state laws, thereby hampering public transparency (D_DGC14; FON 2011a, 2011b, 2011c; Qie 2011a). To further increase transparency, the project's name should clearly identify the project as incinerator, they stated. Moreover, according to the speakers at the meeting, the size of the planned incinerator was much larger than necessary for the Beijing government's official 2009 plans to reach a waste treatment ratio of 4:3:3 for incineration to biochemical treatment to landfilling by 2015 (cp. Huo 2009; Wang 2009). They thus regarded the project as not in line with official Beijing city plans and hence illegitimate (D_DGC14; FON 2011a, 2011b).

Again, the organizations reached out to the media and Friends of Nature published a lengthy opinion statement about the project on their website, followed by several further press releases in the next weeks (FON 2011a, 2011b, 2011c). Zhao Zhangyuan also directly approached the media to publish his expert opinions about the flaws he had identified in the EIA report, further certifying the organizations' claims as a widely known anti-incineration expert (ibid.; interviews Zhao 8-11-12, NGO NU2 4-5-13, NGO FON1 19-10-12; Zhang 2011b). A few Beijing media outlets with personal relationships with the organizations again reported about the 'no burn' community's activities (Guo 2011; Qie 2011a; Zhang 2011a).[104] After the meeting, Friends of Nature submitted their opinion statement to the units responsible for the project, including the EIA unit CAMS, the construction unit Beijing Lühai, and the Beijing EPB. Green Beagle sent a letter to the Haidian district government and Beijing Lühai on 27 and 28 May. The Friends of Nature statement primarily summarized the main points of critique raised at the two meetings and cited their 'house' experts, including Zhao Zhangyuan (D_DGC14; FON 2011a, 2011b, 2011c). Overall, the organizations' central claims primarily revolved around: (1) the project's environmental and health hazards, particularly challenging the siting decision; (2) procedural flaws, such as a lack of communication with the affected residents during the EIA process; (3) planning issues, which they claimed rendered the project illegitimate; and

104 Media reporting about the issue at this stage was from the primarily the state-led but previously project-critical *Beijing News*, the commercial *China Business News*, and the state-owned *Legal Daily* – the latter two regular outlets for the organizations' activities, including in the Panguanying case.

(4) a fundamental critique of incineration as a waste treatment strategy, thus bridging the frame of the letters with that of the broader anti-incineration community (interviews NGO NU1 4-6-13, 1-7-15, NGO NU3 25-7-13, FON1 6-6-13, Xia 30-7-13, Zhao 8-11-12; D_GDC5, D_DGC14; FON 2011a, 2011b; Guo 2011; Qie 2011a; Zhang 2011a). These are the main lines of argument employed in other cases, including the simultaneous Panguanying case. The letter from Green Beagle further questioned whether the construction unit Beijing Lühai met the legal requirements for building and operating the incinerator and criticized the lack of public bidding procedures for the project (D_DGC5). As before, the public and media response remained limited at this stage and the submitted statements were left without a response.

Fragmented Local Contention: Suspected EIA Fraud, Another Site Visit, and Establishing Local Contacts

Thus far, opposition to the incinerator had primarily come from the Beijing 'no burn' community. This was quite different from the Asuwei and Panguanying cases described in the last two chapters, where the local communities were the main forces of opposition, particularly in the initial stages, and had then reached out to intermediaries. In the Dagong case, the intermediaries were first to obtain information about the planned project and understand its nature and implications due to their expertise in the issue field. By June 2011, however, the different social groups in the Dagong area had become alert to the planned project and started to mount their own forms of contention. Around this time, all of the affected social groups in the area started to communicate with the engaged environmentalists, either through the organizations' renewed site visits (for the villagers and the residents of the Jinyu Relocation Buildings) or by reaching out to the environmental organizations directly (for the Dajue Temple and military staff) (interviews DGC2 23-7-13, DGC5 29-5-13, 23-7-13, DGC9 22-7-13, NGO NU1 7-11-12, NGO NU2 4-6-13, NGO NU3 25-7-13, NGO FON1 19-10-12). During the high tide of the Dagong struggle in summer and fall 2011, staff members from the three engaged environmental organizations – Green Beagle, Friends of Nature, and Global Village of Beijing – were the central communication nodes where the threads of contention by the diverse social groups came together. In this phase, the organizations tried their best to unify the different groups and initiate a convergence of activities in line with their own mode of action.

On 1 June 2011, the abridged EIA report was finally published on the Beijing EPB website and another phase for the solicitation of public opinion was set to last until 15 June (D_DGC9 to D_DGC11, D_DGC20, D_DGC21;

Jin 2011a). Whether this was a response to the environmentalists' critique about lacking public transparency or their request that the report be published online is hard to assess. On 11 June, the Beijing EPB announced that a total of 500 questionnaires had been distributed to the members of the residential communities and other units within 2.5 kilometres of the project site during the public announcement phase in May as part of the EIA measures. According to the EPB, 91.4 percent of the participating individual residents from the surrounding villages and the Jinyu Relocation Buildings, as well as 100 percent of the approached 'groups' (团体, *tuanti*) (including Dajue Temple and some of the nature parks) supported the construction (D_DGC3, D_DGC6, D_DGC8 to D_DGC11, D_DGC20, D_DGC21; Jin 2011a, 2011b; Qie 2011d, 2011e; Yan 2011).

After their experience with the EIA unit CAMS in Panguanying village, where CAMS had fabricated the majority of the alleged EIA questionnaires, these results immediately raised the suspicion of the intermediaries. At the same time, the numbers also gave them hopes that, should they discover a similar fraud, it would provide them with a valuable resource to use against the incinerator project, similar to that which had just won the villagers and their supporters a victory in Panguanying. Moreover, the Panguanying case, for which the Hebei EPB had withdrawn the project's EIA ratification on 27 May – just a few days earlier – based on the fake EIA questionnaires, provided the organizations with a neat success story. They thus hoped to use the Panguanying case to mobilize the local residents in Dagong, who had thus far kept alarmingly quiet in the intermediaries' eyes. The 'no burn' community members were well aware that the eventual success of their intervention would likely depend on locally based contention by the affected communities. The main engaged intermediaries – notably the three environmental organizations, professor Zhao Zhangyuan, and the lawyer Xia – had learned plenty from the Panguanying struggle and were intent on using their accumulated knowledge in the Dagong case (interviews NGO NU1 7-11-12, NGO NU2 4-6-13, 26-3-14, 1-7-15, NGO NU3 25-7-13, Zhao 8-11-12, Xia 6-11-12, 30-7-13). As a Friends of Nature staff member put it:

> No matter whether it is Green Beagle or us, we can only provide support with regards to knowledge, techniques and resources. But a very stern problem is that if no local residents come out to express their opposition, then we are also powerless. [...] For us people from the outside, if we don't have support from the local people then we are very weak. [...] For waste incineration [activities] it is still really important that the local people fundamentally oppose it. [...] Moreover, we can't do anything the local

> residents don't do. Because if they go and oppose their incinerator, then they are relatively legitimate (比较合法, *bijiao hefa*). But if we organizations from the outside come and oppose it, the risks we undertake are extremely large. So in the Dagong case it was really important that the local *laobaixing* also oppose the incinerator. (Interview NGO NU1 19-10-12)

On 13 June, two days after the results of the questionnaires were published, staff members of the three organizations paid another visit to the Dagong area to investigate the local situation, ask whether questionnaires had indeed been distributed, conduct their own survey of the surrounding residents to produce comparative results, and establish closer contacts with the different social groups. During their inspection trip they visited all residential units within 2.5 kilometres of the project site as listed in the EIA report, including the Dagong and Xugezhuang villages, the Jinyu Relocation Buildings, Dajue Temple, and the military base (interviews NGO NU2 4-6-13, 1-7-15, NGO NU3 25-7-13; D_DGC6, D_DGC7, D_DGC19 to D_DGC21; FON 2011a, 2011b, 2011c). At the project site, the organization members noticed that the construction company Beijing Lühai had already started the early stages of construction work, such as fencing off the land, even though the EIA report had not yet been officially ratified by the Beijing EPB (ibid.).

In the villages, the organization staff discovered that, while some public notices about the pending project had been posted, the majority of villagers still had a very poor understanding of the planned project; most reported that they had not seen any questionnaires or been personally informed of the plans by local or municipal government officials (ibid.).[105] Moreover, while most of the approached villagers expressed opposition to the project and having to leave the village, the majority regarded any resistance as futile and expressed their willingness to relocate should compensation payments be sufficient (interviews NGO NU2 4-6-13, 1-7-15, NGO NU3 25-7-13, NGO FON1 19-10-12; D_DGC6 to D_DGC8, D_DGC21; FON 2011b; Jin 2011a). During the visit, some individual villagers exchanged contact details with organization members for later communication and potential activities. Overall, however, the organizations saw little potential for joint action due to the villagers' primary focus on compensation claims (ibid.).

105 This was confirmed during the field visits in summer 2013, when the interviewed Dagong villagers claimed that no questionnaires had been distributed and that they had not been informed about the project in any way other than having notices posted in the village. According to the villagers, no personal meeting was held by local or municipal officials (interviews DGC1 23-7-13, DGC2 23-7-13, DGC3 23-7-13, DGC4 23-7-13).

More responsive were the residents of the Jinyu Relocation Buildings. During their visit, the organization staff learned that some of the residents had already started petitioning (上访, *shangfang*) the Haidian district and municipal governments. In one instance this had taken the form of a mass petition, including a small-scale protest outside the municipal government where they had demanded to see the mayor to express their opposition to the project. However, these attempts yielded no results. According to the angered residents, 40 questionnaires had indeed been distributed in the buildings, and all participants had opposed the project. Several dozen other residents who wanted to participate in the survey were prohibited from filing out the forms. Moreover, while the gift of a toothbrush and toothpaste were promised to all participants, residents who did not support the project were later denied them (interviews DGC5 29-5-13, 1-6-13, NGO NU2 4-6-13, 1-7-15, NGO NU3 25-7-13, NGO FON1 19-10-12; D_DGC6 to D_DGC8, D_DGC21; FON 2011a, 2011b; Jin 2011b; Qie 2011d, 2011e; Yan 2011; Zhang 2011b). The organizations saw greater potential for joint action with the residents of the Relocation Buildings and exchanged contact information so they could invite the residents to future activities.

During their visit to Dajue Temple, the organizations found out that the temple staff fiercely opposed the project and had not returned their questionnaire – thus challenging the report's claim that 100 percent of the surrounding units supported the project. The report's claim that the temple had no permanent residents and would not suffer any harm from the incinerator despite its proximity also turned out to be incorrect. In fact, 100 people were permanently living at the temple, the temple's staff reported (ibid.; interview DGC9 22-7-13). During their visit, organization members talked with the temple's secretary, who outspokenly opposed the project; he also expressed his views in media interviews at that time (interviews DGC9 22-7-13, DGC5 29-5-13, 1-6-13, NGO NU2 4-6-13, 1-7-15, NGO NU3 25-7-13, NGO FON1 19-10-12; D_DGC21; Jin 2011a; Qie 2011e). Again, the organization staff exchanged contact information so they could keep the temple staff informed about further actions and coordinate joint activities.

Shortly after their visit, Green Beagle staff were contacted by a leading member of the temple staff. Through this contact, the organization learned that, like the residents of the Jinyu Relocation Buildings, individual temple staff members had started to create their own means of opposing the planned incinerator (interviews DGC9 22-7-13, DGC5 29-5-13, 1-6-13, NGO NU2 4-6-13, 1-7-15, NGO NU3 25-7-13, NGO FON1 19-10-12). Having learned about the project from a friend who greatly valued the temple and was worried about the incinerator project's implications for the cultural site and its inhabitants, this leading member of the temple staff had conducted personal research about the project and come across Zhao Zhangyuan's online articles, blogposts, and media statements, which alerted them to the harms of incineration and dioxin. Through their internet research, the temple staff member also found out about the prior resistance against incinerators in other parts of Beijing, further increasing their sense of alarm and providing them with a feeling of entitlement to mount an action against the incinerator in the name of culture, the temple's permanent residents, and the area's larger residential community. Relying on their (high-ranking) personal relations both within and outside the temple's political system (系统, *xitong*) under the lead of the Ministry of Culture, this leading temple staff member had started to solicit elite allies and political support to oppose the facility in June 2011 and continued to openly express their opposition to the plant in media interviews (interviews DGC9 22-7-13, NGO NU2 4-6-13, 1-7-15, NGO NU3 25-7-13, NGO FON1 19-10-12; Jin 2011a; Qie 2011e). The friend who first alerted the temple staff member to the pending project had further suggested contacting Green Beagle as a helpful supporter of anti-incinerator contention. In the weeks after the organizations' field visit, communication with the temple staff consolidated and the organization's additional information about incineration and its harms, as well as the flaws in the EIA process, further strengthened the temple staff member's 'cognitive justification' for the mounting resistance (interviews DGC9 22-7-13, NGO NU2 4-6-13, 1-7-15, NGO NU3 25-7-13, NGO FON1 19-10-12).

The organization staff's last stop on their field tour was the nearby military base. During the first meeting, the residents were hesitant to talk with the visitors due to the sensitivity of the issue. However, while the military personnel could not talk openly with the organization staff, their family members who also lived on the base could talk with the environmental organizations. Through them, the organizations learned that the residents of the military base had opposed the incinerator ever since they had learned about the project during the first public announcement phase in November 2010. After hearing about the completion of the EIA report in May, they had requested that they be provided with a copy and had demanded to be included in the EIA's public participation measures as the largest affected residential unit in the vicinity. Both requests had been declined, however, since Beijing's regulations regarding a project's public participation process only stipulated the solicitation of public opinions within two kilometres of the project site – thereby excluding parts of the military base, the closer barracks of which were located about two kilometres from the site. This angered the military residents and in June 2011 they had started to oppose the project through reports to higher levels within the military *xitong*, hoping to solve the issue through contention within the system (interviews NGO NU2 4-6-13, 1-7-15, NGO NU3 25-7-13, NGO FON1 19-10-12, DGC9 22-7-13, DGC5 29-5-13, 1-6-13; D_DGC8 to D_DGC11, D_DGC21; Qie 2011e). The organizations' field visit lay the foundation for further communication over the course of the next weeks.

Stepping Up Environmentalist Action: Continued Complaints and a National-Level Issue Campaign

As in the case of Panguanying village, the strong indication of fraud practiced by CAMS with regards to the EIA public participation measures for the Dagong incinerator gave the Beijing 'no burn' community a good ground on which to step up their action against the project. Together with some further flaws in the abridged EIA report that the engaged intermediaries had discovered through another close reading of the report, they now had good evidence to show that CAMS had not fulfilled its responsibilities as an EIA unit and therefore to fundamentally challenge the project. Moreover, they now knew they could build on some local support to oppose the facility.

One day after their field visit, on 14 June 2011, Green Beagle sent a letter to the Beijing EPB, the Haidian district government, construction unit Beijing Lühai, and EIA unit CAMS to report the findings of their investigation and renewed scrutiny of the EIA report (D_DGC6). Explicitly referring to the collected expert advice to provide certification, the organization again challenged, among other points: (1) the qualification of the construction unit Beijing Lühai and the lack of public bidding procedures for the project; (2) the claim that slag residue would be treated in a specific waste treatment facility in Haidian district, which, according to the organization, did not exist; (3) the claim that waste entering the incinerator would be pre-sorted at waste sorting stations, none of which were fully functional in Haidian district, according to the organizations; (4) the findings from their field visit; and (5) the siting decision as a whole, based on the plant's close proximity to Beijing's two main water systems and the surrounding natural parks (ibid.). In both their letter and a parallel information disclosure request submitted to the Beijing EPB, Green Beagle demanded the release of the full EIA report, particularly the public participation section including the EIA questionnaires (D_DGC22; FON 2011b).

This time, the Haidian district government reacted promptly to the organization's letter. Only one day later, on 15 June 2011, the Haidian district government invited Green Beagle and Friends of Nature staff to a meeting with Haidian district government and BMAC officials, as well as representatives of the construction unit Beijing Lühai, EIA unit CAMS, and the Beijing EPB, to discuss the abridged EIA report (interviews NGO NU2 4-6-13, NGO NU3 25-7-13, FON1 19-10-12, 6-6-13; D_DGC6, D_DGC7, D_DGC19 to D_DGC21; FON 2011b, 2011c; Jin 2011b; Zhang 2011b). The project proponents also invited three experts to the meeting, all well-known incineration advocates, including professor Nie Yongfeng from Tsinghua University, who defended the project's siting and EIA measures and claimed that all procedures had been conducted according to the law and that no major environmental and health impacts would be caused by the plant. The opposing side had also brought their own experts, however, including professor Zhao Zhangyuan, who presented the 'no burn' community's collected points of critique, and the lawyer Xia, who challenged the project from a legal perspective. According to the intermediaries, none of the 'no burn' community's questions were sufficiently answered by either the responsible units or their experts, and the opponents' repeated request to disclose the full EIA report including the 500 EIA questionnaires – which they hoped would provide them with conclusive evidence of EIA fraud, as in the Panguanying case – was not met (ibid.).

The 'no burn' community then decided to jointly approach the ongoing Dagong and Panguanying cases, hoping to give more weight to the claims of the individual struggles by demonstrating their parallels. In Panguanying, the potential for renewed EIA measures was looming in the summer of 2011 despite the villagers' success in obstructing the local incinerator by revealing the EIA fraud. The 'no burn' community hoped that direct references to the successful Panguanying case and the prior fraud practiced by CAMS there would significantly further their claims against the Dagong project. The environmental organizations decided to make the two cases the basis for a broader campaign against fraud practiced during EIA measures, which they had also come across in other localities (interviews NGO NU1 7-11-12, NGO NU2 4-6-13, NGO NU3 25-7-13, NGO FON1 19-10-12; FON 2011b). Over the next weeks, the engaged intermediaries started to send a multitude of letters and information disclosure requests to different institutions and tried to initiate a media campaign about both the individual cases and EIA implementation failures more generally. They also actively attempted to engage the local communities in the Dagong area in their activities.

On 17 June, five environmental organizations[106] under the lead of Green Beagle sent another letter to the Haidian district government,

106 Apart from Green Beagle, these environmental organizations were Global Village of Beijing, EnviroFriends Institute of Environmental Science and Technology (环友科学技术研究中心,

the construction unit Beijing Lühai, and the Haidian District Municipal Administration Commission (海淀区市政市容管理委员会, *Haidian qu shizheng shirong guanli weiyuanhui*). They again reported the suspected fraud practiced by CAMS in the Dagong EIA measures, also detailing the prior fraud committed by CAMS in the Panguanying case and the subsequent withdrawal of the EIA approval by the Hebei EPB (D_DGC7; FON 2011b, 2011c). The letter concluded that the EIA unit should be changed in the Dagong case and that the EIA work should be repeated to avoid a loss of investment as experienced in Panguanying or significant environmental impacts. By changing the EIA unit, the organizations argued, the government could demonstrate its dedication to environmental responsibility and win the support of both the public and environmental organizations. The same content was also published by the organizations in a press release distributed online and through their media networks (D_DGC2, D_DGC23; FON 2011a, 2011b). Again, the state-led Chinese-language newspapers *Beijing News* and *Legal Daily* reported about the organizations' claims and activities (Jin 2011b; Qie 2011d, 2011e). The letter received no response (interviews NGO NU2 4-6-13, NGO FON1 19-10-12; FON 2011b).

A few days later, on 20 June, the same five environmental organizations joined professor Zhao Zhangyuan to openly petition the MEP to revoke CAMS' grade-A qualification as an EIA unit based on the fraud practiced in the Panguanying and Dagong cases. The organizations requested that all EIA work conducted by CAMS should be investigated and any ongoing EIA reports halted. Moreover, CAMS should be fined for breaching the law (interviews NGO NU1 7-11-12, NGO NU2 29-10-12, 4-6-13, NGO NU3 25-7-13, NGO FON1 19-10-12; D_DGC 8; FON 2011a, 2011b; L. Liu 2013b; Mao 2013; Qie 2011c; Yan 2011; Zhang 2011b). Taking the issue to a national level, the organizations demanded that the country's EIA units be more strictly controlled and that any fraud practiced during the implementation of EIA measures be more thoroughly punished. Moreover, to counter the frequent occurrence of such problems, the organizations suggested that the MEP change the method of appointing EIA units. Rather than the construction unit commissioning and paying the EIA unit, the responsibility should be handed to the government to avoid direct and fraud-prone relations between construction and EIA units. The organizations further pleaded for more extensive public participation in the EIA process, including more

Huanyou kexue jishu yanjiu zhongxin), Institute of Public & Environmental Affairs (公众环境研究中心, *Gongzhong huanjing yanjiu zhongxin*), and Green Earth Volunteers (绿家园志愿者, *Lü jiayuan zhiyuanzhe*).

encompassing information disclosure, mandatory public hearings, and the inclusion of environmental organizations in the process (ibid.).

The organizations also tried to approach the case from another angle. On 22 June, they sent a letter to the Ministry of Housing and Urban-Rural Development, requesting that the Ministry investigate the qualification of the construction unit Beijing Lühai to build and operate the Dagong incinerator (D_DGC23; FON 2011b). According to the organizations, the company did not possess the mandatory registered capital of 100 million RMB as required by state law for the operation of municipal solid waste incinerators and was hence not eligible to conduct the operation. Moreover, the Haidian district government and Beijing Lühai had by this time admitted that the waste treatment facility to treat the slag residue as described in the EIA report did not yet exist. The organizations thus suggested in their letter that Beijing Lühai did not have the equipment or capacity to operate the incinerator safely and demanded a thorough investigation of the issue. Again raising the case to a more general level, the organizations demanded that the Ministry investigate all of China's current MSWI construction units to avoid corporations without qualifications entering the incineration business amid the major push into the high-profit incineration sector (ibid.; FON 2011b). On 24 June, Green Beagle followed up on this issue with an information disclosure request to the Beijing Municipal Administration Commission, which asked whether Beijing Lühai had the mandatory license for municipal solid waste management (城市生活垃圾经营性处置服务许可证, *chengshi shenghuo laji jingying xing chuzhi fuwu xukezheng*) (D_DGC24). Again, both letters were left without response (FON 2011b).

The 'no burn' community accompanied this campaign with another round of public and media outreach. On 23 June 2011, Green Beagle held a public symposium on the general problems of EIA implementation and the Dagong case in particular. Again, members of the Beijing 'no burn' community, including staff members from the involved environmental organizations, professor Zhao Zhangyuan, and the lawyer Xia came together to voice their concerns about the project. A small wave of media reporting about the organizations' allegations about CAMS, their challenges to the Dagong EIA report, and their general EIA concerns followed in the usual media outlets, with several of the articles mentioning Panguanying as a parallel case.[107]

107 As before, media reporting about the issue was still largely limited to media outlets with direct contacts with organization members. Articles about the issue were primarily published in the *National Business Daily* (*National Business Daily* 2011; Z. Li 2011), *Legal Daily* (Qie 2011a, 2011b, 2011c), *The Beijing News* (Guo 2011a) and the *China Business News* (Zhang 2011a, 2011b). As part of

While this campaign did not produce major responses from the MEP or other institutions at that time, it took up new momentum in late 2012, as outlined in the previous chapter on the Panguanying case.

On 28 June 2011, the Beijing EPB officially ratified the Dagong project's EIA report despite the ongoing controversy, much to the dismay of the engaged intermediaries (interview NGO NU2 4-6-13; China Solid Waste Net 2011; NGO Information Web 2012). On 30 June, Green Beagle again submitted two information disclosure requests with the Beijing EPB, demanding the disclosure of the full EIA report and the official approval document (D_DGC25, D_DGC26). While the Beijing EPB did reply to this request at the beginning of July, they referred the organization to the construction unit Beijing Lühai, from which they were told to request the release of the full EIA report (D_DGC27).[108] Successive information disclosure requests submitted by Green Beagle staff to the Beijing Municipal Development and Reform Commission (北京市发展与改革委员会, *Beijing shi fazhan yu gaige weiyuanhui*) – requesting whether the commission had officially ratified the Dagong project and demanding the official ratification notice (D_DGC28) – and to the Beijing Municipal Planning Commission (北京市规划委员会, *Beijing shi guihua weiyuanhui*) – requesting the disclosure of the Dagong project's 'social stability and risk evaluation report' (社会稳定风险评价报告, *shehui wending fengxian pingjia baogao*) (D_DGC29) – produced no significant results.

On 6 July 2011, the organizations made a last attempt to oppose the Dagong project via official complaints. The five environmental organizations under the lead of Green Beagle sent another joint letter, now to the MEP North China Environmental Protection Supervision Center (华北环境保护督查中心, *Huabei huanjing baohu ducha zhongxin*) (D_DGC20). The letter again summarized all points raised against the project by the organizations, including: (1) that it was not in line with the Beijing plans and hence was illegitimate; (2) that fraud was practiced during the EIA public participation section and that there were other flaws in the report; (3) that the construction

broader reports about problems in the implementation of the EIA Law, articles also appeared in *Southern Metropolis Daily* (X. Yang 2011), which was alert to the issue due to incineration cases in Guangzhou including Panyu, Likeng, and Huadu, and the *Southern Weekend* (He and Shi 2011). The English-language Party-affiliated *Global Times* also published a detailed report about the issue (Yan 2011).

108 According the EIA Law, the MEP and environmental protection bureaus can refuse the publication of full EIA reports if their publication touches upon the construction unit's corporate interests or would reveal corporate secrets. They can thus refer the authors of information disclosure requests to directly get in touch with the construction unit.

unit Beijing Lühai lacked the required qualifications; and (4) that the Beijing EPB's ratification of the project was not legitimate, since the EPB did not respond to the organizations' allegations and did not conduct an investigation into the matter. Building on these points, the organizations demanded that the Center investigate the EIA fraud, the construction unit's qualifications, and whether the EPB ratification was compatible with proper procedures (ibid.). The letter further included the detailed 'list of incidents of violations against regulations by Chinese incinerators' to the MEP, including Dagong, Panguanying, and several other projects (D_DGC1).

The Supervision Center replied to the organizations in mid-August, reporting that it had conducted a thorough investigation of the matter, including a site visit, a meeting with the EIA and construction units, an investigation of the EIA public participation materials, and a phone survey among the respondents (D_DGC3). According to their investigation, the construction unit had not yet commenced the construction work and there were relocation plans for Dagong, Nan'anhe, Beianhe, Zhoujiaxiang, and Xugezhuang villages – the latter four as part of the town's new village transformation plans – as well as for the Jinyu Relocation Buildings, from which relocation would be on a voluntary basis. Relocation work would begin before 2013. Regarding the other points, the Center claimed that: (1) the project was in line with all necessary plans; (2) the slag treatment waste facility would be built in 2012 with an investment by the Haidian district government; (3) the EIA public participation measures were carried out according to the requirements – of the 42 survey participants who had not agreed with the project, 40 were from the Jinyu Relocation Buildings, while all other participants had supported the project; while Dajue temple staff had not returned their questionnaire, they had orally agreed to the project during a meeting, which had been recorded – (4) there was no problem with the operating qualifications of the construction unit Beijing Lühai; and (5) the Beijing EPB was in charge of ratifying the project and hence no problem existed with regards to the ratification procedures (ibid.). Without further engagement by the local communities, with this reply the organizations felt that they had reached the limits of their capacity to oppose the Dagong incinerator.

Losing Momentum: Failed Attempts to Unify Resistance, Emerging Cleavages, and Fading Contention

Since their field visit, the five organizations' – particularly Green Beagle's – communication with the different social groups had intensified. Individual residents from the Jinyu Relocation Buildings and the military base, as well as leading temple staff had all started to actively seek the organizations' advice to gain a better understanding of the project and its harms, and their rights as affected resident communities. Based on this communication and the residents' growing awareness, the organizations hoped to actively engage the different groups in their own activities.

On the invitation of Green Beagle, who took on a brokering function between the different social groups in the hopes of producing a joint action, several of the residents from Dagong village, the Jinyu Relocation Buildings, and the military base attended the public EIA symposium on 23 June. Some of them also joined a Haidian district government-led inspection trip of the Liulitun landfill, as promised by Haidian district Party secretary Zhao Fengtong in January 2011 (Wen 2011; Yi 2011a). When the organization staff and participating residents asked about the proposed incinerator during this excursion, Haidian government officials revealed an alarming ignorance of incineration technology, organization staff recalled (interview NGO NU2 4-6-13).

During these meetings, the environmental organizations and particularly the Green Beagle staff tried to persuade the different groups affected by the Dagong project of the benefit of undertaking joint activities (interviews NGO NU1 7-11-12, NGO NU2 4-6-13, NGO NU3 25-7-13, NGO FON1 19-10-12, Zhao 8-11-12, Xia 6-11-12, DGC5 29-5-13, 23-7-13, DGC9 22-7-13). The organizations and professor Zhao Zhangyuan provided the groups with further information and materials about incineration, the related environmental laws and regulations, and the Dagong project procedures. They also introduced the residents to the details of the Panguanying case as a successful predecessor in which where local (rural) opposition had succeeded in obstructing a planned incinerator based on similar problems with the same EIA unit. Together with the lawyer Xia, the organizations also offered the residents legal advice and urged them to follow the example of the Panguanying villagers in using to legal means (ibid.).[109] Throughout their campaign, Green Beagle staff prepared several draft letters similar to those written by the organizations themselves, which were meant to be sent by the affected communities to the various responsible units, including the Beijing EPB, the MEP, the Haidian district government, and Beijing Lühai. The draft letters requested, among others, that the project's EIA report be revoked, an investigation into the

109 Among the distributed Panguanying materials were the villagers' court file including a list of all the mistakes discovered in the EIA report and its public participation section; the villagers' letter to the MEP Pollution Prevention Division reporting in detail about the discovered EIA flaws and malpractice conducted by the EIA unit CAMS, as well as their legal actions and its consequences; and the verdict by the Shijiazhuang City Qiaoxi District People's Court confirming that the Hebei EPB had withdrawn their EIA approval (D_PGY41, D_PGY42, D_PGY48).

EIA problems be conducted, and a public hearing be held about the issue (D_DGC9 to D_DGC11). The intermediaries also urged the residents to file information disclosure requests to get the full EIA report (interviews NGO NU2 4-6-13, NGO NU3 25-7-13, NGO FON1 19-10-12; D_DGC30).

While communication with the different groups was close in the first weeks after the organizations' field visit and the residents were initially open to the intermediaries' advice and eager to take action, these relations soon cooled off. To the frustration of the engaged intermediaries, the different groups did not actually follow through with taking the organizations' advice, nor did they intensify the newly established relations with each other. Rather, they insisted on pursuing individual courses of action. The Jinyu Relocation Building residents continued to focus on petitioning and quickly lost confidence after the Beijing EPB had officially ratified the project at the end of June. The military and temple staff were certain that the issue could best be resolved by seeking higher-level support within their respective systems (*xitong*) and building on personal relations. Because of the rather high political sensitivity of the case – due to the very high-level political actors supporting the project – they regarded this means of contention as the only (politically) feasible option, as they were members of residential units that were under state jurisdiction. Throughout the summer and fall of 2011, both military and Dajue temple staff continuously reported about the project, related problems, and their opposition to it to higher-ranking individuals within their respective *xitong* and tried to rely on (high-ranking) personal relations to mobilize support and elite allies who would oppose the project (interviews NGO NU1 7-11-12, NGO NU2 4-6-13, NGO NU3 25-7-13, NGO FON1 19-10-12, Zhao 8-11-12, Xia 6-11-12, DGC5 29-5-13, 23-7-13, DGC9 22-7-13). As a Green Beagle staff member recalls about their failed attempts to unify resistance:

> Some of the Mentougou [Jinyu Relocation Buildings] residents stood up, they started petitioning the Haidian district government and municipal governments. They all adopted *shangfang* measures. We told them at the time that *shangfang* doesn't get you anywhere. That we also hope that you residents write some other things. If the residents do what we do, it might produce a different outcome. But from beginning to end they didn't want to do this kind of action. I don't say *shangfang* is bad, it just doesn't solve the issue. At most you make the government know, make them say ok, we will pay attention to this issue. But if you really want to solve this, you need to do a lot of things, not only *shangfang*. Write detailed reports, raise some questions. How you organize your action is really important

> for how much influence you can gain in this issue. [...] Also the military and Dajue temple people wanted to pursue their individual actions. [...] So Dagong was a very peculiar project. There was little *laobaixing* and lots of NGO [sic.] engagement. [...] There was no united strong force. At the time there were a lot of people, but they stepped up in a very fragmented fashion and never joined forces. But such a thing needs a strategy, you have to move together. But when we tried to overcome that problem at the time, they stubbornly thought that what they were doing themselves, this and that, was right. But in fact there was no encompassing strategy. This was a very cumbersome thing. (Interview NGO NU2 4-6-13)

These different approaches soon led to cleavages between the environmentalists and local groups and contact with several of the residents was broken off. Without stronger local community engagement, the intermediaries felt that they had come to the end of their capacities. While continuing to keep an eye on the project's developments, they stopped investing much energy in the issue for the time being and turned to more urgent cases featuring more active resident engagement (interviews NGO NU1 7-11-12, NGO NU2 4-6-13, NGO NU3 25-7-13, NGO FON1 19-10-12). Towards the end of 2011, the local groups' individual means of action also hit dead ends and the contention against the incinerator slowly trailed off. The military base residents' complaints had not produced any results, and they had been warned about conducting further activities. A leading Dajue temple staff member who had openly opposed the project and attempted to seek high-level political allies was removed from his post and transferred to a different unit within the culture *xitong* – a transfer that was officially labelled a promotion. Since contact with the Jinyu Relocation Buildings residents broke off rather abruptly, there were rumours amongst the intermediaries involved in the case that the residents had been offered a compensation deal by the government and had therefore withdrawn from any further activities (interviews NGO NU1 7-11-12, NGO NU2 4-6-13, NGO NU3 25-7-13, NGO FON1 19-10-12, Zhao 8-11-12, DGC5 29-5-13, 23-7-13, DGC9 22-7-13).

Minor Recurring Contention and the Villagers' Perspective: The Beginning of Construction, a Road Blockade, and another Failed Mobilization Attempt

A small peak of contention recurred in spring 2012. In March that year, construction work started on the incinerator without prior relocation of the residents (interviews NGO NU2 4-6-13, 1-7-15, DGC1 22-7-13, DGC2 23-7-13,

DGC4 23-7-13, DGC5 29-5-13, DGC6 1-6-13, DGC7 1-6-13, DGC8 1-6-13, DGC9 22-7-13; D_DGC19; *Beijing News* 2011b). This stifled some acts of contention by the surrounding residents, which were, however, less in opposition to the project and more of a request for timely relocation and suitable compensation. There were rumours, for example, that some individuals from the Jinyu Relocation Buildings had kicked over some construction fences shortly after the construction began in March (interviews NGO NU2 4-6-13, 1-7-15, DGC2 23-7-13, DGC5 29-5-13).

A few months later, in June 2012, a few dozen Dagong villagers – mostly old women[110] – blocked the access road to the construction site for several days. While contacts between the environmental organizations and the villagers had become sparse, a Green Beagle staff member still visited the protesters during this phase to observe the situation, urge the villagers to avoid violent conflict, and give them advice if needed (interviews NGO NU2 4-6-13, 1-7-15; D_DGC19). According to the villagers, there were several reasons for this act of protest. First, the construction workers' cars and trucks had blocked the only road to the village and cut off the villagers' water supply for several days. The workers had also carelessly destroyed some of the environment along the road and surrounding the construction site, interrupting the villagers' daily life (interviews DGC1 22-7-13, DGC2 23-7-13, DGC3 23-7-13, DGC4 23-7-13, DGC5 29-5-13, NGO NU2 4-6-13). What infuriated them even more was that construction had started without any further communication about either the project details or relocation measures and compensation rates. Angered that their opinions had gone unheard, they decided to 'make a little trouble' (闹一下, *nao yixia*) to voice their discontent about being side-lined and draw attention to their claims, which – since they felt that that they had no political efficacy and that resistance against the plant was futile – still mainly centred on compensation rates and a timely resettlement (interviews DGC1 22-7-13, DGC2 23-7-13, DGC3 23-7-13, DGC4 23-7-13). As one of the villagers put it during the field visit in July 2013:

> They did a survey in 2011 and asked for the villagers' opinions. But then they don't listen to the villagers' opinions. The Party says they rely entirely on farmers, but in reality the farmers have nothing to say. [...] So in 2012 when they started construction, the *laobaixing* knew nothing and they still started construction. We didn't know what was going on at all, no

110 This is standard practice in Chinese contention, since old women are less likely to encounter physical attacks by public security staff and signal docile resistance (cp. Hung 2011) – in addition to having enough spare time to sustain lengthy blockades.

> one had talked with us. So the *laobaixing* made a little trouble. [...] This [project] is not good for Beijing because of the wind. I don't know what the cadres are thinking. But opposition is in vein (反对也白反对, *fandui ye bai fandui*). There are still people who oppose this, but it is in vein. The Party doesn't listen to the opinions of the *laobaixing*, they have no power. They did this very clever, you know. In Liulitun there were many people, here we are very few. So they don't make trouble and it's easy to get things done. [...] If there were cadres living here, they wouldn't build the project here. But the people living here are all *laobaixing*, so no one cares. (Interview DGC2 23-7-13)

The overall sense of disempowerment and political distrust was also confirmed by other villagers:

> Until they don't dismantle, we won't leave (不折不走). But relocation is certain, there is no way to oppose that. It's a state decision. [...] There haven't been any cadres, no government has talked with us. Not one single time. Not until today. We have only been informed via notices. Where are the cadres? They don't listen to our opinions, our opinions about that waste plant, our opinions about relocation. But then, even if they talked with us, we wouldn't believe them anyway. [...] They shouldn't build an incinerator here, the environment is very good. They should rather build old people's homes. But the *laobaixing* can't organize (组织不了, *zuzhi bu liao*). They don't have this capacity (没有这个能力, *meiyou zhege nengli*). (Interview DGC4 23-7-13)

When asked whether they had done any petitioning, the same group of villagers laughed at the thought of taking *shangfang* measures and explained that they knew someone who had previously tried petitioning on a different issue, entirely without results. As in the Asuwei case, the villagers' road blockade was eventually dissolved by public security forces with one villager detained, scaring the larger villager community away from any further action. The villagers were promised a timely relocation to nearby relocation homes and adequate compensation payments by Haidian government officials (interviews DGC2 23-7-13, DGC4-23-7-13, NGO NU2 4-6-13, 1-7-15; D_DGC19).

In the summer of 2012, there was also another mobilization attempt after Mrs. Ma, a villager from Nan'anhe village who mostly lived in Beijing city, heard about the project from another villager when the construction started. Mrs. Ma was shocked by the news that a waste incinerator was to be built in this beautiful stretch of land (interviews DGC5 29-5-13, 1-6-13,

23-7-13). While not entirely certain about the harms of waste incineration, she was aware of the prior Asuwei campaign and knew that incineration was somehow linked to harms to the environment and human health. She started to conduct some internet searches and discovered that almost no information about the Dagong project was to be found online. What she did find was plenty of information about the harms of incineration and the ongoing public and media debate about dioxin, as well as more information about the Asuwei and Liulitun campaigns. She also found out that Friends of Nature and Green Beagle were involved with the issue of waste incineration as local environmental organizations. While she at first considered directly contacting the media about the project, she decided to notify the environmental organizations in a first step (ibid.). Excited to hear about a potential new basis for opposition within the villager community, the organizations were glad to provide her with more information and invited her to speak at a one of their public weekend lectures in mid-July, which they dedicated to the Dagong case (ibid.; interviews NGO NU2 4-6-13, 1-7-15; D_DGC19, D_DGC31). As Mrs. Ma recalled:

> At that time I didn't know anything about waste incineration, I knew absolutely nothing. I only knew it was not a good thing, because if you burn waste, with only a little bit of knowledge about chemistry, you know what comes out is hazardous. And I knew about Asuwei, because that issue was in the Beijing newspapers. Beijing reporting about that issue was tremendous. That project was halted because of the residents' protest (抗议, *kangyi*). Then I searched a lot on the internet and that's how I gained some knowledge. [...] The first means I thought of was to use the media to do some things. But when I thought about it, I figured contacting the NGOs [sic.] would be best, because the NGOs can help us do many things. At that time, Beijing had two large environmental organizations, one was Friends of Nature, one was Green Beagle. So I went online to find information and just right, they were concerned with this issue. Online was Chen Liwen's [a Green Beagle staff member] mobile number so I called her and told her about this issue. She said she knew about it early on. And she said: Just right that you care about this, don't you want to come and hold a meeting with us? So that's what I did. [...] The Green Beagle people also explained a lot of things to me. That's how I slowly, slowly started to understand more about the issue myself. (Interview DGC5 29-5-13)

When she asked the organizations for advice, Green Beagle staff members urged her to link up with and mobilize the local villagers (interviews DGC5

29-5-13, 23-7-13, NGO NU2 4-6-13, 1-7-15). Mrs. Ma thus visited Nan'anhe, Dagong, and several of the other surrounding villages to inquire about the villagers' opinions and call for collective action against the project. Much to her frustration, she discovered that, while the residents of the surrounding villages expressed a variety of opinions about whether they personally wanted to be relocated, all of them regarded relocation as mandatory and resistance as futile. Like the organization members before her, she had to accept that the villagers' claims centred mainly on receiving timely relocation and sufficient compensation rates (interviews DGC5 29-5-13, 1-6-13, 23-7-13). In the words of Mrs. Ma:

> It's all villages, there are no capacities at all. Last year, the Green Beagle people told me you have to link up with the villagers. That was the main advice they gave me, to unite the local villagers. Just like Liulitun in the beginning, how they united and made trouble. But after I talked with a bunch of them [villagers] I realized that their way of looking at this issue is completely different than mine. Most of the villagers don't care about this thing. They absolutely don't care about environmental problems or about the problem of losing land. The only thing they care about is economic compensation. Because they have no way of doing resistance (去抗争, *qu kangzheng*). In the cities they might win, it's easier to win there, but not in the villages. In the villages they say: I would actually quite like to move [...]. Whether I cultivate land or move, I will make money either way, so moving is easier, I can take five million at once.[111] I really couldn't get through to them. If the people who live here, not only the farmers, if they don't have any knowledge, if they don't care, then there really is no use in talking with them. [...] What really made me lose hope at the time was that although they all really like living in the village, [...] all they really think about are those five million RMB. But you can't really blame them. Ten years ago I might also have thought five million is a huge amount of money. (Interview DGC5 29-5-13)

The difficulty of mobilizing collective action among the villagers was confirmed during my field visit in July 2013, when I accompanied Mrs. Ma to visit several Nan'anhe villagers she had not previously contacted so she could demonstrate to me that mobilizing the villagers was virtually impossible. The visits included a villager who had represented several of the Nan'anhe villagers in a prior lawsuit related to other land issues (interviews

111 The amount of five million RMB refers to a hypothetical sum.

DGC6 1-6-13, DGC7 1-6-13, DGC8 1-6-13). While all of the visited villagers said they had been informed about the planned waste facility in 2011 and knew about the relocation plans early on, they had a very vague understanding of the project details and little to no knowledge about its related harms. When informed in more detail about the dangers of incineration by Mrs. Ma – with reference to the environmental organizations and the prior resistance in Asuwei and Liulitun as certification – the villagers reacted with initial anger against the government and responsible units for not having informed them about this aspect in a better way. Throughout the course of the conversations, however, the villagers declared that while most of them regretted having to leave the village, there was no point in taking any action, since resistance to a state project was futile and they would be relocated anyway. Moreover, as *laobaixing* they had no knowledge or awareness of such things and thus no capacity to take action (interviews DGC6 1-6-13, DGC7 1-6-13, DGC8 1-6-13).

Various villagers also highlighted that since the plant was located on Dagong village land, it should be the Dagong villagers to oppose the project – simultaneously pointing out that some resistance had indeed been mounted by the Dagong residents (interviews DGC6 1-6-13, DGC8 1-6-13). Moreover, the Nan'anhe villagers explained, the Dagong villagers were 'of one heart and mind' with regards to opposing the plant, while they themselves were torn, with some who wanting to be relocated and some not wanting to leave the village, making unified resistance impossible (interviews DGC6 1-6-13, DGC7 1-6-13, DGC8 1-6-13). A prior fruitless lawsuit related to the appropriation of some of the village's collective land, which many of the Nan'anhe villagers had immersed themselves in for several years, was considered proof that legal action was futile and that *laobaixing* had no means of fighting for their rights (interviews DGC6 1-6-13, DGC8 1-6-13). There also appeared to be rumours in the village that the project had been scrapped and that they were building old people's homes instead (interview DGC8 1-6-13) – as had been a popular wish among the Dagong villagers (interviews DGC1 23-7-13, DGC2 23-7-13, DGC4 23-7-13).

Faced with the difficulty of mobilizing villager action in the summer of 2012, Mrs. Ma went back to her initial idea of reaching out to the media. Having worked in a media environment for years, she had good contacts with some of Beijing's large state-owned media, including CCTV, *Beijing Media Network* (北京人民广播电视台, *Beijing renmin guangbo dianshitai*), and *The Beijing News*. While one journalist accompanied her on a site visit to the Dagong area, the other contacted journalists told her that, due to the mass of environmental issues occurring in Beijing every day, they were unable to report about all of them. Moreover, even if they wanted to report about something like this, the news would immediately 'be killed' by their superiors. The report written by the journalist who did visit the area was also not published. Frustrated, Mrs. Ma turned to social media as a last means of public communication, knowing that this had been the main communication and organization platform for several other anti-siting campaigns, including Asuwei. However, her social media attempts to mobilize the public or surrounding residents failed (interviews DGC5 29-5-13, 1-6-13).

> They [the journalists] all told me that there was no way the news would get out. So in the end I thought of social media means, that's the only way. Weibo. I sent plenty of posts, but no one looks at my weibo. I have no followers. I sent weibo posts, I sent weixin posts. But I didn't get any reactions. You know, every *xiaoqu* has an online forum. I also went on the forums of some close-by residential compounds because I thought they might oppose (反对, *fandui*), because they had invested some money in the apartments there, including some wealthy people from Beijing. But I tell you, most people still don't know [about the project] until today. I also posted on these real estate websites, to tell the people not to buy apartments there. I tried to alert them. But when I went on the forums, I very fast discovered that no one was discussing this issue. There wasn't any reaction to my posts. So in the end I thought: There really is no way. (Interview DGC5 29-5-13).

While the organization staff had provided Mrs. Ma with the contact details of some of the Liulitun and Asuwei campaign leaders and urged her to seek their advice and assistance, she saw no point in contacting community members from other struggles where the initial conditions had been very different. After her frustrating experiences throughout the summer of 2012, she eventually gave up and went back to her daily routine, convinced that if the environmental organizations had not succeeded in mobilizing the local residents or obstructing the project, she as an individual was powerless (interviews DGC5 29-5-13, 1-6-13, 23-7-13).

> They [the environmental organizations] are doing so much. They are fighting so many waste incinerators. They had already tried everything here. Legal means, legal procedures, writing complaints, sending letters and petitions. But they lost. People just turned around and said the project is legitimate. So what was left to do? They had done it all, called them all. Oh dear. [...] She [a Green Beagle staff member] told me there are

> also successful cases. Asuwei and Liulitun, they won. She told me to get in touch with them. At the time I also considered to contact some of these people, to link up with them. But in the end I didn't. It's because they had so many people [that they won]. A couple hundred people all united together. How could they help us here? There are no people. And no one cares. If I had these kinds of resources, I would definitely still do this. But as it is, I just let it go. Maybe I gave up halfway, I don't know. (Interview DGC5 29-5-13)

After this last small peak of contention, the Dagong struggle had virtually come to a halt and the incinerator continued to take shape unimpeded at the foot of the Yangtai Mountain range.

The Aftermath: Relocation and the Beginning of Operation

In March 2013, the engaged environmental organizations paid another site visit to the Dagong area to see the progress at the construction site and examine the situation. Even though they hadn't been able to obstruct the project, they were still observing its development and hoped to monitor the plant once it started operating (interviews NGO NU2 6-4-13, 1-7-15, NGO NU3 25-7-13, FON1 6-6-13). During my field visits in July 2013, Dagong had just started its official relocations, slated for 20 July to 18 August 2013, according to the notices plastered on exterior walls across the village. The road leading to Dagong village past the half-finished incinerator was also flanked with official banners urging an orderly and timely relocation with slogans such as 'Early relocation for early profit, early signature of the agreement for selecting a good apartment' (早搬迁早收益，早签协议选好房, *zao banqian zao shouyi, zao qian xieyi xuan hao fang*), 'Relocation compensation – transparent policies – just and fair – establishing trust between the cadres and the people' (安置补偿 – 政策公开 – 公平公正 – 取信干民 , *anzhi buchang – zhengce gongkai – gongping gongzheng – quxin gan min*), and 'Bid the mountain district farewell, march towards a new life' (告别山区， 迈向新生活, *gaobie shanqu, maixiang xin shenghuo*).

At the time of my field research there were ongoing cleavages between the government departments responsible for this relocation and the villagers, who were discontent with both the amount and the modalities of compensation. While earlier the Haidian district government had promised to provide housing in nearby relocation buildings for the villagers, they were now expected to find housing themselves using their compensation payments which the villagers regarded as too little

Figure 5.2 Relocation notice in Dagong village, July 2013

to buy suitable apartments. The villagers had thus refused to relocate and announced that they would continue their resistance should the compensation modalities not be changed (interviews DGC1 22-7-13, 23-7-13, DGC2 23-7-13, DGC3 23-7-13, DGC4 23-7-13). For the other villages and the Jinyu Relocation Buildings, no concrete timetables for relocation had been announced as of summer 2013, but the residents had heard news that 2015 was the prospective year for their relocation (interviews DGC1 22-7-13, DGC5 29-5-13, 1-6-13, DGC6 1-6-13, DGC7 1-6-13, DGC8 1-6-13, DGC9 22-7-13).

Despite the cleavages in the summer of 2013, the relocation of Dagong and Zhoujiazhuang villages eventually took place in the first half of 2014. When Green Beagle staff returned for a site visit in June 2014, the villages were already demolished. The other villages and the Jinyu Relocation Buildings were still inhabited. When organization staff came in October 2014 to check on the incinerator's construction process, the plant was nearly completed. While there were rumours that the incinerator started operation in 2015, there was no official news about the start of the facility's operation – which the organizations suspected to be a deliberate concealment of information by the municipal government (interviews NGO NU2 26-3-14, 1-7-15; continued email communication with Green Beagle staff).

Figure 5.3 Roadside banner advocating relocation in Dagong village, July 2013

Note: The banner ironically reads 'Relocate for the sake of a more beautiful environment'.

Analysis: The Role of Horizontal and Vertical Linkages in the Dagong Struggle

The case of Dagong village shows the limitation of linkages and diffusion processes in unfavourable local conditions. In this case, a demobilizing local environment for the villagers, fragmented local interests, a restrictive local political setting, and emerging cleavages between the local residents and engaged environmentalists prevented the emergence of unified or sustained collective action. Even the early involvement of environmental organizations and other intermediaries and their active attempts to unify and assist the different local communities could not overcome these obstacles. The case shows that even if horizontal and vertical linkages exist, diffusion effects are highly dependent on the local context and can only unfold if they are rooted in sustained local contention.

The Role of Horizontal Ties

Despite its geographic proximity to the urban Beijing Liulitun and Asuwei campaigns and its strong parallels with the simultaneous Panguanying case, horizontal linkages played only a minor role in the Dagong case. Individual members of most of the social groups in the Dagong area had some vague previous knowledge of the Asuwei struggle due to the extensive reporting in Beijing media, which contributed to their initially heightened sense of alarm when they first heard about the project (as seen for the Jinyu Relocation Buildings residents, the Dajue temple and military staff, individual Dagong and Nan'anhe villagers, and Mrs. Ma). For some social groups, further information about the Liulitun and Asuwei campaigns found online or provided by the intermediaries in the early stages of the struggle raised their awareness about the harms of incineration and their 'cognitive justification' for taking action (for the Jinyu Relocation Buildings residents, the military and Dajue temple staff, and Mrs. Ma). Some of the villagers also mentioned the Liulitun and Asuwei campaigns. This knowledge of the earlier campaigns did not, however, appear to have any impact on their claims or actions, which mainly centred on compensation payments due to their lacking sense of political efficacy and their belief in the unavoidability of their relocation within the broader urbanization process – both discourses that are actively fostered by the government.

Despite the brokerage role played by the intermediaries at the Beijing meeting in June 2011, where individual Dagong villagers and Jinyu Relocation Building and military base residents were brought together with some of the urban campaigners to discuss the project, personal ties were not intensified by the different social groups. Since they preferred to pursue their own courses of action, which differed significantly from the strategies adopted by the urban contenders, they saw no use in intensifying these contacts. Mrs. Ma also refrained from getting in touch with the Liulitun and Asuwei residents since she regarded the urban conditions as so different from the situation in the Dagong area that she saw no use in contacting the communities – particularly after her local mobilization attempts had failed. Due to their diverging interests, the local villagers had no incentive to learn from or contact the urban campaigners.

While the Liulitun and Asuwei community members knew about the Dagong project early on, as demonstrated by their participation in the Friends of Nature meeting in November 2010, they did not play an active role in the struggle apart from their participation in the meeting and the lawyer Huang's repeated public and media statements both against incineration in general and the Dagong project in particular. Members of both the Liulitun and Asuwei campaigns regarded their own struggles as successfully concluded and did not pay much attention to the new location, feeling that it did not personally concern them and that they could not engage in another locality's issues without major political risks (interviews ASW1 31-7-13, NGO NU2

4-6-13, NGO NU3 25-7-13, Zhao 8-11-12, Xia 30-7-13; Wen 2011). Apart from spurring the awareness process of some of the local groups and providing some 'cognitive cues' for taking (fragmented) action, the urban Beijing campaigns thus played no major role in the Dagong struggle despite their geographic proximity.

In a similar vein, the Panguanying case, which the intermediaries introduced to the different Dagong area groups, also failed to have a major impact on the Dagong struggle. While the environmental organizations presented the Panguanying campaign as a successful case of legal action against similar EIA problems and the same EIA unit, the local groups preferred to pursue their own courses of action within the restrictive local political framework. For this reason, the Panguanying case was thought to have no noteworthy implications for their local activities.

The Role of Vertical Ties

More pronounced than horizontal ties was the role of vertical linkages for the Dagong struggle. The engaged intermediaries – mainly the Beijing-based environmental organizations Green Beagle, Friends of Nature, and Global Village of Beijing, as well as professor Zhao Zhangyuan, the lawyer Xia, and some other individual experts and journalists – were the first forces actively opposing the project and remained some of the facility's fiercest opponents throughout the struggle. Apart from their own actions, which centred mainly on official complaints, letter-writing, and information disclosure requests, as well as public and media outreach, the organizations made great efforts to mobilize the different local groups, unify their fragmented activities, and align them with their own mode of action – via both their personal communication with the individual social groups and by bringing them together at the Beijing meeting in June 2011.

During some of the local groups' initial awareness processes, the intermediaries played an important role by providing detailed information about the project and about incineration and its harms. Both the Dajue temple staff and Mrs. Ma gained a better understanding of the harms of incineration through Zhao Zhangyuan's online articles and media presence. These nonrelational ties were later complemented with the organizations' direct communication with the different local communities through both site visits and the intensified direct communication with some of the groups in the summer of 2011 (and with Mrs. Ma in 2012). At the peak of the struggle in the summer of 2011, the organizations functioned as a communication node and central information source for several of the local communities and actively tried to shape their mode of action (for the Jinyu Relocation Building residents, the Dajue temple and military staff, individual Dagong villagers). After the intermediaries failed to initiate unified local action – owing both to the different groups' diverging interests and the restrictive local political setting, which prompted the temple and military staff to keep action restricted to their own *xitong* – cleavages between the environmentalists and the social groups emerged and communication eventually broke off. Here, the Dagong case shows the sometimes 'uneasy alliance' between environmentalists and local communities if their priorities and modes of action diverge.

Without sustained local engagement and particularly after the activities of the temple and military staff were curtailed in the second half of 2011, the intermediaries' actions hit a dead end. This case shows that vertical linkages and diffusion processes can be a significant enabling factor for local contention, but if they are not anchored in sustained local activities because of unfavourable local conditions, their impacts remain limited and diffusion effects fail to unfold. This was also confirmed by several members of the Beijing 'no burn' community, who expressed their feeling of powerlessness as outsiders in local struggles where their activities could not build on autochthonous local contentions.

Notes on Scale Shift

As in the previous cases, the findings of this case point to the limited prospects of scaling up local contention based on alliances or collaborative action between different affected communities in China. No closer personal ties were established by the Dagong area residents with either the geographically near urban Beijing cases or the parallel Panguanying case, despite the brokerage efforts of the involved environmental organizations. In the Dagong case, this can be attributed to the local communities' strong reliance on their own modes of action and diverging interests, the restrictive local political setting, which significantly limited the means of contention that were deemed to be feasible, and the perception that the initial conditions of the (particularly urban) other cases were significantly different and therefore irrelevant to the developments in Dagong. Even among the different social groups in the Dagong area, which were all affected by the same incinerator and hence prone to cooperate – particularly in light of the intermediaries' brokerage attempts – no collaborative or unified action emerged for similar reasons.

As in the Panguanying case, the intermediaries' engagement in the Dagong case nonetheless furthered the cause of the Chinese 'no burn' community. The Dagong case helped the intermediaries draw greater public and media attention to both the hazards related to incineration and to broader regulatory failures, such as the frequent failure to implement EIA measures at the local level. Tackling the Dagong and Panguanying cases in a combined fashion enhanced the environmental organizations' public and media campaign about EIA failures and helped them exert stronger pressure on the MEP and related institutions – urging them to tackle this problem at a higher political level, and pleading for stricter control of the EIA and construction units, different methods for appointing EIA units, stronger public participation in the EIA process, and greater transparency. The (in this case cleavage-burdened) linking up of local affected communities and the national 'no burn' community can thus strengthen a national-level advocacy network calling not only for more sustainable Chinese waste treatment strategies but also drawing attention to broader environmental issues – even if the individual local struggle remains unsuccessful.

6 Conclusion

Networked Contention: No Longer Fragmented, Not Yet a Movement

This chapter outlines the network of incineration-related contention that has emerged under Hu Jintao based on the findings from both the three case studies presented in the previous chapters and the broader case spectrum. It then turns to the prospects and limitations of scaling up local contention and forming an issue-specific movement and, moving beyond the issue field of waste incineration, the implications for Chinese environmental activism more broadly. In the conclusion, the chapter comes back to the social movement literature and discusses what the findings of this study reveal about networked contention, the dynamics of diffusion, and scale shift in the context of a restrictive political regime.

Environmental Actors Linking Up: Implications for Local Anti-Incineration Contention in China

Contrary to the scholarly literature's widespread assumption that contentious local communities in China are isolated from both each other and supra-local environmentalists, the affected communities in this study are by no means secluded entities. Facilitated by the liberalizing media landscape and frequent reporting on social and environmental issues under Hu Jintao, by the spread of the Internet and particularly social media during the same period, and by the growing local engagement of intermediaries – often functioning as brokers between different localities – at the time of research, the affected communities under study have established a dense network of ties with each other and with members of the Chinese 'no burn' community. This network permits diffusion processes and learning effects across geographic space. The nature of linkages established by local communities and their role in individual struggles do, however, depend strongly on the local context. As in the previous chapters, this section first discusses the nature and role of horizontal linkages before turning to vertical ties.

Horizontal Ties: Linkages Among Local Contentious Communities

As shown in the previous case studies, local communities affected by planned or operating incinerator projects in their neighbourhood are significantly impacted by the actions of their predecessors in other localities. The contentious communities in Beijing have established relationships not only within their cluster, but also with communities in other parts of the country and across different issue fields. These linkages are not restricted to urban cases, but also link rural and peri-urban communities to a sprawling network of ties between contentious communities across the country.

This finding is supported by the larger spectrum of case studies collected for this research. Figures 6.1 to 6.4 illustrate that all cases in the sample are embedded in the country-spanning network of linkages, with each community linked to at least two of the other communities in the sample. In the figures, a line between two cases indicates that at least one of the two communities has experienced some kind of influence from the other case (with earlier cases impacting later ones, in the case of non-simultaneous contentious actions).[112] The linkages are separated into the categories of nonrelational ties; direct relational ties based on one-time face-to-face contact, such as at meetings and conferences; and more sustained relational ties based on either online communication or intensified face-to-face contacts.

Within this network, some cases have emerged as important role models for other local struggles and can be regarded as 'initiator' or 'spin-off' movements, as those terms are used in social movement theory. As outlined in Chapter Three, the urban cases of Liulitun and Asuwei in Beijing as well as the urban Panyu case in Guangzhou have significantly impacted many of the other communities. Among these, the Asuwei campaigners have established linkages with every other case in the sample (see Figures 6.1 to 6.4). The high impact factor of these cases has several reasons. First, their respective community members were very active in disseminating information online, thus making it easy for other affected communities to access information about their activities via the Internet and social media. Second, all three

112 The sample cases can broadly be distinguished into two phases, with about half experiencing a peak year in 2009 and 2010 (Auwei, Dagong, Panguanying, Huadu, and Likeng), and the other half between 2011 and 2013 (Huangtutang: 2011, Songjiang: 2012; Huiyang and Fengqiao: 2013). However, most local struggles spanned several years and some experienced various phases of contention (e.g., Asuwei, Panguanying, and Huadu, where renewed siting procedures triggered second phases of contention). Thus, most cases had some overlap in time. The figures display all of the linkages found in the data, irrespective of the phase of contention when they were established. Further linkages may exist that were missed during the data collection process.

Figure 6.1 Nonrelational linkages among sample cases

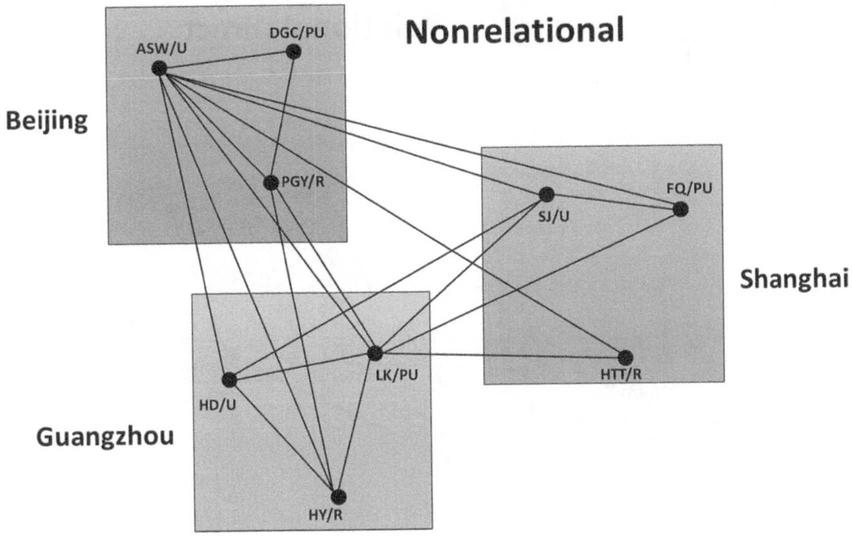

Notes: Acronyms before the slash indicate the case; for full case names and further information see Appendix IV. Acronyms after the slash indicate the setting: U = Urban, PU = Peri-Urban, R = Rural. LLT, GAT and PY are central background cases not included in the sample.
Source: Own compilation. I thank Tobias Scholz for assistance with generating these figures.

Figure 6.2 Direct relational linkages among sample cases (face-to-face; one-time)

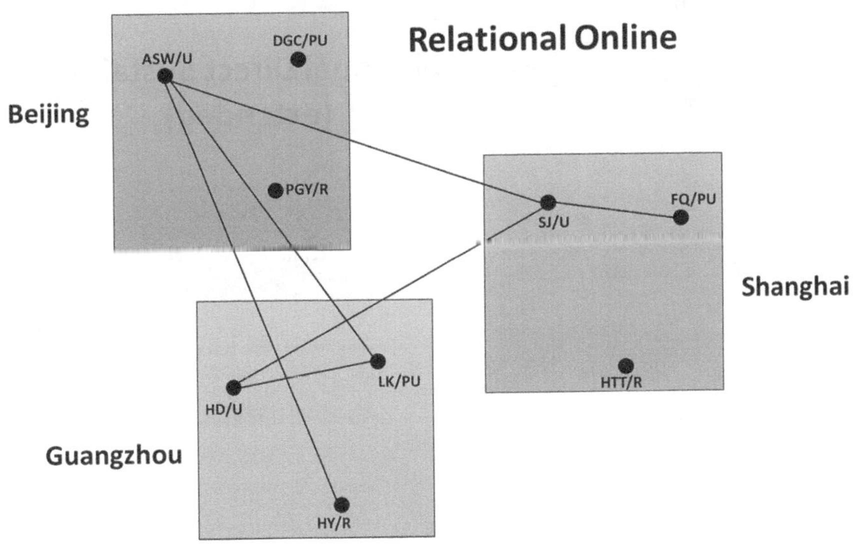

Notes: Acronyms before the slash indicate the case; for full case names and further information see Appendix IV. Acronyms after the slash indicate the setting: U = Urban, PU = Peri-Urban, R = Rural. LLT, GAT and PY are central background cases not included in the sample.
Source: Own compilation. I thank Tobias Scholz for assistance with generating these figures.

Figure 6.3 Online relational linkages among sample cases

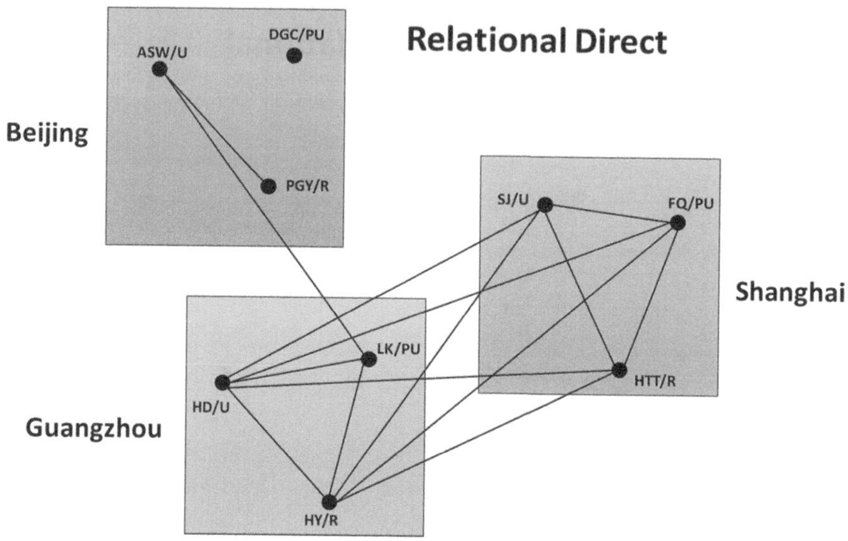

Notes: Acronyms before the slash indicate the case; for full case names and further information see Appendix IV. Acronyms after the slash indicate the setting: U = Urban, PU = Peri-Urban, R = Rural. LLT, GAT and PY are central background cases not included in the sample.
Source: Own compilation. I thank Tobias Scholz for assistance with generating these figures.

Figure 6.4 Sustained direct relational linkages among sample cases (face-to-face; sustained)

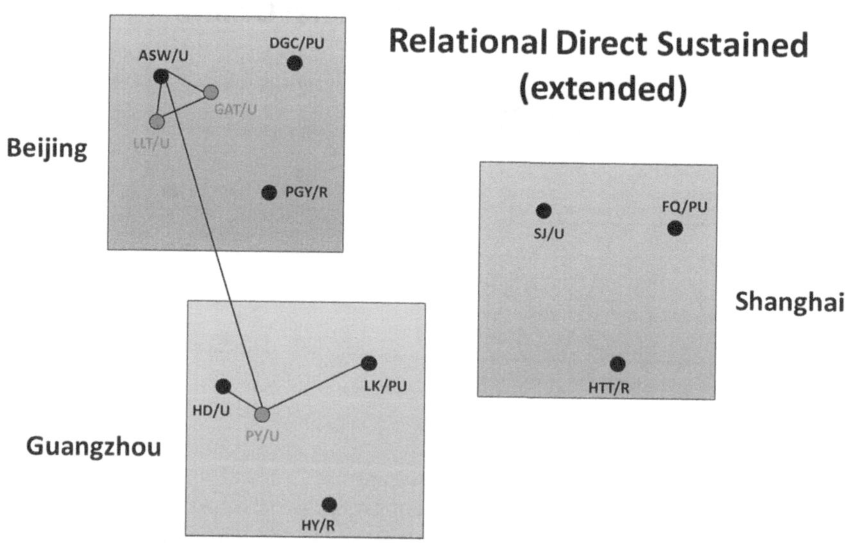

Notes: Acronyms before the slash indicate the case; for full case names and further information see Appendix IV. Acronyms after the slash indicate the setting: U = Urban, PU = Peri-Urban, R = Rural. LLT, GAT and PY are central background cases not included in the sample.
Source: Own compilation. I thank Tobias Scholz for assistance with generating these figures.

communities fully immersed themselves in the issue of waste incineration and compiled comprehensive materials about the technical aspects and risks of incineration that offered important information and critical frames for other (particularly resource-poorer) communities (cp. Chapters Three and Four).

Third, the three cases featured prominently in the national media, where they were hailed as positive examples of successful interaction between local governments and an aggravated public. As described in Chapter Three, the Asuwei campaign leader turned into a media star after turning into a full-time environmentalist and founding his own environmental organization. A similar career can be observed in the case of the main Panyu campaign leader and has also been seen in other issue fields (Cai and Sheng 2013). Both this media prominence – which ensured that information about the cases would reach less Internet-connected rural communities, such as Panguanying, via the mass media – and the campaigns' (temporary) success in obstructing the plants added to their radiant power. Another case that reached national media prominence and was well-known among other affected communities was the case of Likeng in Guangzhou, which first brought the health-related risks of waste incineration (particularly cancer) to the attention of the public and media and served as an important warning signal and awareness trigger for many other communities.

The emergence of initiator or 'spin-off' cases was further facilitated by the strategic dissemination of information by 'no burn' community members, who pointed other local communities to the positive examples of Asuwei, Panyu, and Panguanying and promoted their use of 'rational' and legal means rather than disruptive and potentially violent action. This included disseminating the comprehensive research report compiled by Asuwei campaigners in the action 'toolkit' compiled by Nature University (cp. Chapter Three) and promoting the Panguanying struggle as an example of a successful case of environmental litigation (cp. Chapters Four and Five).

The nature and depth of the linkages between different contentious communities vary and permit different kinds of diffusion effects. The Beijing cluster cases show that horizontal linkages are largely restricted to nonrelational ties without establishing online or offline personal relations. Closer relational ties were only developed between the urban homeowner cases in Beijing (Liulitun, Gaoantun, and Asuwei) and, across clusters, with the homeowners in Guangzhou's Panyu district. The rural and peri-urban cases in the Beijing cluster (Panguanying and Dagong) showed little interest in establishing closer personal linkages with any other cases either in the cluster region or in other parts of the country despite repeated opportunities.

This pattern is also visible in the larger case spectrum. As shown in Figure 6.1, the communities in the sample are linked primarily via nonrelational linkages. Here, the Internet and social media functioned as the main channels for the diffusion of information both within clusters and across time and space. Social media in particular played a crucial role in most of the cases as an important source of critical and less strictly censored information. Affected communities could follow the activities of other communities almost in real-time via microblog services, weblogs, or public online forums, or could learn about prior actions without having to establish personal contact. The availability of such online information played a major role in the onset and development of the majority of the cases under study. While the urban communities in the sample were more acquainted with using the Internet to strategically disseminate information about their own claims and activities, the Internet was also used as an important source of information by peri-urban and rural communities. Internet access by individual community members sufficed to make full use of the web as a valuable source of information (such as in Panguanying, as described in Chapter Four, or in Fengqiao and Likeng). For the dissemination of information about their own case, rural communities mainly relied on intermediaries acting as communication nodes to the outside world – particularly when media reporting about their case was blocked or limited.

A second major channel for the nonrelational transmission of information was reporting about waste incineration and related local contention in the mass media – which was also mostly available online. Reporting about waste incineration projects became more frequent after the risks related to incineration started to attract public and media attention in 2009. Nonetheless, supra-local and Chinese-language reporting about local anti-incinerator struggles remained limited to a handful of cases independent of which cluster region the cases were located in. As outlined above, the cases that featured prominently in the mass media were mainly Liulitun, Asuwei, and Panyu as positive examples and Likeng as the first incinerator-related cancer case. A particularly important role was played by the CCTV program broadcast in the context of the 2009 International Dioxin Symposium in Beijing, which prominently featured the Liulitun struggle and helped to unfold its impact as a 'spin-off' movement (impacting, among others, the Asuwei and Panguanying cases in the Beijing cluster as well as Panyu and Huadu in Guangzhou).

Apart from the Internet and mass media, deliberate brokerage efforts by members of the Chinese 'no burn' community and by individual activists such as the Asuwei and Panyu campaign leaders also fostered nonrelational diffusion processes between the different communities. As outlined above, the intermediaries strategically informed affected communities about the activities and success or failure of other communities in order to increase issue awareness and expectations of success and to actively influence the mode of action used to oppose the incinerators.

While the nonrelational linkages illustrated in Figure 6.1 span all clusters and also link peri-urban and rural cases, it is noticeable that linkages among the Shanghai cluster cases are more limited than in the other two cluster regions. This can mainly be attributed to the rather isolated villager community of Huangtutang, who established few ties with other communities, and those only after a violent clash with public security forces had already occurred. Moreover, since Huangtutang villagers had limited access to the Internet and media reporting of the case was blocked, other communities did not learn about the case unless directly informed by intermediaries.

Direct relational ties based on online or face-to-face communication were significantly more limited, as reflected in the Beijing cluster cases. This was not due to a lack of opportunity: as shown in Figure 6.2, many of the communities in the sample, particularly those in the Shanghai and Guangzhou clusters, had the opportunity to establish direct relational linkages during meetings and conferences. However, only a few of them established more sustained communication either online (Figure 6.3) or based on face-to-face relations (Figure 6.4) – even in cluster regions where geographic proximity would have permitted regular offline meetings.

Like in the Beijing cluster, the majority of relational ties (Figure 6.2) were deliberately brokered by members of the Chinese 'no burn' community with the aim of bringing together affected communities in different localities. Most importantly, members of several local communities were present at a two-day symposium on the 'public supervision of municipal solid waste incinerators' (生活垃圾焚烧厂公众监督研讨会, *shenghuo laji fenshao chang gongzhong jiandu yantaohui*) organized by Nature University in Shanghai in June 2013. This symposium had the explicit goal of linking affected communities, lawyers, and national and international experts and environmental organizations to promote local contention against incinerator projects and to foster a national issue network.[113] Present at the symposium were, among others, members of all of the Shanghai cluster sample cases (Songjiang, Huangtutang, and Fengqiao – the first two more or less concluded at the time of the conference), campaigners from the

113 A similar symposium was again organized by Nature University and Wuhu Ecology Center – with the support of GAIA – in Shanghai in April 2014 under the title 'Community Training on the Impacts of Waste Incineration' (垃圾焚烧影响社区培训, *Laji fenshao yingxiang shequ peixun*).

ongoing Huadu and Huiyang cases, and members of the ongoing cases in Hai'an county (Jiangsu Province), and in Taizhou city and Hangzhou city's Binjiang district (both in Zhejiang Province).

Most of the other relational ties displayed in Figure 6.2 were based on deliberate brokerage efforts by intermediaries. Most communities in the Beijing cluster (Liulitun, Gaoantun, Asuwei, Dagong, and Panguanying) were brought together at symposiums and public forums by the Beijing-based environmental organizations in the attempt to spark more direct communication between the communities. In a similar vein, the Guangzhou cluster communities had personally met at meetings organized by Panyu campaigners and their newly founded organization or during simultaneous petitioning activities. Some cross-cluster relational ties were established during the shooting of a Phoenix TV program on the pros and cons of incineration. Campaigners from the Beijing homeowner cases (Liulitun, Gaoantun, and Asuwei) and members of the Panyu and Likeng communities were invited to present their cause during this program (see Chapter Three). This was not necessarily intended as an opportunity for community members and intermediaries to meet personally, but did have that effect.[114]

Despite these personal meetings, the communities developed few intensified online or face-to-face relational ties. After the Shanghai conference, only the Huadu and Fengqiao communities established online contact with Songjiang community members, who they asked for advice (Figure 6.3). None of the sample communities present at the conference developed more sustained relational ties based on face-to-face communication, even within clusters (Figure 6.4). The Phoenix TV recording initiated some online and face-to-face communication between the Asuwei and Likeng communities (with the latter asking the Asuwei campaign leader for advice and active assistance). The Songjiang and Huiyang communities also used social media and email to contact the Asuwei campaign leader for help due to his media and online prominence (Figure 6.3). Sustained relational ties based on face-to-face communication were only established among the homeowner cases in Beijing and Guangzhou – with regular cross-cluster meetings between Asuwei and Panyu campaigners – and among the cases within Guangzhou municipality, with Panyu acting as the central communication node (Figure 6.4). Overall, more online ties were established by urban communities, as was to be expected due to their better access to and familiarity with the Internet.

114 A similar phenomenon was also observed by Zhu (2017) in Maoming, where activists were provided with the unintended opportunity to establish networks at an official press conference.

The overall reluctance to establish sustained relational ties among the communities can partly be attributed to the restrictive political setting, in which closer relations between different contentious communities still pose a significant risk. Collaborative action across localities or engagement in another community's struggle are still a political taboo and can lead to serious repercussions, as pointed out by several community members from different cases. As outlined in Chapters Three and Five, this perception of political risks prevented the Beijing homeowner cases from establishing collaborative action or getting actively engaged in other cases in the area (such as Dagong). The Asuwei campaign leader, a trained lawyer, declined the request to function as a legal representative for the Likeng villagers for similar reasons. In a similar vein, contentious action among the Guangzhou cluster cases was deliberately disconnected despite the simultaneity of the campaigns. The contentious communities in Guangzhou municipality (Panyu, Huadu, and Likeng) reported that they had formed an 'alliance of support' similar to the homeowner communities in Beijing, but had refrained from any active collaboration due to the related political risks. An exception was some joint petitioning activities by the Panyu and Likeng communities in 2009 and early 2010.

Also in part attributed to the political risk, all of the urban communities under study who had successfully obstructed an incinerator in their neighbourhood (Liulitun, Panyu, Huadu, and Songjiang) refrained from becoming involved in the new localities even though the projects were moved to nearby rural areas within the same district. The prevalent explanation for this lack of engagement given in interviews was that (certain kinds of) environmentally-related contentious action were tolerated by the regime if they were based on one's own grievances. Once contentious action was conducted in the name of another community, however, it was regarded as highly political and likely to meet with repercussions – as personally experienced by several community members who had attempted to assist resource-poorer communities around the new project sites.

Apart from the apparent political risks, another reason for the lack of closer relational ties was just as evident: simple lack of interest on the part of many communities. The rural and peri-urban communities in particular – but also the majority of urban community members – showed little interest in actively engaging with other communities. As reflected in the Panguanying and Dagong cases (Chapters Four and Five), many communities expressed doubts that they could learn from other communities via direct relational ties because they either had already gathered all of the information they deemed necessary via nonrelational ties and did not see any added value in direct communication, or regarded their own cases as significantly different from other cases. Only urban communities sought to learn from each other via direct exchange, while rural and peri-urban communities preferred to rely on intermediaries for assistance. Several of the communities also regarded other cases as 'bad examples', which was a strong disincentive for establishing closer ties (such as the Asuwei campaigners' negative assessment of other 'NIMBY' cases or the Panguanying villagers' critical evaluation of Gaoantun and Dagong).

Most importantly, many communities simply took little interest in other communities' grievances or broader waste and environmental issues beyond their personal concerns. Except for individual urban campaigners, the communities' concerns and activities largely revolved around their own well-being, without shifting from a Not-In-My-Backyard to a Not-In-Anybody's-Backyard attitude. Whenever their struggles came to an end (deemed successful or not), most community members returned to their daily lives without continuing to give attention to the issue. This includes a widespread lack of interest in the nearby and mostly rural new sites of relocated projects among (urban) community members, which contributed to an environmental justice debate among intermediaries and some of the (particularly rural) communities.

The lack of interest beyond individualized local concerns significantly limits the prospects for scaling up local contention based on collaborative action or a heightened sense of solidarity among affected communities. The exact reasons for this lack of interest and the extent to which they are related to China's (political) culture, such as its Confucian heritage, are hard to pin down. A similar lack of interest in other communities and broader environmental issues has also been observed in other countries. However, it is likely that in China this is at least enhanced by the political risks related to engaging with other contenders.

An exception were the campaign leaders and individual community members of the resource-rich Asuwei and Panyu communities, who continued to care for the issue of waste incineration beyond their own localities and founded their own environmental organizations dedicated to promoting more sustainable waste policies, as outlined in Chapter Three. Here, a shift from not only NIMBY to NIABY took place, but also from anti-incineration to pro-alternative waste management approaches. Both shifts are necessary for broader policy changes in the issue field. Individual Songjiang community members also showed a continued interest in waste issues – with one of them planning to establish a local waste recycling project – and attempted to assist the villager communities around the new site. These attempts to assist were curtailed by both political pressure and a lack of awareness and interest on the part of the villagers.

Apart from the linkages established among the communities affected by incineration, the local campaigners were also strongly influenced by contentious communities from other issue fields. Most communities referred to the anti-PX protests in Xiamen as an important point of reference. As outlined in Chapter Three, the Liulitun community had actively learned from the almost simultaneous contention in Xiamen and also spread these learning effects to other contenders. Similarly, Songjiang campaigners reported that they had learned a lot from the previous struggle against the extension of the city's magnetic levitation (maglev) train. Through close personal ties with several of the maglev campaigners (some of whom had moved to Songjiang), Songjiang community members received important advice on protest activities and how to circumvent observation and online censorship by public security forces. Due to the repercussions they had previously experienced, however, the maglev campaigners declined to get actively involved against the planned incinerator.

Several of the communities – particularly the more tech-savvy urban communities that could access a wider array of (English-language) information and surmount the Chinese Firewall to enter the World Wide Web to find uncensored information – were also influenced by anti-incinerator struggles in other parts of the world, thus extending the network of linkages to a transnational scale. None of the communities developed direct relational ties with campaigners abroad, except for Panyu, where the community members met some Taiwanese anti-incineration activists who were visiting Guangzhou for a media event.

Local Impacts: The Role of Horizontal Linkages for Anti-Incineration Struggles

Despite the overall limitation to nonrelational linkages, the emerging ties still permitted diffusion processes among the affected communities, which had a major impact on both individual struggles and the overall development of anti-incineration contention in China. Since active engagement in other localities was largely taboo, the content exchanged between communities was mainly restricted to information and did not encompass other types of (more manifest) resources. Local factors (primarily the communities' resource structure and the local political and socio-economic setting) played a more important role in the development of linkages and diffusion effects than regional factors did.

Amid the restricted access to reliable information, critical information – particularly about the details and hazards of incineration, contentious actions in other parts of the country, and possible modes of action – passed between the communities played a crucial role in the awareness process of most cases under study. Such information provided important cognitive cues and alternative frames for many communities, initiating the first sense of alarm, contributing to a collective interpretation of the situation, and fostering the formation of a contentious consciousness and cognitive justification for taking action. However, as described by Jeffrey Broadbent (2003: 222-226) in the case of environmental protests in Japan, not much 'cognitive liberation' was needed for most of the communities under study. External cognitive cues about hazards or regulatory failures related to the waste facilities generally fed into broader ambivalences towards modernization and resonated with the overall sense of risk and distrust of experts and governments (cp. also Lora-Wainwright 2013d, 2017). Most communities had prior experiences of political deprivation (such as prior corruption issues, land grabs, or a lack of participatory structures) or, particularly in urban cases, a more fundamental sense of repudiation of many aspects of the political system. Moreover, most communities had experienced prior environmental grievances such as pollution and health issues, since many of the incinerator sites in the sample were home to other industry or waste facilities such as landfills that had already burdened the local environment for years (such as in Liulitun, Gaoantun, Asuwei, Songjiang, Panguanying, Likeng, Fengqiao, Huangtutang, and Hai'an).

These prior experiences and broader concerns generally facilitated the mobilization of action unless they had already led to a sense of lacking political efficacy and resignation, as in the case of Dagong described in Chapter Five. Many homeowners in urban communities had attempted to escape the environmental downturns of urbanization and invested large sums of money to move to cleaner and calmer suburban areas or neighbouring districts, which were again being threatened by planned incinerator plants (Asuwei, Huadu, Panyu, Huiyang). This often agitated the communities and fostered the onset of contentious action, particularly when a lack of transparency about the planned projects had obscured the possibility for making an informed decision about where to invest to find a clean living environment. Hence, while more detailed information was central in later stages of the awareness process and for taking action, not much was needed to trigger the first sense of alarm and injustice that caused many communities to activate their own resources and actively search for further information – a process that in urban areas generally lead to a substantial amount of lay expertise. This was particularly the case in later struggles, when waste incineration had already become a major public and media issue and thus resonated with many of the communities.

After the first stage of initial awareness, externally provided information and cognitive cues helped many of the local communities pierce the 'information haze' caused by the highly contested and technical issue of waste incineration in combination with restricted access to reliable information and a prevailing distrust of the claims of party-state officials and (government) experts. Such external information helped certify alternative frames and interpretations that deviated from those promoted by party-state representatives and companies. This was relevant for all cases in the sample irrespective of their setting. Information about or compiled by other communities (such as the Liulitun 'opinion booklet' or the Asuwei research report) proved to be an important resource for campaign leaders across the country, allowing them to certify their concerns and issue interpretations both internally vis-à-vis their own communities – thus functioning as mobilization resources, as described in the Panguanying case – and externally towards the larger public and project proponents (such as government and company representatives and official experts). Here, many communities referred to other cases to frame themselves as one in a lengthening line of communities standing up for their rights and against misguided waste policies carried out on the back of the *laobaixing*, thus adding legitimacy to their claims and activities.

Knowledge about other contentious communities also impacted many of the campaigners' perceptions of threat and opportunity, as suggested in the literature on contentious politics. Knowledge about successful cases raised several communities' expectations of success and contributed to their decision to resort to contentious action. Information about the success or failure of other cases also had a significant impact on the mode of action employed. The perceived higher chances of success for nondisruptive modes of action led to a significant emulation of tactics and strategies and a convergence of methods across the cases. Due to the prominence of communities that had (temporarily) succeeded in obstructing projects via 'rational' and legal means (such as Liulitun, Asuwei, Panyu, and Panguanying, but also cases from other issue fields such as the anti-PX struggle in Xiamen), these were the preferred methods adopted by the communities in the case sample. The convergence of tactics was further reinforced by 'no burn' community members' repeated urging to resort to peaceful and legal means. All of the urban cases under study employed 'peaceful strolls' and mass petitions – thus emulating the Xiamen and Liulitun communities and their successors such as Asuwei and Panyu. The communities actively adopted means such as writing 'opinion booklets', printing (similar) slogans on paper sheets or T-Shirts for use during the strolls – since the shouting of slogans was deemed politically unfeasible – and using similar symbols, such as gas masks (prominently introduced by the Panyu campaigners in 2009). Moreover, imitating the prominent cases and guided by 'no burn' community members, most urban, rural, and peri-urban communities included legal means – such as letter-writing, information disclosure requests, administrative redress applications, and environmental litigation – in their portfolio of activities.

Information about other cases also contributed to the formation of a collective (contentious) identity in several cases. Across the case spectrum, the communities' primary identity frame centred on the protection of their rightful claims and interests (as captured by O'Brien and Li's (2006) notion of 'rightful resistance'). However, as outlined in the Asuwei and Panguanying cases, this self-perception often broadened to a wider identity frame of being members of a Chinese *weiquan* community, which spread across the cases. Most communities depicted themselves as but one in a country-spanning wave of affected communities standing up not only against waste incinerator projects in their neighbourhood, but also against environmental degradation more generally. Several of the rural and peri-urban communities linked this with an 'injustice frame' (Futrell 2003), criticizing how resource-poorer and more vulnerable villager communities were expected to bear the burdens of the growing amounts of waste produced by urban residents and the downturns of urbanization and modernization more generally, thus feeding into a broader environmental justice debate. Some of the communities also presented themselves as members of a (trans-)national 'no burn' community. However, this alignment of local self-perceptions with broader identity frames was predominantly used to further localized claims and did not result in a larger community of solidarity or alliances across the cases.

Last but not least, the information transmitted via horizontal linkages changed the resource structure of many of the communities under study, thereby enabling or facilitating the mobilization of contentious action. As outlined above, knowledge about (successful) modes of action employed in other localities and state responses to those actions that indicated the boundaries of the politically permissible were important resources for most communities. This included legal knowledge and information related to the modes of action that were possible within the Chinese political framework. In particular, information about contentious action in other parts of the country helped resource-poorer rural and peri-urban communities who might otherwise not have been able to conduct or proceed with contentious action (as outlined in the Panguanying case in Chapter Four). As in Panguanying, information about the activities of other contenders also pointed several of the communities in the sample toward intermediaries active in the issue field (such as Zhao Zhangyuan, the lawyer Xia, or Nature University), who could then be approached for assistance by the affected communities.

Both resource-rich urban and resource-poorer rural and peri-urban communities benefitted from the action-related technical and issue-specific information transmitted via horizontal ties. Externally provided information about the details and hazards of incineration, related regulations and standards, and mandatory procedures – such as those related to the EIA process – provided the basis for contentious action in most communities in the sample. Such issue-specific information also enabled the local communities to adopt broader issue interpretations and align their frames with the arguments and interpretations of the (trans-)national 'no burn' community, such as portraying the Chinese context as unsuited for waste incineration or regarding incineration as a misguided and unsustainable waste policy. As in the case of identity frames, this was predominantly used to add weight to the communities' localized claims rather than fostering broader claims.

As demonstrated in the Dagong case described in Chapter Five, diffusion effects strongly depend on the local context: they can only unfold if anchored in sustained local action. In cases where fragmented local interests or a sense of lacking political efficacy demobilized local communities, the existence of strong horizontal or vertical ties was not sufficient for mobilizing unified local action and diffusion processes failed to take effect. Apart from Dagong, this was also observable in the peri-urban Shanghai communities of Fengqiao and Yuqiao, where an overall sense of the futility of contentious action and lack of political efficacy by self-perceived *laobaixing* who were without education and resources stopped large parts of the communities from taking sustained or coordinated action despite active mobilizing attempts by both community members and intermediaries. A similar resignation tempering pollution-related action despite an awareness of its negative effects is also described by Lora-Wainwright (2017) in her anthropological investigation of pollution-affected rural communities. Like in the communities studied by Lora-Wainwright, a long-term sustained strategy that could have united the different social groups in Dagong was impeded by their overall sense of lacking political efficacy and the inevitability of urban and economic development. In Fengqiao, their sense of the futility of action was further coupled with the villagers' economic dependency on a nearby industrial park. While they recognized that economic and industrial development was the cause of pollution, they regarded it as inevitable since they depended on it for their livelihood. This seemed to also affect their attitude towards the planned MSWI project.

The impacts of external linkages were also hampered where local communities preferred relocation to fighting against incinerator plants in their neighbourhood. Again, this was mainly the case in peri-urban neighbourhoods, particularly if the villager communities no longer cultivated their own land. The defence of land was a major mobilizing force for several rural communities (such as for the farmers in Panguanying, Huangtutang, and Huiyang). Without an attachment to land, several of the communities turned to relocation claims in the hopes of improving their livelihood through compensation payments or because they had already suffered from previous pollution in the area which they hoped to escape (such as the villager communities in Asuwei and in later stages also Likeng). Such relocation or compensation claims were complex and can not necessarily be understood as the communities' preferred path of action. In several cases such claims were based on the above-described feeling of resignation and lack of political efficacy or were a strategic reframing of claims where other demands and contentious action had failed (such as in Dagong and Likeng). Most of the communities in this study were all but unified when it came to attitudes towards relocation (as described in the Dagong case, where different preferences contributed to stifling collective action either against the plant or for relocation).

Repressive party-state reactions can also hamper diffusion effects and disrupt contentious action at any point in the struggle, as was the case in Dagong, Gaoantun, Likeng, and Huiyang. In this study, urban communities seem more immune to political pressure. This can be attributed to the larger number of campaigners, better access to the Internet and social media as mobilizing devices and for the dissemination of information, and the less easily discernible leadership structure of such groups.

The Two Facets of Chinese Environmental Activism Linking Up: Vertical Ties Between Local Communities and Environmentalists

Apart from the horizontal ties amongst local communities, this study has also revealed a dense network of linkages with supra-local intermediaries active in the issue field. In contrast to the widespread assessment in the literature that Chinese environmentalists are shunning direct engagement at the local level, leaving local communities isolated from intermediary support (e.g., Johnson 2010, 2013a; Matsuzawa 2012; Tang 2012; van Rooij 2010; D. Zhao 2007), the cases under study show that environmental organizations, experts, and lawyers engaged in the field of waste incineration were becoming increasingly active at the local level at the time of research. At least under Hu Jintao, the 'two facets' of Chinese environmental activism – grievance-based local contention and environmentalist action – were linking up in the field of waste incineration and forming an alliance for the mutual benefit of both sides.

As suggested by the literature, the environmental organizations contacted by contentious communities in the early years of Chinese anti-incineration contention did indeed decline to become actively engaged in the local struggles (such as in the cases of Liulitun, Gaoantun, and Asuwei described in Chapter Three). This greatly changed throughout the course of the research period. The growing numbers of local anti-incinerator campaigns after 2009 started to attract the organizations' attention and – in combination with major transnational influences – several environmental organizations soon started to get actively involved in the issue field (see Chapter Two). As in other (Western) countries, it was grievance-based local action that brought the issue of waste incineration to the attention of national-level environmental organizations. After the foundation of the Beijing-based organization Green Beagle in 2009 as the first Chinese environmental organization with the explicit aim of supporting local communities affected by environmental degradation, several other Chinese organizations dedicated to waste issues started to actively assist local communities living in the surroundings of planned or operating incinerators. They approached local communities on their own initiative (as in the case of Dagong) or became involved when contacted for advice and assistance (such as in Panguanying). This network of organizations was institutionalized in 2011 with the foundation of the Chinese 'Zero Waste Alliance'. Under the lead of Green Beagle (and its later split-off organization Nature University) and with the help of transnational organizations such as GAIA and IPEN, the organizations in this study not only supported local communities and impacted their local struggles, but also used these local engagements to further their own higher-level advocacy work for more sustainable waste policies and against regulatory failures in the environmental realm.

Via their joint engagement in some of the local cases, the environmental organizations in the issue field closed ranks with environmental experts and a handful of lawyers and legal associations such as the Beijing-based Center for the Legal Assistance to Pollution Victims, who are dedicated less to waste issues and more to furthering environmental litigation in China (cp. Chapter Two). Together, these three main groups of actors – environmental

Table 6.1 Vertical linkages between intermediaries and local communities

Intermediaries \ Cases	HD/U	PY/U	LK/PU	HY/R	ASW/U	LLT/U	GAT/U	DGC/PU	PGY/R	SJ/U	FQ/PU	HTT/R
Zhao Zhangyuan (BJ)	NR	NR RD RO RDS	NR	NR RO	NR RD RO RDS	NR RD RO RDS	NR RD RO	NR RD	NR RD RDS	NR RO	NR	NR RD RDS
Guangzhou expert (GZ)	NR RD RO RDS	NR RD RO RDS	NR RD RO RDS									
Individual Beijing lawyers (BJ)			NR	NR RD RO RDS	NR RD	NR RD RO RDS	NR RD RO RDS	NR RD	NR RD RDS		RD	RD RDS
Center for the Legal Assistance to Pollution Victims (BJ)									NR RD RDS			RD
Nature University / Green Beagle (BJ)	NR RO	NR RD RO	NR RD RO RDS	NR RD RO RDS	NR RD RO RDS	NR RD RO RDS	NR RD RO RDS	NR RD RO RDS	NR RD RDS	NR RD RO	NR RD RO	NR RD RO RDS
Friends of Nature (BJ)	RD RO RDS	NR RD RO	NR		NR RD RO RDS	NR RD	NR RD	NR RD	NR		NR	RD RDS
Global Village of Beijing (BJ)					NR			NR	NR			
Aifen Huanbao (SH)												
Wuhu Ecology Center (Wuhu, Anhui)				NR RO				NR	NR			
Zero Waste Alliance (national)								NR	NR			

Intermediaries \ Cases	HD/U	PY/U	LK/PU	HY/R	ASW/U	LLT/U	GAT/U	DGC/PU	PGY/R	SJ/U	FQ/PU	HTT/R
Global Alliance of Incinerator Alternatives (GAIA) (transnational)	RD	RD	RD	RD	NR	NR	NR	NR	NR	RD	NR RD	RD
International POPs Elimination Network (IPEN) (transnational)		NR		RD	NR				NR	RD	RD	RD
Other international experts and NGOs	NR	NR		NR	NR	NR	NR			NR	NR	

Notes: Acronyms before the slash indicate the case; for full case names and further information see Appendix IV. Acronyms after the slash indicate the case setting: U = Urban, P = Peri-Urban, R = Rural. Acronyms after the intermediary names refer to their home base: BJ = Beijing, GZ = Guangzhou, SH = Shanghai. LLT, GAT and PY are central background cases not included in the sample. The fields in the matrix indicate the types of linkages established between the intermediaries and local communities: NR = Nonrelational, RD = Relational Direct (based on one-time face-to-face interaction), RO = Relational Online (direct online communication), RDS = Relational Direct Sustained (sustained direct relations based on face-to-face communication).

organizations, experts, and lawyers – formed what in this study is captured as the Chinese 'no burn' community. While the number of actors actively engaged at the local level remains limited, they have nonetheless played a major role in the majority of the cases under study. Table 6.1 lists the main intermediaries involved in the sample (and some background) cases, and indicates the nature of the linkages they established with the respective communities. As in Figures 6.1 to 6.4, these linkages are separated into non-relational ties, direct relational ties based on one-time face-to-face contact, such as at meetings and conferences, and more sustained relational ties based either on online communication or intensified face-to-face contacts.

Across the case spectrum, none of the cases in this study was fully isolated from intermediary support (see Table 6.1). Due to the growing prominence of both the issue of waste incineration and individual intermediaries (cp. Chapter Two), an increasing number of communities actively established contacts with the 'no burn' community during the time of research. However, as pointed out by several 'no burn' community members, many cases of local anti-incinerator contention (particularly in rural areas) are still unknown to both the public and intermediaries, and thus remain isolated from potential intermediary support.

As shown in Table 6.1, the core Chinese 'no burn' community members actively engaged in local struggles across the country were professor Zhao Zhangyuan, the environmental organizations Green Beagle/Nature University and Friends of Nature, and a handful of lawyers – all of them based in Beijing. These intermediaries had direct contacts with all (Zhao Zhangyuan and Nature University) or most (Beijing lawyers and Friends of Nature) of the affected communities under study, regardless of their cluster region.

Zhao Zhangyuan, a retired researcher from the Chinese Research Academy of Environmental Sciences, turned into a fierce and outspoken critic of waste incineration after conducting his own research into the issue in 2001 (see Chapter Two). In the following years, he played a similar role for Chinese anti-incineration contention as the professors Paul Connett and Barry Commoner have played for the anti-incineration and environmental justice movements in the United States. After his personal engagement as a critical expert in the urban Beijing campaigns (Liulitun, Gaoantun, and Asuwei), he reached national media prominence by outing himself as a fierce opponent of China's waste incineration policies during a meeting in Guangzhou in 2010 (cp. Chapters Two and Three). Due to his prominent appearance in the media, his publication of online articles and blogposts, and his active engagement in the early local campaigns, all of the affected communities in this study were impacted (through nonrelational linkages) by his critical assessment of incineration technology. Many of the communities also directly contacted him for advice and assistance (such as the urban Beijing cases, Panguanying, Panyu, Songjiang, and Huiyang). Other intermediaries such as Nature University or the Beijing lawyers also told local communities to contact professor Zhao and actively engaged him in local struggles (such as in Huangtutang and Likeng). While in earlier years Zhao paid site visits to several communities across the country, in later years his declining health and the growing political pressure did not permit him to undertake local visits.

A country-wide role was also played by the Chinese environmental organizations active in the issue field under the lead of Green Beagle/ Nature University and Friends of Nature. Starting with active engagement in the larger Beijing area, they soon expanded their activities to other parts of the country, facilitated by the growing number of local organizations active in the Zero Waste Alliance that could be mobilized to reach out to local communities in other regions. During the research period, Nature University – who was both present in the media and easy to find online – in particular gained significant prominence among the affected communities and was actively contacted by most communities in the sample, albeit at different stages of contention. Other Beijing-based organizations such as Global Village of Beijing were primarily engaged via nonrelational ties, such as participating in the advocacy campaigns led by Nature University and Friends of Nature without establishing direct contact with the affected communities (cp. the Panguanying and Dagong cases in Chapters Four and Five). These case-related issue campaigns were joined by the Anhui Province-based environmental organization Wuhu Ecology Center and other members of the Zero Waste Alliance (see Table 6.1).

The local engagement of Beijing-based lawyers and the legal association Center for the Legal Assistance to Pollution Victims primarily focused on cases in the Beijing cluster area. However, after they acted as legal representatives in cases that succeeded in obstructing planned facilities, such as Liulitun and Panguanying, and that received wider attention, they were also contacted by communities or were actively involved by other 'no burn' community members in cases in other parts of the country (such as in Likeng, Huiyang, and Huangtutang).

Beyond Beijing, the number of supra-local intermediaries engaged in more than one struggle is very limited. In Shanghai, only the environmental organization Aifen Environmental Protection Science and Technology Service Council Center (爱芬环保科技服务咨询中心, *Aifen huanbao keji fuwu zixun zhongxin*, hereafter referred to as Aifen) collected some information and communicated with Shanghai community members (Caolu and Yuqiao, both not included in the sample). As members of the Zero Waste Alliance, Aifen staff introduced members of Nature University to the two communities during their visit to the Shanghai area. In the Guangzhou cluster, a social science researcher working on social conflicts at Guangzhou's Sun Yat-sen University (中山大学, *Zhongshan daxue*), who established close ties with members of the contentious communities in Guangzhou municipality (Panyu, Huadu, and Likeng) functioned as an important expert in and information broker between these cases. Moreover, in both Beijing and Guangzhou, the Asuwei and Panyu campaign leaders and their newly founded organizations (Green House and Eco Canton) became important members of the 'no burn' community. They are not included in Table 6.1 because they were mainly contacted by other communities in their function as successful and knowledgeable contentious community members and because their organizations are primarily dedicated to promoting sustainable municipal waste policies and are less directly focused on incineration (cp. Chapter Three).

Further, the domestic Chinese 'no burn' community was joined by members of the global anti-incineration movement. In particular, the transnational organizations GAIA and IPEN joined ranks with the Chinese environmental organizations, supporting and participating in their case-related issue campaigns, certifying their claims and activities, and otherwise embedding them into the transnational anti-incineration movement (cp. the Beijing cluster cases in Chapters Three to Five). This included the participation of GAIA and IPEN staff in the Shanghai conference for affected communities (see above), which permitted the present local communities to establish direct relational ties with members of the transnational network. As in the case of horizontal linkages, however, none of these ties were intensified after the conference (see Table 6.1). Under the lead of Wuhu Ecology Center and Eco Canton, GAIA staff also paid site visits to the affected communities in Guangzhou (Panyu, Huadu, and Likeng).

Other international experts and organizations, such as the central figures of the US anti-incineration and environmental justice movements Paul Connett and Barry Commoner, as well as transnational organizations such as Greenpeace and the World Health Organization, impacted the local communities in this study via nonrelational information flows. Here, urban communities with access to the World Wide Web and the ability to read English-language information played an important brokerage and translation role for less Internet-acquainted local communities – such as via the information in the Liulitun 'opinion booklet' or the Asuwei research report, which were distributed to the rural and peri-urban communities (cp. Chapters Three and Four) .

Local Impacts: The Role of Vertical Ties for Anti-Incineration Struggles

Despite the limited number of supra-local intermediaries who became directly engaged at the local level, the established vertical linkages had a major impact on the spread and development of local anti-incineration struggles. Many of the 'no burn' community members' case-related activities were embedded in or the starting point for broader issue campaigns, as described in the Panguanying and Dagong cases (Chapters Four and Five). Overall, strong learning effects are discernible on the part of the intermediaries throughout the research period. Through their engagement in different local struggles, they expanded and consolidated both their individual repertoires and their collaborations with each other. Their activities at the local level thus became more knowledgeable, professionalized, and effective over time. Across the case spectrum, all intermediaries developed their own core set of activities which they offered to the affected communities, irrespective of their location or local situation. As in the case of diffusion effects based on horizontal linkages, however, the actual role played by intermediaries was strongly dependent on the local context.

Professor Zhao Zhangyuan played an important role for most communities in this study as trusted expert who provided hard-to-obtain 'neutral' information on waste incineration and its harms within the overall 'information haze'. As in the cases of Asuwei and Panguanying (Chapters Three and Four), the information provided by Zhao played an important role in the awareness process of most communities in the sample, either by consolidating a first sense of alarm, like in Panguanying, or by functioning as cognitive cues and alternative frames, like in both Panguanying and Asuwei. This helped foster the communities' collective interpretation of the situation and the formation of a contentious consciousness, as suggested in the social movement literature.

As a renowned expert on incineration, Zhao also played an important role for the certification of critical information – both internally, by substantiating campaign leaders' warnings of environmental and health effects emanating from the plants vis-à-vis their own communities, and externally, by legitimizing the local communities' claims and activities towards the public and project proponents. He presented his views in media articles and interviews, online publications, and in direct communication with party-state institutions, companies, and government-commissioned experts. Here, information provided by Zhao was often more trusted than that provided by environmental organizations, since several (particularly urban) communities suspected 'NGOs' to be irrational, radical, or tendentious – as expressed by members of the Asuwei, Songjiang, and Fengqiao communities. A similar 'Achilles' heel' of environmental organizations, who tend to be regarded as emotive, is also described by Mertha (2008: 146).

Zhao assisted several communities by offering them advice and information that was helpful for mobilizing action (such as in the urban Beijing cases, in Panguanying, Panyu, Huiyang, Songjiang, and Huangtutang). This included providing the communities with: technical information related to incineration; legal knowledge and information about the mandatory procedures related to MSWI projects, including the collection of evidence related to procedural flaws such as in the EIA process; and advice related to other modes of action, such as peaceful strolls and seeking media attention. In several cases, Zhao also served as a broker between different communities and with other 'no burn' community members. Further, he functioned as a central diffusion node between domestic communities and the transnational anti-incineration movement by providing Chinese-language translations of transnational materials that enabled the local communities' (identity) frame to align with that of the transnational 'no burn' community – although, as outlined above, this was mainly used to promote localized claims.

Similar roles were played by the environmental organizations active in the issue field. The organizations' activities and related impacts on local contention outlined in the Beijing cluster cases encompass the standard repertoire of activities employed by organization staff across the wider case spectrum. Like professor Zhao, the organizations provided critical issue-specific information on waste incineration and its risks, as well as on the activities of other communities across the country. This fostered communities' awareness processes, promoted the formation of a collective contentious identity, raised communities' expectations of success, and certified their claims and activities both within the community and towards the public. The provision of information was standardized in the 'toolkit' specifically compiled by Nature University staff to assist affected communities, which was distributed to all community members who sought the organization's advice and was also available for public download from an organization member's Sina microblog account starting in 2012 (cp. Chapter Three).

The organizations also certified the local communities' claims and activities through media outreach activities and by inviting local community members to public forums and symposiums targeted at providing them with a public platform to disseminate their claims and grievances. By disseminating information about the cases via their personal and online networks, the organizations functioned as important communication nodes with the outside world, particularly for less Internet-acquainted rural and peri-urban communities (like in Panguanying, Likeng, Huangtutang, and Fengqiao). In some cases, this generated external attention included a protective function for the local campaigners, as described in the Panguanying case (and in Huangtutang, Likeng, and Huiyang).

The organizations also actively assisted the communities in the mobilization of contention by providing both relevant information and further assistance and support – thus filling resource-gaps and impacting the development of contentious action. This was particularly relevant for resource-poorer communities in peri-urban and rural areas, but also during the later stages of urban struggles (such as in Asuwei, Huiyang, and Songjiang). Like Zhao, the organizations actively promoted legal and peaceful means and warned contentious communities against taking disruptive and potentially violent action – thus fostering a convergence of strategies across the different cases. As outlined in the case studies, the organizations' mobilizing support included the provision of: detailed technical information about incineration; information regarding environmental laws and regulations and the mandatory environmental procedures related to MSWI projects, including active assistance in finding evidence of procedural flaws (such as EIA fraud); assistance with submitting information disclosure requests, filing administrative redress applications or environmental lawsuits, and finding legal representation; and mobilizing advice and information related to other modes of action (such as strolls and petitions); as well as representing local communities during negotiations with government institutions or companies, including letter-writing campaigns such as to the MEP. In Panguanying, Nature University staff also took on an important protective function for the contentious villagers and acted as election observers during the local village elections (see Chapter Four).

The organizations (particularly Nature University) also deliberately attempted to broker direct horizontal communication, learning processes, and alliances between the different communities. Before 2013, they mainly invited community members to symposiums and public forums in Beijing and distributed contact information among the communities. Since 2013, the organizations have attempted to promote horizontal linkages on a national scale – such as in-person via the Shanghai conferences mentioned above, or online via the establishment of issue groups on the messaging service WeChat (微信, *Weixin*) and other social media platforms. These brokerage efforts include attempts to provide local communities with access to (trans-)national experts, lawyers, and media representatives by inviting them as speakers and participants to both case-related symposiums and the Shanghai conferences. While intensified horizontal linkages based on these meetings remained limited, as outlined in the previous section, the symposiums did help several communities establish valuable contacts with intermediaries and journalists who proved helpful in later stages of contention (such as the media contacts established by Panguanying villagers or contact with the Beijing lawyers established by the Huangtutang and Huiyang communities). Like professor Zhao, the organizations also functioned as important diffusion nodes between domestic communities and the transnational 'no burn' community, hence enabling local communities' alignment of (identify) frames with the transnational anti-incineration movement.

The third main actor group in the Chinese 'no burn' community, the mainly Beijing-based lawyers and legal associations, primarily facilitated the communities' access to justice by providing legal advice and representation and assisting in the collection of evidence (such as in the case of EIA flaws). The active engagement of lawyers and their success in some of the earlier cases contributed to the convergence of strategies across the case spectrum, with many communities including legal means in their contentious repertoire due to their heightened expectations of success for this mode of action. As described in the case of Panguanying (Chapter Four), some of the lawyers also played additional functions such as the dissemination and certification of the communities' claims in the media, through their own publications, and at symposiums and conferences. They further acted as representatives of some of the communities in negotiations with party-state representatives or companies. Irrespective of the density of linkages between local communities and lawyers, the success of legal activities was strongly dependent on whether the advocates could find a legal entry point in the cases. Despite active attempts to legally assist the communities in Huangtutang or Huiyang, for example, this was not possible in these cases.

While the core repertoire of activities offered by the 'no burn' community members in this study is similar across the case spectrum, the extent to which local communities made use of this engagement varies significantly. This is clearly visible across the Beijing cluster cases described in Chapters Three to Five. Like the Asuwei campaigners, the other urban communities in this study were largely self-reliant in their initial problem awareness, search for information, and mobilization of action. They mainly fell back on their own resources and extensive social networks (including to other contentious urban communities). In this context, the major role of intermediaries was played by trusted domestic experts (such as Zhao Zhangyuan) or international experts who provided critical information and interpretive frames that helped the communities pierce the overall 'information haze' and come to a collective interpretation of the situation.

More divergence can be seen in the case of peri-urban and rural communities. Overall, these resource-poorer communities were more dependent on intermediaries and many actively sought their assistance. Here, the case of Panguanying is at one end of the case spectrum: the villagers made full use of the assistance and advice provided by intermediaries (see Chapter Four). The Dagong case stands at the other end of the spectrum: no unified and sustained contentious action emerged despite several 'no burn' community members' active attempts to promote collective activities (see Chapter Five).

As outlined in the previous section, horizontal and vertical linkages fail to take effect if they are not rooted in local contention. 'No burn' community members repeatedly pointed out that their efforts to get engaged at the local level were in vain if the local communities did not take sustained or coordinated action themselves – apart from Dagong, this was also observable in Nangong, Fengqiao, and Yuqiao, as well as in Lanzhou city (兰州市, *Lanzhou shi*) in Gansu Province (甘肃省, *Gansu sheng*) and Luodai town (洛带镇, *Luodai zhen*) in Sichuan Province (四川省, *Sichuan sheng*). In these cases, the intermediaries soon reached their limit and often turned to more promising cases. This was particularly the case when local communities did not believe in the possibility of success through contentious action, lacked a sense of political efficacy, or were scared away from further action by repressive party-state responses at any point in the struggle. Moreover, if the priorities and modes of action significantly diverged between the local communities and intermediaries – if local communities decided to opt for relocation rather than opposing the incinerator project, like in the case of the Dagong villagers, for example – it became evident that collaborations between local communities and environmentalists were still be somewhat of an 'uneasy alliance' (McAdam and Boudet 2012: 135) as described in the literature.

Beyond Localized Struggles: Networked Contention and the Prospect for a Broader Movement

As outlined in the previous section, the environmental actors in this study were far from isolated from each other; rather, they were connected via an extensive network of linkages across sites and social groups. In the social movement literature, such ties are regarded as a promising basis for a potential scaling up of local contention, i.e., a shift in the scope and scale of contentious issues from the local to regional, national, or transnational levels (Givan, Roberts, and Soule 2010; McAdam, Tarrow, and Tilly 2001; McAdam and Boudet 2012; Tarrow 2010). Within China's restrictive political setting, however, the prospect of scaling up a local contention based on horizontal linkages is limited. While the network of contention found in this study has played a significant role in the spread and development of local anti-incineration contention, the political framework has hampered collaborations across localities and the formation of alliances that are active beyond individual settings. In particular, resource-poorer peri-urban and rural communities have avoided fostering direct relations with other communities or becoming active beyond their own localities. While urban homeowner communities within individual municipalities have formed loose 'alliances of support' (such in the urban Beijing and Guangzhou cases, and also across issue fields like between the Songjiang and maglev campaigners in Shanghai), more formal alliances or collaborative action are regarded as politically taboo. In a similar vein, repercussions were feared or experienced by urban homeowners who attempted to assist resource-poorer (rural) communities in their region – such as in the case of relocated incineration plants.

Horizontal collaborations were impeded not only by the political setting, but also by the communities' overall lack of interest in other communities' grievances and broader waste-related or environmental issues. This lead to a widespread lack of interest in intensified relations with other communities not only on the part of rural and peri-urban contenders, but also by the majority of urban community members. As outlined above, most affected communities were primarily concerned with their own localized claims and grievances. Rural and peri-urban communities in particular did not see the potential for much added value in establishing closer ties with other contenders. In most cases, contentious action receded soon after the local struggles were concluded. This was also the case for urban campaigns that lead to the relocation of incinerator projects to nearby rural areas. While a few individual campaigners tried to assist the resource-poorer successor communities, most community members returned to their daily lives and regarded the new projects as beyond their concern – feeding into an environmental justice debate among members of the 'no burn' community. While this localized attitude is likely enforced by the political hazards related to getting involved elsewhere, it also reflects a broader tendency to show little personal interest beyond individual concerns, as observed in other studies (e.g., Jian and Chan 2016). The extent to which this is influenced by the political context or is part of the Chinese (political) culture is hard to discern. In Western countries, many local struggles against hazardous construction projects also tend to revolve mainly around localized grievances and do not produce broader claims or activities (Futrell 2003; McAdam and Boudet 2012; Schively 2007; Walsh, Warland, and Smith 1997). It is likely that in China this is at least enhanced by the political risks related to engaging with other contenders.

Apart from direct alliances and collaborations, communities across the case spectrum aligned their issue and identity frames both with each other and with the (trans-)national 'no burn' community. Most of the communities under study regarded themselves as part of a broader *weiquan*-community or the (trans-)national 'no burn' community and actively aligned their issue frames with both other contenders and the 'no burn' actors, but this alignment was primarily used to further the communities' localized claims and activities and did not produce collective claims beyond the individual localities. An actual shift from NIMBY to NIABY claims as observed in Western countries – albeit also limited in those contexts (Futrell 2003: 377; McAdam and Boudet 2012: 135) – only occurred on the part of the resource-rich urban communities in Asuwei, Panyu, and Songjiang, where it was limited to individual community members. The campaign leaders of Asuwei and Panyu have not only turned to focus on broader waste and environmental issues and attempted to impact municipal waste policies, but have also founded their own organizations dedicated to promoting more sustainable waste policies at higher political levels. Here, a shift of claims and grievances well beyond the local level has occurred and in some cases has produced effective policy changes, as described in Chapter Three (cp. also Johnson 2013a, 2013b; Wong 2016). While such developments have been limited to individual members of what has been termed the Chinese 'middle class', these cases nonetheless show that local struggles can scale up and produce results at higher political levels in China.

While the prospects for scale shifting based on horizontal networks remain limited, collaborations between local contenders and the Chinese 'no burn' community, i.e. between the 'two facets' of Chinese environmental activism, have significantly furthered the interests of the supra-local actors in this study. Vertical cooperation between the two groups has strengthened a national-level advocacy network that tackles not only waste-related issues but also broader environmental problems. As has been observed in other world regions, the growing number of local struggles against incinerator projects in China has helped bring the issue of waste incineration onto the agenda of environmental organizations and promoted the formation of a Chinese 'Zero Waste Alliance' (cp. Chapter Two). Moreover, joint engagement in individual local struggles has intensified the collaborations between different actor groups in the Chinese 'no burn' community, thereby strengthening their advocacy activities.

As outlined in these case studies, many of the 'no burn' community members' local activities were embedded in advocacy campaigns targeting national-level political institutions, which increased public and media attention to not only the risks associated with waste incineration and China's national waste policies, but also the broader regulatory failures reflected in the issue field. These include the lax implementation of EIA procedures or other environmental laws and standards at the local level, weaknesses in China's environmental litigation system, lacking public participation and transparency in the environmental realm, and local corruption issues more generally. The resistance of local communities has significantly enhanced national-level actors' struggle for greater awareness and policy change related to these issues. In some cases, locally-rooted national advocacy activities have produced manifest results, such as the MEP's nation-wide critical evaluation of EIA units or the national media campaign about EIA fraud at the local level (as described in Chapters Four and Five). By aggregating localized claims and grievances into broader public demands and long-term advocacy campaigns, supra-local intermediaries have not only lifted disparate local claims and grievances to the national level, but have also transformed the mostly short-lived local struggles into more sustained policy advocacy.

For the time being, it is unlikely that these developments will consolidate into an issue-specific or environmental movement. Within China's restrictive political framework, networked contention against waste incinerators has stalled at a meso-level of contention between fragmented activism and a full-grown movement. Particularly in the current political climate, it is unlikely that the Chinese 'no burn' community will continue a trajectory of growth and institutionalization. However, the existence of networked contention as a meso-level social phenomenon is a significant development in and of itself in the Chinese political context, strengthening the supra-local actors in the issue field and facilitating their assistance in local conflicts. At the same time, the actors stay loosely organized enough to not prompt repression by the party-state. Networked contention may thus hold significant advantages under the current adverse political conditions in China.

Beyond Waste Incineration: Linking Up as a Broader Trend in Chinese Environmental Contention

The findings from this study are limited to the issue field of waste incineration and cannot simply be transferred to Chinese environmental activism at large. The field of waste incineration can be regarded as a most likely case for the emergence of linkages among the actors in China's green sphere, particularly because the close relationship between the issue of waste incineration and broader waste policies permits the alignment of local grievances with broader claims and has attracted the attention of a large number of environmental organizations. This has facilitated the emergence of vertical linkages and national-level policy advocacy when compared to other siting disputes, such as those against PX plants, in which local grievances tend to remain more localized since they are less easily embedded into the activities of supra-local actors with a broader political agenda. In addition, the presence of a vivid global anti-incineration and 'zero waste' movement and the close linkages between the Chinese and transnational 'no burn' communities have promoted Chinese anti-incineration contention and the emergence of a Chinese issue network in the shape of the 'Zero Waste Alliance'. Similar to the role of professors Connett and Commoner in the US anti-incineration and environmental movements, the prominent role of Zhao Zhangyuan as an anti-incineration expert with national influence can be regarded as somewhat exceptional. His position as a retired researcher facilitated his outspoken stance against incineration, since he did not have to fear retaliation linked to his employment situation (interview Zhao 8-11-12).

The local engagement of the large number of supra-local actors in the waste realm has significantly fostered the emergence and spread of local anti-incinerator conflicts. The (media) prominence of many of the 'no burn' community members has greatly facilitated local communities' access to information and external assistance. This was enhanced by the public and media debate about incineration after 2009, which made information not only about the risks of incineration but also about numerous local anti-MSWI struggles publicly available. Media information about these cases could then be complemented with online information. This availability of critical information and the rise of a rather clear-cut and outspoken community of opponents have facilitated the assessment of risks for local communities and their access to external assistance, especially when compared to other issue fields and types of polluting plants in China.

Moreover, most opposition to incinerator projects in China has taken the shape of siting disputes against planned or not yet completed facilities. Siting disputes are a specific form of environmental contention that differ from the conflict in cases where pollution has already occurred. The perception of anticipated risks and particularly health effects in often highly technical siting disputes tends to rely heavily on knowledge, information, and expertise. In many of the anti-incinerator struggles under study, this high initial threshold was overcome by building on ongoing conflicts about already existing waste facilities such as landfills in the vicinity. In these cases, the clearly discernible smell from the operating facilities had already produced grievances that rendered opposition to another – even more harmful – waste facility more likely. This was the case with the opposition to several operating incinerator plants such as in Likeng or Gaoantun, where it was the stench or dust from the waste combustion that first angered the residents, who only found out about the more hidden environmental and health hazards associated with incineration when looking deeper into the matter. This is different from other types of hazardous facilities such as nuclear plants, where environmental and health hazards are less palpable, and mobilization relies on a sense of alarm related to hidden risks even in the initial stages of awareness. Moreover, when compared to other polluting industries, contention against waste incinerators is facilitated by the rareness of economic dependencies on or (expected) benefits from the plants for the local population, which might otherwise hamper open opposition to polluters (cp. Deng and Yang 2013; Lora-Wainwright 2017; McAdam and Boudet 2012; Tilt 2010).

While the siting of other types of hazardous facilities, particularly PX plants and other waste facilities, has also produced significant local resistance in China, waste incineration can thus be regarded as particularly favourable for the emergence of local conflicts and the development of networked contention. This was confirmed by several of the intermediaries interviewed for this study. The construction of nuclear power plants, which in other countries has produced strong local reactions and national-level anti-nuclear movements (Kitschelt 1986; Markham 2005; Sherman 2011a; Walsh, Warland, and Smith 1997), have thus far triggered relatively little resistance at either level in China (Sheng forthcoming; Sheng 2014; Grano and Zhang forthcoming). This is interesting particularly in light of the nuclear disaster in Fukushima in 2011 and China's expansive nuclear strategy (Dorfmann 2015; He 2013; *Probe International* 2011; Wang 2013). Moreover, while some supra-local engagement in local contention has been observed and national-level alliances have formed in anti-dam disputes (Büsgen 2006; Mertha 2008), the environmental organizations, other intermediaries, and Chinese social science researchers interviewed for this study have regarded anti-dam contention as significantly more disparate and parochial than anti-incineration contention. The specific reasons for these limitations and a structured comparison across issue fields that would permit a more comprehensive assessment of the transferability of the findings of this research to other issue areas is an interesting entry point for future research.

Nonetheless, the emergence of networked contention among the different actors in China's environmental arena has also been observed in recent studies investigating the resistance to other types of industrial and infrastructure projects and Chinese environmental contention more broadly. On a horizontal scale, research has found that diffusion processes and 'spill-over' effects via Internet- and mass media-based learning processes between affected communities have driven the nation-wide protest wave against PX plants that has spread across China since the first major anti-PX protest in Xiamen in 2007 (Zhu 2017). Horizontal ties have, to some extent, also played a role in the small number of Chinese anti-nuclear protests (Grano and Zhang forthcoming). This suggests that the new communication channels enabled by the pluralizing media landscape and the spread of the Internet, particularly social media, under Hu Jintao have also fostered communication and diffusion processes between contentious communities in other areas.

On a vertical scale, recent studies have observed intensifying linkages between local street protesters and environmental organizations or other supra-local policy advocates in other environmental issue fields (Steinhardt and Wu 2015; Sun, Huang, and Yip 2017; Tang forthcoming; Wright 2018). Wright (2018: 115-37) argues that a central factor in the relative success of recent environmental protests compared to other types of popular protests is the engagement of environmental organizations, which provide aggravated local groups with information, advice, and support. Sun, Huang, and Yip (2017) show that supra-local intermediaries – including regional and national-level environmental organizations, lawyers, and journalists – played an important role in the development of the local struggle against a planned PX plant in Kunming in 2013. To assist the local community there, these intermediaries drew on a repertoire of activities similar to that observed in this study, including media and public outreach activities, the promotion of information disclosure requests, environmental litigation, and letter-writing campaigns. Moreover, and again similar to the national issue campaigns described in this study, environmental organizations' engagement in the Kunming episode and the occurrence of similar problems in other localities prompted them to conduct a thorough and in-depth investigation of disputed PX and other chemicals projects across the country. This not only added legitimacy to the local contenders, but also resulted in broader policy advocacy.

In a similar vein, Steinhardt and Wu (2015) in a study of four environmental protests against different construction projects[115] and Steinhardt

115 The four environmental protests under scrutiny are the anti-PX protest in Xiamen (2007), the anti-incinerator protest in Panyu (2009), a tree-saving campaign related to the construction

(forthcoming) in a follow-up study of 25 cases of environmental contention find that Chinese environmental organizations were increasingly active at the local level during the observation period (between 2007 and 2016) and that a mutual reinforcement of street mobilization and policy advocacy was taking place. A similar finding is reported by Tang (forthcoming) in her study of 20 environmental protests. She argues that environmental protests in today's China are characterized by a 'multifaceted advocacy dimension': different social actors, including affected communities and supra-local actors such as social organizations and lawyers, employ different strategies and together invoke the necessity of policy change for environmental governance.

While their case studies are all related to environmental contention, Steinhardt and Wu (2015) point to similar developments in other areas and regard this as part of a new repertoire for popular contention in China that signals broader developments in Chinese contentious politics. While more research about the linkages between different sites and actor groups in the Chinese environmental realm is needed and while these developments might have been curtailed under the current leadership, the developments described in this study and the emergence of networked contention point to a broader trend in Chinese environmental activism, at least under the government of Hu Jintao.

Beyond China: Networked Contention, Linkages, and Diffusion in a Restrictive Political Setting

The potential for the formation of an environmental or other social movement is significantly hampered in the context of a restrictive political regime. This study shows that networked contention in the form of a meso-level phenomenon between fragmented activism and a full-blown movement can nonetheless develop under adverse political circumstances and foster both local and higher-level contention. Within the limits imposed by a restrictive setting, this loosely organized form of resistance can hold significant advantages for contentious actors by strengthening supra-local policy advocacy and facilitating the engagement of supra-local actors in local conflicts without attracting too much attention.

The findings from this study speak to the literature on the diffusion of contention and contentious politics more broadly. They shed light on

of a new subway station in Nanjing (2011), and the anti-PX struggle in Kunming (2013) also investigated by Sun, Huang, and Yip (2017).

the exact ways different types of contentious actors may link up under a restrictive political regime and how these linkages can foster both local and higher-level contention in such a context. The channels and mechanisms that drive contention-related diffusion and the scaling up of local contention to higher levels have mainly been investigated in the context of Western democratic countries. Only few studies have looked at diffusion or scale shift in authoritarian settings where the dynamics of contention tend to diverge widely from those observed in liberal Western regimes – and most of those studies have only tackled certain aspects of the phenomenon (Lohmann 1994; McAdam et al. 2010; Osa 2003; Osa and Schock 2007).

The horizontal and vertical linkages behind networked contention are shaped by the political context. In regimes where close relationships among contentious actors are associated with high personal risks, it is unlikely that horizontal linkages between affected communities will take the shape of formal supra-local alliances, collaborations, or new organizational forms that could be the basis for scaling up local contention to regional or higher-level movements, as suggested by the social movement literature based on research in democratic contexts. As this study shows, however, this does not mean that contenders do not link up under a restrictive framework. Instead, diffusion processes are based more on informal networks and loose 'alliances of support' (see the Beijing and Guangzhou cases) as also observed by Osa (2003: 78). As suggested in the literature (Bennett and Segerberg 2012; Diamond 2010; Diamond and Plattner 2012; Earl and Kimport 2011), information and communication technologies, particularly social media, play a crucial role in both enabling relational diffusion across geographically dispersed localities and allowing the broadcast dissemination of information. Overall, diffusion processes in such restrictive contexts are particularly reliant on nonrelational ties via the Internet and mass media, since the formation of offline and online relational linkages is hampered by associated political risks. For the resource-poor, less internet-savvy communities in this study, vertical relational linkages with intermediaries remained a major diffusion channel. Where open support is risky, it may take the form of informal advice and the online provision of precious information.

The nature of these linkages renders the scaling up of local contention or formation of a broader movement based on horizontal ties among affected communities unfeasible. As this study shows, however, the diffusion processes enabled by networked contention can nonetheless play a particularly important role in the spread and development of local contention under restrictive conditions and produce 'protest waves' or 'cycles of contention' in an adverse political climate. Overall, the same basic mechanisms that drive the diffusion of contention in Western democratic regimes are also at play under restrictive conditions, although there are specific features in how these mechanisms intertwine and what role they play for local contention within a restrictive political framework.

First, the information disseminated via horizontal and vertical linkages is of particular importance for the awareness process of affected communities, and hence for both the onset and development of local contention, in settings where access to uncensored and critical information – particularly to the highly relevant technical and medical information that is needed to assess the nature and risks of environmentally hazardous projects such as incinerators, PX and nuclear plants, or dam projects – is limited (compare with Mertha 2008). This includes fostering the first awareness of related risks, a collective interpretation of the situation, and a cognitive justification for contentious action. As this study shows, 'neutral' experts, (transnational) organizations, and social movements in other world regions are particularly important sources of information, while supra-local intermediaries function as important 'translators' and diffusion nodes between the transnational contentious community and local contenders.

Second and closely related, these above sources of information play a particularly important role in the certification of information within the overall 'information haze'. Due to the highly contested nature of environmental issues, particularly health-related ones, the certification of the communities' claims and activities is not only important vis-à-vis the public and project proponents, but also internally vis-à-vis the affected communities if collective action is to be mobilized despite the risks associated with contentious activities.

Third, as outlined in the literature on diffusion processes in authoritarian contexts (Allagui and Kuebler 2011; Lohmann 1994; Lynch 2011; Mueller 1999), the transmitted information plays a central role in changing the affected communities' perceptions of threat and opportunity. Against the backdrop of potentially repressive party-state responses and the blurred and continuously shifting boundaries of the politically permissible, knowledge about other contentious communities, their activities, and related party-state reactions can have a significant impact not only on the onset of contention but also on the mode of action that is employed. In this study, knowledge about other communities' activities not only helped the adopting contenders assess the risks related to certain kinds of activities, but also increased or decreased their belief in the chances of success of certain types of action – such as legal actions – within the restrictive political opportunity structures.

Fourth, diffusion scholars have pointed to the role of information flows for 'identity formation', i.e., the formation of a shared contentious identity within the local context or as part of a broader contentious struggle or movement. As this study demonstrates, the self-perception of affected communities as part of a broader (national or transnational) wave of contention can constitute a strong mobilizing force and help community members overcome their reluctance to resort to contentious means in the context of high political risks and insecurities associated with contentious action. However, it is unlikely that a scaling down of identity frames to the local level leads to broader claims, alliances, or communities of solidarity in this context. Rather, the affected communities in this study adopt identity and other frames primarily to further their own localized claims.

Fifth, diffusion processes may also impact the dynamics of contention by changing the adopting communities' resource structure, thus enabling or facilitating the mobilization and organization of action. Under a restrictive political framework with limited access to critical issue-specific and action-related information, such information plays a particularly important role in the onset and development of contention. This includes the above-mentioned information about specific tactics, strategies, and organizational forms that have been applied in other sites of contention and that can be emulated by the adopting communities. This also includes information about the opportunities provided within the legal and political system, as well as hard-to-obtain issue-specific information, expertise, and evidence that is critical for taking action in often highly technical environmental disputes. The resource-poorer communities in this study were particularly dependent on the provision of more manifest resources for mobilizing action, including financial support, legal advice, and representation in negotiations with project proponents. Here, intermediaries can play a major role, particularly in rural struggles, by filling in resource gaps that would otherwise render contentious action unfeasible.

With regards to changing the resource structure of affected communities, this study has observed another aspect that is not covered in the literature on the diffusion of contention. In a repressive context where contentious action is often met with repercussions and entails high personal risks, particularly for community leaders, networked contention can play an important protective function, thus enabling a broader contentious repertoire. This is best demonstrated in the Panguanying case, but also played an important role in several other cases in this study. For example, the materials provided by other communities or the state media can be distributed without making the distributors vulnerable to legal repercussions as the authors of the material. Intermediaries can also be called on as experts or eyewitnesses to refute official claims, allegations, or charges, thereby protecting local campaigners from repressive party-state actions. In several of the cases in this study, external attention from intermediaries, journalists, and researchers (including myself) and the public dissemination of information about ongoing events helped to raise the costs for state repression. In the case of Panguanying, this included the role of organization staff and journalists as election observers and as eyewitnesses in official charges against a community leader.

Networked contention can also have a significant impact at higher political levels in a restrictive setting. In a political context where the range of activities by supra-local actors and particularly social organizations that is tolerated by the regime is limited, collaboration with local communities helps broaden these actors' repertoire of action and can strengthen a national issue network that also tackles broader (environmental) problems. It permits them to address larger issues or regulatory failures that manifest at the local level – such as lacking implementation of environmental regulations or adverse policy effects – and adds weight to their higher-level political claims. As shown in this study, local action can be embedded in or the starting point for long-term national-level advocacy campaigns that increase public and media attention to the issues at hand and place pressure on policymakers. While the emergence of a full-blown movement might be restricted, networked contention can thus help lift disparate local claims and grievances to the national level and transform the mostly short-lived local struggles into broader public demands and more sustained policy advocacy. At the same time, the actors themselves stay small and loosely organized enough to remain under the radar of state forces. As such, networked contention can hold significant advantages in a restrictive political setting and merits attention in its own right.

Appendices

Appendix I List of interviews

ASW1 17-10-12	Interview with Asuwei campaign leader 1, Beijing, 17 October 2012.
ASW1 31-5-13	Interview with Asuwei campaign leader 1, Beijing, 31 May 2013.
ASW1 31-7-13	Interview with Asuwei campaign leader 1, Beijing, 31 July 2013.
ASW2 27-5-13	Interview with Asuwei campaign leader 2, Beijing, 27 May 2013.
BMAC 7-11-12	Interview with BMAC waste expert and NPC member, Beijing, 7 November 2012.
CL 22-5-13	Interview with Shanghai Caolu campaigner, Shanghai, 22 May 2013.
DGC1 22-7-13	Interview with Dagong villager 1, Dagong village, 22 July 2013.
DGC1 23-7-13	Interview with Dagong villager 1, Dagong village, 23 July 2013.
DGC2 23-7-13	Interview with Dagong villager 2, Dagong village, 23 July 2013.
DGC3 23-7-13	Interview with group of Dagong villagers 1, Dagong village, 23 July 2013.
DGC4 23-7-13	Interview with group of Dagong villagers 2, Dagong village, 23 July 2013.
DGC5 29-5-13	Interview with Nan'anhe villager 1, Beijing, 29 May 2013.
DGC5 1-6-13	Interview with Nan'anhe villager 1, Nan'anhe village, 1 June 2013.
DGC5 23-7-13	Interview with Nan'anhe villager 1, Nan'anhe village, 23 July 2013.
DGC6 1-6-13	Interview with Nan'anhe villager 2, Nan'anhe village, 1 June 2013.
DGC7 1-6-13	Interview with Nan'anhe villager 3, Nan'anhe village, 1 June 2013.
DGC8 1-6-13	Interview with group of Nan'anhe villagers, Nan'anhe village, 1 June 2013.

DGC9 22-7-13	Interview with person knowledgeable about Dajue Temple developments, Beijing, 22 July 2013.
EnvJourn 16-10-12	Interview with journalist for *China Dialogue* and the environmental magazine *Green Youther* (青年环境评论, *Nianqing huanjing pinglun*), Beijing, 16 October 2012.
EnvJourn 23-4-13	Interview with same *China Dialogue* journalist, Beijing, 23 April 2013.
FQ1 22-6-13	Interview with Fengqiao village resident 1, Shanghai, 22 June 2013.
FQ1 14-7-13	Interview with Fengqiao village resident 1, Fengqiao village, 14 July 2013.
FQ2 14-7-13	Interview with group of Fengqiao village residents, Fengqiao village, 14 July 2013.
GAT 1-11-12	Interview with Gaoantun campaign leader, Beijing, 1 November 2012.
GAT 15-6-13	Interview with Gaoantun campaign leader, Beijing, 15 June 2013.
GAT MSWI 9-11-12	Interview with public relations representative of Beijing Chaoyang Circular Economy Industrial Park, Beijing, 9 November 2012.
GZ expert 1 30-6-13	Interview with Guangzhou expert 1, who is working on anti-incineration contention in Guangzhou with personal ties to members of the affected Guangzhou communities, Guangzhou, 30 June 2013.
GZ expert 2 2-7-13	Interview with Guangzhou expert 2, who is working on anti-incineration contention in Guangzhou with personal ties to members of the affected Guangzhou communities, Guangzhou, 2 July 2013.
HD1 2-7-13	Interview with Huadu campaigner 1, Guangzhou, 2 July 2013.
HD2 6-7-13	Interview with Huadu campaigner 2, Guangzhou, 6 July 2013.

HD2 8-7-13	Interview with Huadu campaigner 2, Guangzhou, 8 July 2013.
HD3 7-7-13	Interview with Huadu campaigner 3, Guangzhou, 7 July 2013.
HZ 23-6-13	Interview with villager from Hangzhou's Binjiang district, Shanghai, 23 June 2013.
HY1 23-6-13	Interview with Huiyang campaigner 1, Shanghai, 23 June 2013.
HY2 11-7-13	Interview with Huiyang campaigner 2, Shenzhen, 11 July 2013.
LK 4-7-13	Interview with Likeng villagers, Yongxing village, 4 July 2013.
LK 9-7-13	Interview with Likeng villagers, Yongxing village, 9 July 2013.
NGO Aifen 15-5-13	Interview with Aifen Environmental Protection Science and Technology Service Council Center staff, Shanghai, 15 May 2013.
NGO Boell 25-10-12	Interview with Boell Foundation's Beijing office staff, Beijing, 25 October 2012.
NGO CLAPV 18-10-12	Interview with Center for the Legal Assistance to Pollution Victims staff, Beijing, 18 October 2012.
NGO ECO1/PY1 1-7-13	Interview with Panyu campaign leader and founder of Eco Canton, Guangzhou, 1 July 2013.
NGO ECO1/PY1 10-7-13	Interview with Panyu campaign leader and founder of Eco Canton, Guangzhou, 10 July 2013.
NGO ECO2 17-6-13	Interview with Eco Canton staff 2, Guangzhou, 17 June 2013.
NGO ECO2 10-7-13	Interview with Eco Canton staff 2, Guangzhou, 10 July 2013.
NGO FON1 19-10-12	Interview with Friends of Nature staff 1, Beijing, 19 October 2012.
NGO FON1 6-6-13	Interview with Friends of Nature staff 1, Beijing, 6 June 2013.
NGO FON2 21-10-11	Interview with Friends of Nature staff 2, Beijing, 21 October 2011.

NGO FON3 27-9-11	Interview with Friends of Nature staff 3, Beijing, 27 September 2011
NGO FON3 25-10-12	Interview with Friends of Nature staff 3, Beijing, 25 October 2012.
NGO GAIA 23-6-13	Interview with GAIA staff, Shanghai, 23 June 2013.
NGO GAIA 4-7-13	Interviews with GAIA staff, Guangzhou, 4 July 2013.
NGO GCB 5-10-12	Interview with Green Camel Bell staff, Lanzhou, 5 October 2012.
NGO GreenAnhui 27-9-12	Interview with Green Anhui staff 2, Beijing, 27 September 2012.
NGO GreenAnhui 23-10-12	Interview with Green Anhui staff, Hefei, 22 October 2012.
NGO GVB 18-10-11	Interview with Global Village of Beijing staff, Beijing, 18 October 2011.
NGO IPE 13-10-11	Interview with Institute of Public and Environmental Affairs staff, Beijing, 18 October 2011.
NGO IPEN1 22-6-13	Interview with IPEN staff 1, Shanghai, 22 June 2013.
NGO IPEN2 23-6-13	Interview with IPEN staff 2, Shanghai, 23 June 2013.
NGO NU1 11-10-11	Interview with Green Beagle/Nature University staff 1, Beijing, 10 October 2011.
NGO NU1 7-11-12	Interview with Green Beagle/Nature University staff 1, Beijing, 7 November 2012.
NGO NU2 28-9-11	Interview with Green Beagle/Nature University staff 2, Beijing, 28 September 2011.
NGO NU2 18-10-11	Interview with Green Beagle/Nature University staff 2, Beijing, 18 October 2011.
NGO NU2 14-10-12	Interview with Green Beagle/Nature University staff 2, Beijing, 14 October 2012.
NGO NU2 29-10-12	Interview with Green Beagle/Nature University staff 2, Beijing, 29 October 2012.
NGO NU2 8-5-13	Interview with Green Beagle/Nature University staff 2, Beijing, 8 May 2013.
NGO NU2 4-6-13	Interview with Green Beagle/Nature University staff 2, Beijing, 4 June 2013.

NGO NU2 26-3-14	Interview with Green Beagle/Nature University staff 2, Beijing, 26 March 2014.
NGO NU2 1-7-15	Interview with Green Beagle/Nature University staff 2, Beijing, 1 July 2015.
NGO NU3 25-7-13	Interview with Green Beagle/Nature University staff 3, Beijing, 25 July 2013.
NGO NU3 26-3-14	Interview with Green Beagle/Nature University staff 3, Beijing, 26 March 2014.
NGO NU4 8-11-12	Interview with Green Beagle/Nature University staff 4, Beijing, 8 November 2012.
NGO PacEnv 27-10-11	Interview with Pacific Environment staff, Beijing, 27 October 2011.
NGO PacEnv 21-10-12	Interview with Pacific Environment staff, Beijing, 21 October 2012.
NGO PacEnv 23-4-13	Interview with Pacific Environment staff, Beijing, 23 April 2013.
NGO SEE 14-10-11	Interview with Society of Entrepreneurs and Ecology Foundation staff, Beijing, 14 October 2011.
NGO WEC 23-10-12	Interview with Wuhu Ecology Center staff, Hefei, 23 October 2012.
NGO WEC 23-4-13	Interview with Wuhu Ecology Center staff, Beijing, 23 April 2013.
Nie 7-6-13	Interview with professor Nie Yongfeng, School of Environment, Tsinghua University, and central expert advocating waste incineration in China, Beijing, 7 June 2013.
PGY1 4-11-12	Interview with Panguanying village leader 1, Panguanying village, 4 November 2012.
PGY1 5-11-12	Interview with Panguanying village leader 1, Panguanying village, 5 November 2012.
PGY1 27-7-13	Interview with Panguanying village leader 1, Panguanying village, 27 July 2013.
PGY2 4-11-12	Interview with Panguanying village leader 2, Panguanying village, 4 November 2012.

PGY2 5-11-12	Interview with Panguanying village leader 2, Panguanying village, 5 November 2012.
PGY2 27-7-13	Interview with Panguanying village leader 2, Panguanying village, 27 July 2013.
PGY2 28-7-13	Interview with Panguanying village leader 2, Panguanying village, 28 July 2013.
PGY3 28-7-13	Interview with Panguanying village leader 3, Panguanying village, 28 July 2013.
PY1 1-7-13/NGO ECO1	Interview with Panyu campaign leader and founder of Eco Canton, Guangzhou, 1 July 2013.
PY1 10-7-13/NGO ECO1	Interview with Panyu campaign leader and founder of Eco Canton, Guangzhou, 10 July 2013.
SH official 23-5-13	Interview with Shanghai official with responsibilities in waste management, Shanghai, 23 May 2013.
SJ1 18-5-13	Interview with Songjiang campaigner 1, Shanghai, 18 May 2013.
SJ1 25-5-13	Interview with Songjiang campaigner 1, Shanghai, 25 May 2013.
SJ2 20-5-13	Interview with Songjiang campaigner 2, Shanghai, 20 May 2013.
SJ2 23-5-13	Interview with Songjiang campaigner 2, Shanghai, 23 May 2013.
SJ3 15-7-13	Interview with Songjiang campaigner 3, Shanghai, 15 July 2013.
SJ4 17-5-13	Interview with Songjiang campaigner 4, Shanghai, 17 May 2013.
SJ4 20-5-13	Interview with Songjiang campaigner 4, Shanghai, 20 May 2013.
TZ 23-6-13	Interview with villager from Taizhou city, Shanghai, 23 July 2013.
WH 4-7-15	Interview with campaigners of Wuhan city's Yongfeng township, Beijing, 4 July 2015.
WX 6-5-13	Interview with Huangtutang village leader, Beijing, 16 May 2013.

Xia 6-11-12	Interview with lawyer Xia, Beijing, 6 November 2012.
Xia 30-7-13	Interview with lawyer Xia, Beijing, 30 July 2013.
Xie 22-6-13	Interview with Xie Yong, main plaintiff in the Hai'an county incinerator case, Shanghai, 22 June 2013.
YQ 15-5-13	Interviews with villagers living next to the Shanghai Pudong Yuqiao MSWI, Shanghai, 15 May 2013.
Zhao 8-11-12	Interview with professor Zhao Zhangyuan, retired researcher from the Chinese Academy of Environmental Sciences and central expert opposing waste incineration in China, Beijing, 8 November 2012.

Appendix II List of cited documents

Please note: in the text, the title of each document is prefaced with 'D_' (i.e., D_ASW1) to indicate the kind of source.

ASW1	Research report entitled '中国城市环境的生死抉择 – 垃圾焚烧政策与公众意愿' [*Zhongguo chengshi huanjing de shengsi jueze – laji fenshao zhengce yu gongzhong yijian*; The life and death decision of China's urban environment – Waste incineration policies and public will], compiled by the Aobei residents' Beijing Aobei volunteers research small group, 77 pages, version 4.0, 31 January 2010.
ASW2	Aobei residents' PowerPoint presentation summarizing the situation and their claims against the planned Asuwei incinerator project, entitled '关于阿苏卫垃圾焚烧' [*Guanyu Asuwei laji fenshao;* About Asuwei waste incineration], no date.
ASW3	PowerPoint presentation by the Beijing Aobei voluntaries research small group entitled '中国城市环境的生死抉择 – 大城市大规模兴建垃圾焚烧厂隐患巨大' [*Zhongguo chengshi huanjing de shengsi jueze – da*

chengshi da guimo xingjian laji fenshao guang yinhuan juda; The life and death decision of China's urban environment – The gigantic hidden danger of the large-scale construction of waste incinerators by China's major cities], 3 September 2009.

ASW4 Official notice by construction unit Beijing Huayuan Huizhong Co. Ltd., announcing the planned Beijing Asuwei Municipal Waste Incineration Electricity-Generating Plant construction project, posted on Xiaotangshan town government public notice board, 28 July 2009.

ASW5 Invitation from Phoenix Television for the recording of a program entitled '建垃圾焚烧发电厂在中国是否可行?' [*Jian laji fenshao fadian chang zai Zhongguo shifou kexing?*; Is the construction of waste-to-energy plants feasible in China?] as part of the program '一虎一席谈' [*Yi hu yi xi tan;* Tiger Talk], to be conducted in Beijing on 9 December 2009.

ASW6 Nature University staff's detailed report about the Asuwei case from the villagers' perspective entitled '阿苏卫垃圾焚烧厂建设不能成为搬迁的筹码' [*Asuwei laji fenshao chang jianshe bu neng chengwei banqian de chouma;* The construction of the Asuwei incinerator must not become an impediment for relocation], 2015, no detailed date given.

ASW7 Letter by Nature University staff to the Asuwei incinerator's operating unit Beijing Huayuan Huizhong Environmental Technology Corporation and EIA unit Sinoma Geological Engineering Exploration Academy Limited, requesting a public hearing about the project's EIA publication on 15 December 2014, no date given.

ASW8 Aobei resident representative's detailed protocol of public hearing for the renewed Asuwei incinerator EIA procedures, held in April 2015, no detailed date given.

ASW9 Aobei resident representative's opinions about the above public hearing (D_ASW8), no date given.

ASW10 Open letter to the MEP by Friends of Nature and seven other environmental organizations, including Green Beagle/Nature University, requesting the withdrawal of the EIA qualifications of the Asuwei project's EIA unit Sinoma Geological Engineering, no date given.

ASW11 Detailed list of problems in the Asuwei project EIA, compiled by Green Beagle/Nature University to accompany the above open letter to the MEP (D_ASW10).

ASW12 Aobei resident representative's administrative redress application to the MEP, requesting that approval of the Asuwei project EIA given by the Beijing EPB on 28 April 2015 be revoked, submitted on 23 June 2015.

ASW13 Aobei resident representative's response letter to the MEP's rejection of the residents' administrative redress (D_ASW12), 9 September 2015.

DGC1 Document entitled '中国垃圾焚烧厂违规事件列表' [*Zhongguo laji fenshao chang weigui shijian liebiao*; List of incidents of violations against regulations by Chinese incinerators] submitted to the MEP on 6 July 2011 by the five Chinese environmental organizations Green Beagle Environment Institute, Beijing Global Village Environmental Education Center, EnviroFriends, Institute of Public and Environmental Affairs, and Green Earth Volunteers.

DGC2 Press release published by Green Beagle regarding the suspected fraud practiced by the EIA unit CAMS in the environmental impact assessment procedures for the Dagong incinerator, in the context of the organization's broader campaign against breaches of the EIA Law, 17 June 2011.

DGC3 Letter of the MEP North China Environmental Protection Supervision Center in response to the environmental organizations' letter (D_DGC20), 15 August 2011.

DGC4 Official relocation notice posted in Dagong village announcing the relocation period of 20 July to 18 August 2013, to be implemented by Beijing Lühai, dated 18 July 2013.

DGC5 Letter by Green Beagle sent to the Haidian district government and construction unit Beijing Lühai, 27 May 2011.

DGC6 Letter by Green Beagle sent to Beijing EPB, the Haidian district government, construction unit Beijing Lühai, and the EIA unit CAMS, 14 June 2011.

DGC7 Letter by Green Beagle sent to the Haidian district government, the construction unit Beijing Lühai, and the Haidian District Municipal Administration Commission, 17 June 2011.

DGC8 Joint letter to the MEP by five Beijing-based environmental organizations under the lead of Green Beagle, requesting that CAMS' EIA qualifications be revoked, 20 June 2011.

DGC9 Draft letter to the MEP and Beijing EPB, written by Green Beagle for the residents of the Western Hills Fengjingyuan Jinyu Relocation Buildings and the nearby military site, to request that the EIA report of the Dagong incinerator project be revoked, 24 June 2011.

DGC10 Draft letter to the Haidian district government, construction unit Beijing Lühai, and the Haidian District Municipal Administration Commission, written by Green Beagle for the residents of the Jinyu Relocation Buildings and the nearby military site, requesting that the Dagong project's EIA report be revoked, no date given.

DGC11 Draft letter to the Beijing EPB, written by Green Beagle on behalf of all affected communities, requesting a public hearing on the Sujiatuo project's EIA report, 25 June 2011.

DGC12 Response letter of Beijing EPB to the letter from Green Beagle (D_DGC6), 27 June 2011.

DGC13 Meeting protocol for the public lecture on the Dagong incinerator organized by Green Beagle in Beijing, 23 November 2010.

DGC14 Opinion statement on the Dagong incinerator project by Friends of Nature entitled '自然之友关于《北京市海淀区循环经济产业园再生能源发电厂工程环境影响报告书简本》的意见' [*Ziran zhi you guanyu 'Beijing shi Haidian qu xunhuan jingji chanye yuan zaisheng nengyuan fadian gongcheng huanjing yingxiang baogao shu jianben' de jijian;* Opinions by Friends of Nature on the 'Abridged EIA report on the Beijing City Haidian District Circular Economy Industrial Park Renewable Energy Power Plant Project'], 25 May, 2011.

DGC15 Opinion statement on the Dagong incinerator project by Zhao Zhangyuan entitled '对'苏家坨垃圾焚烧厂项目环评信息公告'的看法' [*Dui 'Sujiatuo laji fenshao chang xiangmu huanping xinxi gonggao' de kanfa;* Standpoint about the 'Public announcement of the Sujiatuo incineration plant project'], prepared for a public lecture in Beijing, dated 22 November 2010.

DGC16 PowerPoint presentation of Mao Da entitled '对苏家坨垃圾焚化炉项目的意见' [*Dui Sujiatuo laji fenhualu xiangmu de yijian;* Opinions about the Sujiatuo refuse incinerator project], presented during a public lecture in Beijing, 23 November 2010.

DGC17 PowerPoint presentation of Zhao Zhangyuan entitled '对'苏家坨垃圾焚烧厂项目环评信息公告' 的粗浅看法' [*Dui 'Sujiatuo laji fenshao chang xiangmu huanping xinxi gonggao' de cuqian kanfa;* Preliminary standpoint about the 'Public announcement of the Sujiatuo incineration plant project'], given at a public lecture in Beijing, 23 November 2010.

DGC18 Comments on the Dagong incinerator project submitted by Green Beagle to the Haidian district government, construction unit Beijing Lühai Environmental Protection Co. Ltd., and the EIA unit Chinese Academy of Meteorological Sciences as part of the Sujiatuo project's first public opinion solicitation phase, 28 November 2010.

DGC19 PowerPoint presentation by Green Beagle staff introducing the Dagong case, prepared for the public lecture on the Dagong case in July 2012, no detailed date given.

DGC20 Joint letter to the MEP North China Environmental Protection Supervision Center by five Beijing-based environmental organizations under the lead of Green Beagle, reporting problems with the Dagong incinerator project's construction and EIA units and requesting an investigation into the matter, 6 July 2011.

DGC21 Case summary written by Green Beagle staff for their symposium on environmental impact assessments in general and the Dagong case in particular, held in Beijing on 23 June 2011.

DGC22 Information disclosure request by Green Beagle staff with the Beijing EPB, demanding the release of the full Dagong incinerator EIA report and public participation questionnaires, 14 June 2011.

DGC23 Joint letter to the Ministry of Housing and Urban-Rural Development by five Beijing-based environmental organizations under the lead of Green Beagle, requesting the investigation of Beijing Lühai's qualification as the construction unit for the Dagong incinerator project, 22 June 2011.

DGC24 Information disclosure request by Green Beagle staff to the Beijing Municipal Administration Commission, requesting whether the construction unit Beijing Lühai has the mandatory license for municipal solid waste management, 24 June 2011.

DGC25 Information disclosure request by Green Beagle staff to the Beijing EPB, requesting the disclosure of the Dagong project's EIA report and official approval document, 30 June 2011.

DGC26 Online information disclosure request by Green Beagle staff to the Beijing EPB, requesting the disclosure of the Dagong project's EIA report and official approval document, 30 June 2011.

DGC27 Beijing EPB's response letter to the information disclosure request by Green Beagle staff, 3 July 2011.

DGC28 Information disclosure request from Green Beagle staff to the Beijing Municipal Development and Reform Commission, requesting whether the commission had officially ratified the Dagong project and demanding the official ratification notice, 14 July 2011.

DGC29 Information disclosure request by Green Beagle staff to the Beijing Municipal Planning Commission, requesting the disclosure of the Dagong project's 'social stability and risk evaluation report' (社会稳定风险评价报告, *shehui wending fengxian pingjia baogao*), 16 August 2011.

DGC30 Draft of an information disclosure request by Jinyu Relocation Buildings resident to the Beijing Municipal Development and Reform, requesting whether the commission had officially ratified the Dagong project and demanding the disclosure of the official ratification notice, 12 July 2011.

DGC31 Announcement and program for a public lecture about the Dagong case organized by Green Beagle and Friends of Nature in Beijing on 14 July 2012.

FON1 'Zero Waste Alliance' progress report, written by Friends of Nature staff in February 2012.

GAIA1 PowerPoint presentation by GAIA staff entitled 'Climate Change and Waste: Zero Waste for Zero

Warming!', invited by Eco Canton, presented in Guangzhou on 6 July 2013.

GAT1 Information booklet about the Beijing Chaoyang Circular Economy Industrial Park entitled '行动在手，绿色在心' [*Xindong zaishou, lüse zaixin;* Action is in the hand, green is in the heart] and distributed by the facility's public relations office during a guided facility tour on 9 November 2012.

GAT2 Information booklet about the Beijing Gaoantun Waste-to-Energy Corporation (北京高安屯垃圾焚烧有限公司, *Beijing Gaoantun laji fenshao youxian gongsi*) entitled 'Environment from Our Hearts/ 环境从心开始' [*Huanjing cong xin kaishi*; Environment starts in the heart] and distributed by the facility's public relations office during a guided facility tour on 9 November 2012.

GAT3 Gaoantun residents' letter entitled '生命的复议' [*Shengming de fuyi*; Reconsideration about Life (residents' own translation)], which summarizes their reasons for asking for an administrative redress to accompany their application to SEPA (document not listed), no date.

GAT4 Gaoantun residents' open letter entitled '百姓的希望在哪里?' [*Baixing de xiwang zai nali*; Where is the common people's hope?], detailing the course of events related to the Gaoantun struggle until the end of September 2008, no date.

GAT5 Reply by the Beijing Municipal Administration Commission regarding Gaoantun resident's information disclosure application (document not listed), 25 December 2008.

GAT6 Reply by Beijing EPB to Gaoantun resident's information disclosure request (document not listed), 23 January 2009.

GAT7 Judgement by Beijing City Chaoyang District People's Court (北京市朝阳区人民法院, *Beijing shi Chaoyang qu renmin fayuan*) regarding Gaoantun resident's lawsuit against the Gaoantun MSWI operator Beijing Jinzhou Anjie Waste Disposal Company (北京金州安洁废物处理有限公司, *Beijing jinzhou anjie chuli youxian gongsi*), 13 April 2010.

GAT8 Administrative lawsuit application by Gaoantun resident to the Haidian District People's Court (北京市海淀区人民法院, *Beijing shi Haidian qu renmin fayuan*) against the Beijing EPB's rejection of the resident's information disclosure request submitted on 7 January 2010, March 2010, no exact date given.

GAT9 Judgement by Beijing Haidian District People's Court regarding the Gaoantun resident's lawsuit application (D_GAT8) against the Beijing EPB's rejection of the resident's information disclosure request, 9 April 2010.

GAT10 Reply by Beijing EPB to the Gaoantun resident's information disclosure request (document not listed), 1 October 2011.

GAT11 Meeting protocol for the internal meeting between Liulitun and Gaoantun residents, as well as professor Zhao Zhangyuan, lawyer Wu, and several NGO members, no date given.

GAT12 Public announcement of an open lecture about municipal solid waste treatment strategies in Beijing organized by Green Beagle together with Global Village of Beijing and Friends of Nature, which featured Zhao Zhangyuan and Gaoantun residents as key speakers, 12 August 2009.

GAT13 Letter by Green Beagle staff to Beijing EPB, submitting comments on the planned second phase of the Gaoantun MSWI, 31 October 2011.

IPEN1 PowerPoint presentation by IPEN staff entitled 'Stockholm Convention Obligations for Wastes and UPOPs', given at the 'Forum on Waste Management Practices and Policies: Building Civil Society Capacity and Participation' organized by Nature University, Shanghai, 22 June 2013.

IPEN2 PowerPoint presentation by IPEN staff entitled 'Burning Issues', given at the 'Symposium on the Public Supervision of Municipal Solid Waste Incinerators' organized by Nature University, Shanghai, 23 June 2013.

LLT1 Liulitun residents' opinion booklet in opposition to the construction of a waste incinerator in Liulitun, entitled '反对在六里屯建垃圾焚烧厂意见书' [*Fandui zai Liulitun jian laji fenshao guang yijian shu;* Opinion booklet in opposition to the construction of a waste incinerator in Liulitun], 31 March 2009.

NG1 Open letter to KfW signed by Friends of Nature, Nature University, and 16 other Chinese environmental organizations, as well as numerous individuals, including Zhao Zhangyuan, 8 August 2012.

NG2 Press release regarding the organizations' open letter to KfW (D_NG1), 8 August 2012.

NG3 Meeting protocol for the meeting between Nature University and KfW staff in Beijing on 4 September 2012.

NG4 Letter to KfW written by Nature University staff regarding the meeting with KfW staff (D_NG3), 7 September 2012.

NG5 Information disclosure request demanding the publication of the Nangong incinerator EIA report submitted by Nature University staff to the MEP, 7 September 2012.

NG6 Information disclosure request demanding the publication of the Nangong incinerator EIA report submitted by Nature University staff to the Beijing EPB, 24 October 2012.

NG7 Information disclosure request demanding the publication of the Nangong incinerator EIA report submitted by Nature University staff to BMAC, 24 October 2012.

NU1 Internal notes by Nature University staff on the three cases of 'homeowner' resistance against incinerator projects in Beijing (Liulitun, Gaoantun, and Asuwei), 2013, no specific date given.

NU2 Compilation of documents specifically compiled by Green Beagle staff to assist affected communities.

PGY1 Project proposal (项目建议书, *xiangmu jianyi shu*) for the planned Qinhuangdao West WtE BOT project, written by Zhejiang Weiming, December 2008.

PGY2 Official statement by the Qinhuangdao City Management Bureau (秦皇岛市城市管理局, *Qinhuangdao shi chengshi guanli ju*) on the distance between the planned Qinhuangdao West WtE and various locations including Panguanying village and the village leaders' houses, based on an inspection trip by bureau staff and village head Qiao, 9 September 2010.

PGY3 Letter by the three village leaders Pan to the MEP to report about the illegal activities by the EIA unit CAMS regarding the environmental impact assessment procedures for the Qinhuangdao West WtE, 8 March 2011.

PGY4 Report by Zhao Zhangyuan regarding the flaws in the Qinhuangdao West WtE MWSI EIA report, no date.

PGY5 Official list by the Panguanying village committee of ten Panguanying villagers who agreed to sell their land for the construction of the MSWI for 34,400 RMB per Mu, 6 May 2009.

PGY6 *Xinfang* letter to the Panguanying village, Liushouying town, and Funing county governments signed by numerous Panguanying villagers to report about the land requisitioning for the MWSI and inquire about the lawfulness of this land use, 2009, exact date illegible due to bad quality of scan.

PGY7 Letter by the Funing County Land and Resources Bureau (阜宁县国土资源局, *Funing xian guotu ziyuan ju*) to the Qinhuangdao City Management Bureau stating its agreement with the project's land use and certifying that the land category of the construction land is 'garden' and not 'basic farmland', 10 April 2009.

PGY8 *Xinfang* letter by numerous Panguanying villagers (signatures and fingerprints attached) inquiring about the lawfulness of the land requisitioning and the project details, 16 April 2009.

PGY9 *Xinfang* request by farmer Pan to the Funing County Land and Resources Bureau reporting the procedural flaws in the land requisitioning, including the illegal change of the land category from 'basic farmland' to 'garden', and earlier misconduct regarding the village's collective land perpetrated by the village committee, 15 July 2009.

PGY10 Verdict by the MEP regarding the villagers' administrative redress request that the project approval given by the Hebei EPB in May 2009 be revoked, 19 December 2010.

PGY11 Clean Development Mechanism Project Design Document Form with detailed project description of the Qinhuangdao West WtE, for the project's application for CDM status, 15 June 2011.

PGY12	Villagers' administrative redress application to the Qinhuangdao City People's Government (秦皇岛市人民政府, *Qinhuangdao shi renmin zhengfu*), 19 November 2010.
PGY13	Supplementary comments by lawyer Wu to the MEP regarding the villagers' administrative redress request that the project approval given by the Hebei EPB in May 2009 be revoked, 15 December 2010.
PGY14	Collected evidence lists given as attachments to the villagers' several administrative redresses and lawsuits. The collected evidence detailed in these lists is on file with the author.
PGY15	Letter by numerous Panguanying villagers (with signatures and fingerprints attached) to the Funing County Land and Resources Bureau requesting a thorough investigation of the situation, 29 September 2009.
PGY16	Letter by the Qinhuangdao City Management Bureau to the Hebei Province Environmental Protection Bureau, stating that all project-related procedures are in line with state laws and regulations, including the project land's categorization as 'garden' land and the project's inclusion in the relevant Qinhuangdao city plans, 15 May 2009.
PGY17	Letter by the Qinhuangdao City Land and Resources Bureau to the Qinhuangdao City Management Bureau stating its agreement with the project's land use, 15 May 2009.
PGY18	Letter by the Qinhuangdao City Planning Bureau (秦皇岛市规划局, *Qinhuangdao shi guihua ju*) to the Qinhuangdao City Management Bureau stating its agreement with the project siting decision, 6 May 2009.
PGY19	Letter by the Hebei Province Department of Land and Resources to the Qinhuangdao City Land and Resources Bureau and the Qinhuangdao City Management Bureau stating its agreement with the project land use, 24 August 2009.
PGY21	Official reply by the Funing County Land and Resources Bureau to the villagers' *xinfang* request from 15 July 2009 (D_PGY9), claiming that the project land's category is 'garden' land and ordering a temporary construction halt based on the other procedural flaws pointed out by the villagers, 15 September 2009.

PGY22 Panguanying villagers' opinion booklet, entitled '反对在潘官营村建垃圾焚烧发电广意见书' [*Fandui zai Panguanying cun jian laji fenshao fadian guang yijian shu;* Opinion booklet in opposition to the construction of a WtE plant in Panguanying village], with attached signature collection and statements and stamps from surrounding village committees, 7 June 2010.

PGY25 Villagers' administrative lawsuit application to the Ministry of Land and Resources, requesting that the Ministry order the Hebei Province Department of Land and Resources to accept their earlier administrative redress, August 2011, exact date illegible due to bad quality of scan.

PGY26 Villagers' administrative redress application with the Hebei Province People's Government (河北省人民政府, *Hebei sheng renmin zhengfu*), requesting that the government revoke its project land use approval, 19 November 2010.

PGY27 Contract between Panguanying villagers and lawyer Xia, entrusting the lawyer as their legal representative, 19 August 2010.

PGY28 Villagers' administrative redress application with the MEP, requesting that the project approval given by the Hebei EPB in May 2009 be revoked, 19 August 2010.

PGY29 Attachment to the villagers' administrative redress application with the MEP (D_PGY28) detailing their charges and the reasoning for the lawsuit, 19 August 2010.

PGY30 Notification by the MEP demanding that the villagers adjust their administrative redress (D_PGY28), 30 August 2010.

PGY31 Letter by Zhejiang Weiming to the MEP, stating that an environmental impact assessment had been conducted by the EIA unit CAMS as required by state law, 7 September 2010.

PGY32 Letter by the Hebei Province Environmental Protection Bureau to the MEP, stating that the project was in line with the relevant laws and regulations and that an environmental impact assessment had been conducted by the EIA unit CAMS as required by state laws, 13 September 2010.

PGY33 Letter by the Ministry of Land and Resources to the villagers, informing them that the Ministry declined their administrative lawsuit application (D_PGY25), 12 September 2010.

PGY34 Notice by the Hebei Province People's Government to the villagers, informing them that the government had declined their administrative redress (D_PGY26), 27 December 2010.

PGY36 Compilation of documents provided to the villagers by Green Beagle staff, encompassing articles and blog posts on incineration and its harms by different experts including Zhao Zhangyuan and the Green Beagle staff member Mao Da; media articles and blog posts about other cases including Beijing Gaoantun, Beijing Asuwei, Guangzhou Likeng, Guangzhou Panyu, Nanjing Jiangbei, and Shenzhen Longgang; as well as the relevant environmental laws and regulations.

PGY37 Contract between Panguanying villagers and the lawyer Xia, entrusting the lawyer as their legal representative for their lawsuit against the MEP's decision to uphold the Hebei EPB's project approval (D_PGY41), 19 August 2010.

PGY38 Contract between Panguanying villager representatives and the three Pans as well as two other villagers, entrusting the latter as the villagers' representatives in the lawsuit against the MEP (D_PGY41), no date.

PGY39 Notes of the villager representative meeting stating that the villager representatives who were present agree that the construction of the plant should be halted and that they support the Pans' legal actions, 2 December 2010.

PGY40 Case report by Green Beagle and GAIA staff about the Panguanying events between April 2009 and November 2012, no date.

PGY41 Court file including the villagers' administrative lawsuit against the MEP's decision to uphold the Hebei EPB's EIA approval with the Shijiazhuang City Qiaoxi District People's Court including the lawsuit application, prior legal documents, e.g., concerning their earlier administrative redresses, collected evidence, and the villagers' extensive report about the situation, 4 January 4 2011.

PGY42 Verdict of the Shijiazhuang City Qiaoxi District People's Court regarding the villagers' administrative lawsuit (D_PGY41), 8 June 2011.

PGY43 Original EIA public participation questionnaires, no date.

PGY44 Meeting notes by the village committee from the villager representative meeting on 23 March 2009, claiming that all 27 villager representatives who were present had agreed with the construction of the incinerator, including the attached signatures of villager representatives, 23 March 2009.

PGY45 Testimony of villager representatives that they had opposed the construction of the incinerator during the 2009 villager representative meeting, that the meeting protocol was falsified, and that the signatures were given to acknowledge their receipt of a participation fee of 10 RMB, no date.

PGY46 Handwritten list by the village leaders detailing all mistakes related to the EIA public participation questionnaires that had been discovered, no date.

PGY47 Original EIA public participation questionnaires with the statements, signatures, and fingerprints of the respective villagers, testifying that they had never before seen the questionnaire, do not know who signed it in their name, and do not agree with the construction of the incinerator, no date.

PGY48 Letter by two of the Pans to the MEP Pollution Prevention Division, reporting in detail about the discovered EIA flaws and malpractice conducted by the EIA unit CAMS and construction unit Zhejiang Weiming, as well as about the details of their legal actions and its consequences, and asking the division to decline Zhejiang Weiming's application for the MEP's environmental ratification to enter the stock market, 14 September 2011.

PGY49 Villagers' administrative lawsuit application to the Beijing Municipality First Intermediate People's Court (北京市第一中级人民法院, *Beijing shi diyi zhongji renmin fayuan*) against the MEP's decision to ratify Zhejiang Weiming's market entry, 18 May 2012.

PGY50 Notification by the Shijiazhuang City Qiaoxi District People's Court that a hearing in the villagers' case (D_PGY41) would take part on 30 May 2011, sent to the villagers on 19 May 2011.

PGY51 Critical comment on the Qinhuangdao West WtE project by Friends of Nature staff, submitted to the Clean Development Mechanism department on 25 July 2011.

PGY52 Critical comment on the Qinhuangdao West WtE project by Wuhu Ecology Center staff, submitted to the Clean Development Mechanism department on 22 December 2011.

PGY53 Print copy of MEP website announcing an environmental evaluation phase for Zhejiang Weiming, 14 September 2009.

PGY54 Notice by Shijiazhuang City Intermediate People's Court (石家庄市中级人民法院, *Shijiazhuang shi zhongji renmin fayuan*) informing the villagers that their lawsuit had been accepted, 4 August 2011.

PGY55 Letter by the Hebei Province Government to the Shijiazhuang City Intermediate People's Court in representation of the Hebei Province Department of Land and Resources, stating that all procedures regarding the project siting and project land requisition had been in accordance with the law, 2 September 2011.

PGY56 Verdict by the Shijiazhuang City Intermediate People's Court regarding the villagers' lawsuit, 3 November 2011.

PGY57 Villagers' administrative redress application to the China Securities Regulatory Commission (中国证券监督管理委员会, *Zhongguo zhengquan jiandu guanli weiyuanhui*) demanding that the MEP's decision to ratify Zhejiang Weiming's market entry be revoked, 18 May 2012.

PGY59 Verdict by the Beijing Municipality First Intermediate People's Court regarding the villagers' administrative lawsuit (D_PGY49), 18 June 2012.

PGY60 Villagers' administrative lawsuit application with the Beijing Municipality Higher People's Court against the Beijing Municipality First Intermediate People's Court's verdict in their administrative lawsuit (D_PGY59), 28 June 2012.

PGY61 Verdict by the Beijing Municipality Higher People's Court regarding the villager's administrative lawsuit (D_PGY60), 12 September 2012.

PGY62 Villagers' *xinfang* letter to the Liushouying town government reporting about the election problems, 14 January 2013.

PGY63 Response letter from the Liushouying town government to the villagers regarding their report about the election problems (D_PGY62), 21 March 2013.

WEC1 Program and related press release of a symposium and press conference organized by Wuhu Ecology Center regarding their information disclosure requests submitted to 122 Chinese MSWIs, held in Beijing on 23 April 2013.

WEC2 Wuhu Ecology Center booklet entitled '垃圾焚烧的真相 [*Laji fenshao zhen xiang*; The real situation of waste incineration] – Stop Burning', which provides information about incineration and its harms, and which was distributed to the public and particularly to the affected communities.

XYLC1 Comprehensive case summary containing original documents, medical reports, photographs, expert statements, and news reports about the Hai'an case, compiled and provided by Xie Yong, no date given.

Appendix III Basic information on collected cases

Cluster	Province/ Municipality	City	Location	Setting	Main Activists	Onset	Peak Year	Major Events (protest or sit-in unless otherwise specified)	Important Dates (*trigger of contention in italics*)	Main Strategies	Project status (at time of onset)	Outcome (at time of research)
BJ	Beijing	Beijing	Liulitun area, Haidian district	Urban	Homeowners	2006	2007	05/06/2007	Landfill operation begin: 1999 First MSWI plans: 2005 *Disclosed by media: 11/2006*	Long-term campaign, protest	Planned	Plant moved to Dagong village
BJ	Beijing	Beijing	Gaoantun area, Chaoyang district	Urban	Homeowners	2007	2008	08/2008	Landfill operation begin: 1999 First MSWI plans: 2004 *Trial run: 08/2008 (during Olympics)*	Long-term campaign, protests, lawsuits	Running	Unsuccessful
BJ	Beijing	Beijing	Asuwei area, Changping district	Urban (rural)	Homeowners (villagers)	2009	2009	08/2009 04/09/2009	Landfill operation begin: 2004 *MSWI announcement: 08/2009 (resident discovers public notice)*	Long-term campaign, Protest	Planned	Temporarily halted, but construction resumed

Cluster	Province/ Municipality	City	Location	Setting	Main Activists	Onset	Peak Year	Major Events (protest or sit-in unless otherwise specified)	Important Dates (*trigger of contention in italics*)	Main Strategies	Project status (at time of onset)	Outcome (at time of research)
BJ	Beijing	Beijing	Nangong village, Daxing district	NA	Environmental organizations	2010	2010		*MSWI announcement*	Petitioning, letter-writing	Planned	Unsuccessful
BJ	Beijing	Beijing	Dagong village, Sujiatuo town, Haidian district	rural	villagers	2010	2012	06/2012	*MSWI announcement: 16/11/2010*	Protest, sit-in	Under construction	Unsuccessful
BJ	Hebei	Shijiazhuang	Dongliangxiang village and Bishui Lanwan residential area, Qiaoxi district	Rural/ urban	Villagers, homeowners	04/ 2009	2009	05/2009	*Operation begin: 05/2009*	Petitions, protest, road blockade	Under construction (near completion)	Pending (plant closed temporarily in 2009 due to proximity to South-North Water Diversion Project)
BJ	Hebei	Qinhuangdao	Panguanying village, Liushouying town, Funing county	Rural	Villagers	04/ 2009	2009	2009 (petitions) 01/2011 (major lawsuit) 11-12/2012 (village election)	*Construction begin (farmers notice land encircling): 04/2009*	Lawsuits and petitions, activist running for village head in village elections	Under construction	Project abandoned

Cluster	Province/ Municipality	City	Location	Setting	Main Activists	Onset	Peak Year	Major Events (protest or sit-in unless otherwise specified)	Important Dates (*trigger of contention in italics*)	Main Strategies	Project status (at time of onset)	Outcome (at time of research)
SH	Shanghai	Shanghai	Jiading district	Urban	Homeowners	2005	2009	09/2009	*Operation begin: 2005→complaints about odor* *Expansion plans revealed: 2009*	Protest	Expansion planned	Expansion stalled
SH	Shanghai	Shanghai	Songjiang district	Urban	Homeowners	2012	2012	27/05/2012 02/06/2012 05/06/2012 14/06/2012	Complaints about landfill since 2005 *MSWI announcement: 18/05/2012*	Protests	Planned	Plant moved within district
SH	Shanghai	Shanghai	Caolu town, Pudong district	Urban	University staff and homeowners	2012	2012		*MSWI announcement*	Long-term campaign, petitions	Planned	Pending (EIA ongoing)
SH	Shanghai	Shanghai	Fengqiao village, Fengxian district	Peri-urban	Individual residents of Fengxian	2013	2013		*Villager discovers MSWI plans in 2013*	Failed attempted mobilization of residents	Planned	Unsuccessful
SH	Jiangsu	Nanjing	Huangyao village , Taishan town, Pukou district	Urban	Homeowners from five residential areas in Pukou district	2006	2009	2006 14/05/2009	MSWI announcement: 10/2006 *End of EIA phase: 15/05/2009*	Protests, petitions, lawsuit	Planned	Project halted
SH	Jiangsu	Nanjing	Shilicun village, Pukou district	Rural	Villagers	2012	2012				Planned	

Cluster	Province/ Municipality	City	Location	Setting	Main Activists	Onset	Peak Year	Major Events (protest or sit-in unless otherwise specified)	Important Dates (*trigger of contention in italics*)	Main Strategies	Project status (at time of onset)	Outcome (at time of research)
SH	Jiangsu	Nantong	Hai'an county	Rural	Villager, various NGOs	10/ 2010	2010	2010 (litigation begin)	*2010: Doctors relate son's disease (cerebral palsy since birth) to pollution*	Individual lawsuits on various levels up to MEP	Running	Out-of-court agreement with MSWI operator, financial compensation
SH	Jiangsu	Suzhou	Pingwang township, Wujiang district	Urban	Residents of Pingwang township		2009	21-22/10/2009	*Trial run: 19/10/2009* *Site opening ceremony: 21/10/2009*	Petitions, protests	Under construction	Construction suspended
SH	Jiangsu	Wuxi	Huangtutang village, Donggang township	Rural	Villagers	2007	2011	13/04/2011	MSWI announcement: 2007 *Construction begin: 2011*	Blockade of construction site, protest	Under construction	Pending (construction halted, deadlock)
SH	Jiangsu	Taizhou		Urban	Residents	2013	2013		*MSWI announcement: 2013*	Planned action	Action planned	
SH	Zhejiang	Jiaxing	Jiaoshanmen village, Daqiao town	Rural	Villagers	2003	2008	08-12/11/2008	Operation begin: 06/2003→ start of complaints *2008 protest trigger unclear*	Protest, blockade of facility entrance	Running	Operation halted for several days, promises of new furnaces
SH	Zhejiang	Hangzhou	Binjiang district	Rural	Villagers	2000s	2013		*High cancer rates in village*	Long-term campaign, protests	Running	

Cluster	Province/ Municipality	City	Location	Setting	Main Activists	Onset	Peak Year	Major Events (protest or sit-in unless otherwise specified)	Important Dates (*trigger of contention in italics*)	Main Strategies	Project status (at time of onset)	Outcome (at time of research)
GZ	Guangdong	Dongguan	Luoma village, Zhangmutou town and Qingxi town (MSWI in between)	Rural/ urban	Villagers, homeowners	2010	2010	28/03/2010 18/05/2010	*MSWI announcement: early 2010*	Protests	Planned	
GZ	Guangdong	Dongguan	Xinwei, Dapaizhai and Henggang villages, Dalingshan town	Rural/ urban	Villagers, homeowners	2010	2012	21/08/2012	MSWI announcement: 2010 *2012 protest trigger unclear*	Petitions, protest	Planned	
GZ	Guangdong	Foshan	Nanhai district; Gaoming district; Xiqiao town	Urban	Residents of Nanhai and Gaoming districts and Xiqiao town	2009	2010	23-24/01/2010	*MSWI announcement (disclosed by media): 13/12/2009*	Long-term campaign, protests	Planned	Gaoming authorities promise to oppose plans
GZ	Guangdong	Guangzhou	Huijiang village, Dashi township, Panyu district	Urban/ peri-urban	Homeowners, villagers	2009	2009	23/11/2009 05/2011	*Announcement (revealed in newspaper): 09/2009* *Construction begin: 08/11/2009*	Long-term campaign, protest	Planned	MSWI moved within district

Cluster	Province/ Municipal-ity	City	Location	Set-ting	Main Activists	Onset	Peak Year	Major Events (protest or sit-in unless otherwise specified)	Important Dates (*trigger of contention in italics*)	Main Strategies	Project status (at time of onset)	Outcome (at time of research)
GZ	Guangdong	Guangzhou	Likeng plant, Yongxing village, Taihe township, Baiyun district	Peri-urban	Villagers	2005	2010	19-20/01/2010 25/01/2010	Operation begin: 2005 Announcement of second phase: 10/2007 Second phase EIA: 04/2009 *Visit by Panyu campaigners: 11/2009* *Explosion in MSWI: 07/01/2010*	Long-term complaints, protests	Expansion planned	Unsuccessful
GZ	Guangdong	Guangzhou	Dagang town and Shawan town, Panyu district	Peri-urban	Villagers of Shawan town, joined by Yongxing village residents	2011	2012	03-05/2012	*Announcement of potential new sites for Huijiang village MSWI: 2011, 2012*	Protest, residents of 4 potential sites mobilize (Shawan town, Dongchong town, Lanhe town)	MSWI site pending	Unsuccessful
GZ	Guangdong	Guangzhou and Qingyuan	Huadu district (Guangzhou); Fensui village, Shiling township (Qingyuan)	Urban	Urban residents from Guangzhou and Qingyuan	2009	2012	12/2009 23/05/2012 10/06/2012	*MSWI announcement: 09/2009* *Siting decision confirmed: 01/04/2012*	Protests	MSWI site pending	Plant moved within district

Cluster	Province/ Municipality	City	Location	Setting	Main Activists	Onset	Peak Year	Major Events (protest or sit-in unless otherwise specified)	Important Dates (*trigger of contention in italics*)	Main Strategies	Project status (at time of onset)	Outcome (at time of research)
GZ	Guangdong	Huizhou	Huiyang district	Rural/ urban	Villagers, homeowners	2013	2013	2013	*MSWI announcement: 2013*	Petitions, protests	Planned	Unsuccessful
GZ	Guangdong	Shenzhen	Qingshuihe district	Urban	Residents	1998	2006	01/2006	Operation begin: 1998 *2006 protest trigger unclear*	Long-term campaign, protest	Running	Plant moved
GZ	Guangdong	Shenzhen	Baoan district	Urban	Residents		2009	05/2009 12/2009	*Construction begin: 02/2009*	Protests	Under construction	
GZ	Guangdong	Shenzhen	Longgang district	Urban	Residents		2009	08/2009		Sit-in	Running (2 MSWIs), under construction (1 MSWI)	
GZ	Guangdong, Hong Kong	Shenzhen, Hong Kong	Shekou, Nanshan district (Shenzhen); Tsang Tsui, Tuen Mun (Hong Kong)	Rural/ urban	Shenzhen residents and villagers from Tuen Mun and	2008	2009	12/2008 04/2009	Operation begin: 2003 *Announcement of expansion: 2008* *Second phase EIA hearing: 04/2009*	Petitions, protests	Expansion planned	

Cluster	Province/ Municipal-ity	City	Location	Set-ting	Main Activists	Onset	Peak Year	Major Events (protest or sit-in unless otherwise specified)	Important Dates (*trigger of contention in italics*)	Main Strategies	Project status (at time of onset)	Outcome (at time of research)
GZ	Hong Kong	Hong Kong	Shek Kwu Chau and Cheung Chau islands	Rural (island)	Cheung Chau residents, staff of rehabilitation center on island	2012	2012	09/04/2012	*MSWI announce-ment: 2012*	Protest, lawsuit by four Cheung Chau residents	Planned	
GZ	Hong Kong	Hong Kong	Soko island, South Lantau			2012	2012		*MSWI announce-ment: 2012*		Planned	
Outlier	Hubei	Wuhan	Yongfeng village, Hanyang district	Urban	Residents	2013	2013	2013	*Significant occurrence of respiratory diseases*	Protests, peti-tions, lawsuits	Running	
Outlier	Guangxi	Guilin	Silian village, Lingtian township	Rural	Villagers	2010	2010	03/07/2010	MSWI announce-ment: 05/2009 *Construction begin: 04/2010*	Protest, mass petition	Planned	Project halted at the end of 2010; government seeking expert advice
Outlier	Guizhou	Guiyang	Wudang district	Urban	Mainly university staff		2010	2010				
Outlier	Sichuan	Chengdu	Luodai town, Longquanyi district	Rural	Villagers, en-vironmental organizations					Mobilization attempts by environmental organizations		Unsuccessful

Cluster	Province/ Municipality	City	Location	Setting	Main Activists	Onset	Peak Year	Major Events (protest or sit-in unless otherwise specified)	Important Dates (*trigger of contention in italics*)	Main Strategies	Project status (at time of onset)	Outcome (at time of research)
Outlier	Sichuan	Nanchong	Jialing district				2009	10/2009	*Celebration of won bid by operator: 21/10/2009*		Planned	
Outlier	Gansu	Lanzhou	Zhongpu district	Urban	Environmental organizations	2012	2012			Attempted mobilization	Planned	

Notes: The cases are sorted by clusters (BJ = Beijing, GZ = Guangzhou, SH = Shanghai) and within clusters peak year. Peak year refers to the first year with a major contention-related event (e.g. protest, environmental litigation, mass petition). The case outcome is based on the latest available information within the research period. "Unsuccessful" indicates that the contested MSWI (or second phase) commenced or continued operation. Where no outcome is specified, the case was still ongoing at the time of field research or no detailed information about the outcome was available.

Appendix IV Brief description of sample and central background cases

Cluster	Setting	Acronym	Location	**Brief Description** [Date: onset/peak year]
Sample Cases				
BJ	Urban	ASW	Asuwei area, Changping district, Beijing municipality	[2009/2009] Project plans in Asuwei landfill vicinity announced in 2009; site is villa area of wealthy homeowners with several old villages around; villagers' earlier complaints against landfill odors unsuccessful; incinerator plans discovered by villa owner in 2009; immediate mobilization, mainly via social media; two motorcades and major protest in front of Agricultural Exhibition Center in Beijing leading to some detentions; thereafter cooperation of main activist with government; development of lay expertise and writing of an extensive research report about harms of incineration; project temporarily halted; wealthy main leader founds own waste-NGO (cp. Panyu case) and comes to some popularity in Chinese media; cleavages between homeowners and villagers, who want to be relocated due to landfill pollution issues; road blockade by villagers about resettlement issues in 2013; project plans and related contention resumed in 2014.
BJ	Peri-urban	DG	Dagong village, Sujiatuo town, Haidian district, Beijing municipality	[2010/2012] New site of obstructed Liulitun incinerator; site is scenic area in the West mountains around Beijing; area characterized by diverse social groups with different interests; villagers around MSWI site see relocation as unavoidable, hence no anti-MSWI contention; close-by Dajue Temple and military base staff take action within their own *xitong*, but are soon faced with repercussions and scared off further action; residents of close-by relocation buildings take some action that soon ebbs down; very early NGO-engagement, who unsuccessfully try to mobilize and unify the local community activities; cleavages between the local communities and intermediaries; MSWI finished and villagers relocated in 2014 and 2015.

Cluster	Setting	Acronym	Location	**Brief Description** [Date: onset/peak year]
BJ	Rural	PGY	Panguanying village, Liushcuying town, Funing county, Hebei Province	[2009/2009] Villagers discover MSWI plans when local government starts encircling collective village land for the construction site; site is major agricultural base; affected farmers immediately mobilize and write unsuccessful petitions to various government departments after discovering severe flaws in the project process; large-scale mobilization in villages around the MSWI site after farmers obtain materials about incineration from a CCTV broadcast and the Liulitun campaign; after unsuccessful petitioning, villagers seek external help; with intermediary support, farmers win environmental litigation case and MSWI construction is halted; due to fears of continued construction, village leader runs for the position of village head in the local village elections in 2012; elections unsuccessful due to local corruption problems; case deadlocked.
SH	Urban	SJ	Songjiang District, Shanghai municipality	[2012/2012] MSWI plans announced in 2012 in existing landfill site; area is new residential district; unsuccessful prior complaints by homeowners about odor from landfill; large-scale protests in 2012, mainly organized via social media; project thus halted; new site announced in close by village area; attempted mobilization of villagers in new site by individual homeowners unsuccessful.
SH	Peri-urban	FQ	Fengciao village, Fengxian district, Shanghai municipality	[2013/2013] Project announced in early 2013; mobilization attempts by one villager who discovered the MSWI plans unsuccessful since villagers regard contention as futile; main activist seeks intermediary help, but no entry points for intermediaries without local mobilization; EIA process ongoing at time of research.
SH	Rural	HTT	Huangtutang village, Donggang township, Wuxi city, Jiangsu Province	[2007/2011] Project announced in 2007; villagers claim they were tricked into agreeing to MSWI by claims that an ancestral temple would be built at the site; in 2011 long-term camp at MSWI construction site after villagers discover MSWI plans; blockage leads to violent clash with public security forces, several arrests and injuries; project temporarily abandoned; intermediaries try to assist villagers, but no legal basis for lawsuit; case deadlocked.

Cluster	Setting	Acronym	Location	**Brief Description** [Date: onset/peak year]
GZ	urban	HD	Huadu district, Guangzhou municipality	[2009/2012] Project plans announced in 2009 at same time as Panyu MSWI; area is residential district with many homeowners; first mobilization in 2009, strongly influenced by ongoing Panyu protests; project temporarily halted; protest site confirmed in 2012; large-scale protests in 2009 and 2012 that fail to attract major media attention; site at border of Guangzhou municipality and Qingyuan city; activists actively engage Qingyuan government; new site selection process after 2012; according to campaigners, this outcome is mainly based on internal negotiations between the two governments; new site selection and renewed contention ongoing at time of research.
GZ	Peri-urban	LK	Likeng plant, Yongxing village, Taihe township, Baiyun district, Guangzhou municipality	[2005/2010] Operation begin in 2005 as first MSWI in Guangzhou in former landfill area; site is peri-urban village area; early contention by villagers against landfill pollution and since 2005 also against MSWI construction; very high cancer rates in the village that are attributed to the MSWI by the villagers; villagers' complaints and petitions unsuccessful; in 2009, Panyu campaigners visit Yongxing village to see example of running MSWI; Panyu campaigners help Yongxing villagers to mobilize further action; thereafter nation-wide media attention on the case as first MSWI-related cancer case; renewed contention after explosion in MSWI in 2012; second MSWI phase commenced operation in 2013; intermediary assistance, including an information disclosure lawsuit, unsuccessful at time of research; relocation of villagers planned and negotiations about compensation payments ongoing at time of research.
GZ	Rural	HY	Huiyang district, Huizhou city, Guangdong Province	[2013/2013] MSWI plans announced in 2013; site is village area with apartment buildings mainly used as weekend resort by Shenzhen residents; homeowners mobilize immediately after site announcement and also activate surrounding villagers; petitions and several protests that led to detentions and repercussions for main activists; homeowners seek intermediary assistance and consider legal action; contention dies down after renewed detentions, however; EIA ongoing at time of research.

Cluster	Setting	Acronym	Location	**Brief Description** [Date: onset/peak year]
Background Cases				
BJ	Urban	LLT	Liulitun neighborhood, Haidian district, Beijing municipality	[2006/2007] Project plans in Liulitun landfill vicinity announced in 2006; first major anti-incinerator campaign that can be regarded as spin-off movement due to major impact on many other cases; site is Beijing residential area; residents of Liulitun neighborhood mobilize long-term campaign; after unsuccessful petitions, legal action and a mass petition in front of the MEP in 2007; project is temporarily halted; renewed contention after new project plans are resumed in 2008; development of lay expertise and writing of an extensive "opinion booklet;" project again halted in 2009 and officially moved to Dagong village in 2011.
BJ	Urban	GAT	Gaoantun neighborhood, Chaoyang district, Beijing municipality	[2007/2008] Construction begin in 2005 in the vicinity of Gaoantun landfill as first MSWI in Beijing; site is newly developed residential area; mobilization after Liulitun mass petition in 2007, including unsuccessful legal action; several large-scale street protests in 2008 after trial run of MSWI during Olympic Games produces major stench; following repercussions for campaigners scare residents off further action; legal action continued by individual residents, among others a young woman who suffers from a respiratory disease she attributes to the incinerator; continued MSWI operation and died down action at time of research.
GZ	Urban	PY	Panyu district, Guangzhou municipality	[2009/2009] Project announcement in 2009; site is residential area; homeowners mobilize promptly, mainly via social media; large-scale street protests in late 2009 that receive major nation-wide media attention lead to negotiations with the Guangzhou city government; project halted in 2010 and site officially moved to rural area within Pany district; due to its major presence in the media who hail the case as positive example of successful negotiations between affected residents and the government, the case can be regarded as second central spin off-movement (with Liulitun); campaign leader founds own waste-related NGO (cp. Asuwei case).

Note: According to the sample matrix introduced in Chapter 1, the cases in this table are sorted by cluster region (BJ = Beijing, GZ = Guangzhou, SH = Shanghai) and setting (urban/peri-urban/rural).

Appendix V Case protocol template

Case	
Location	
1. Overview	
Abstract – *Status of case/outcomes* – *Setting/background* – *Prehistory* – *Brief activist information* – *Major framing/claims/aims* – *Identity* – *Government position/reactions/role* – *Main strategies* – *Vertical ties* – *Horizontal ties* – *Notes on Scale Shift*	**Cluster:** **Status:** **Setting:** **Prehistory:** **Activists:** **Framing/Claims:** **Identity:** **Government:** **Main strategies:** **Vertical ties:** **Horizontal ties:** – **General:** – **Other cluster cases:** – **Other cases:** – **Own broker role:** – **Collaborations:** – **International/transnational:** **Notes on scale shift:**
Duration (onset; ending)	
Status of case	
Evaluation of success *Including likely factors for success or non-success*	
Evaluation of sensitivity	
2. Who's Who (brief overview of engaged actors)	
Core activists	
Participants Brief characterization of engaged parties/ communities	
Government Core individuals Levels of government/institutions (including brief description of role, position, stakes)	
Proponents e.g. MSWI corporations, EIA units etc.	
Engaged other activists/cases (brief description role)	
Engaged intermediaries (brief description role)	
Others e.g. important local experts etc.	
3. MSWI Information	
Name of MSWI Including changes	

Locality	
Prior landfill/facilities	
Date of first proposal	
End date (if applicable) e.g. date cancelled, withdrawn, new siting decision, operation begin, resettlement etc.	
Status at time of onset of contention	
Status at time of field research	
Updates (since 09/2013)	
Important dates related to project First announcement EIA dates Planned construction begin Construction begin End date: Cancelled/built/… Etc.	
Distance of location to respective MSWI	
MSWI **investment and operation** e.g. local government, district government, corporation (including stakes and background)	
EIA unit	
Further MSWI information	
Resettlement plans/information	
4. Broader Context of Case	
Context information Relevant policies (e.g. Guangzhou waste policies), political decisions, level of project, stakes etc.	
Broader context E.g. one of five proposed MSWIs in Beijing, all announced on the same day or: new location of XY, or: part of three closely linked cases ABC, etc.	
Embedment in Cluster	
5. Setting	
Setting Information on setting, including size, population, socio-economic background, economic dependence, industries in area, etc.	
Prehistories/Background Information on prehistories and background relevant for understanding the case (e.g. experiences with contention, corruption, pollution, other facilities etc.; broader background e.g. stories of urbanization/industrialization etc.)	

6. Chronology/Narrative	
Activist Information (Incl. Socio-economic background, education, cleavages between different parties)	
Self-perception Identity	
Framing	
Chronology of Events	
7. Summary: Linkages	
Horizontal ties (with other cases) With whom? Nature of ties, level of engagement	
Vertical ties (with intermediaries) With whom? Nature of ties, level of engagement	
Role of linkages (or lack thereof) Including: Awareness/interpretation Certification Identity formation Threat/opportunity Resource structure Emulation Frame alignment/bridging etc.	
Role for other cases / Influence	
Notes on scale shift (or lack thereof) Broader awareness Frame alignment/bridging Collective claims beyond individual cases Broader collective identity Collaborative action Broader alliances or organizations	
8. Data	
General Notes	
Field Visits	
Facility Visits	
Interviews	
Documents	
Internet Sources	
Media Reports	
Other	
9. Further notes/information	
Misc.	

References

Disclaimer: Interviews and unpublished documents cited throughout this study can be found in Appendices I and II.

Ahlers, Anna. 2014. *Rural Policy Implementation in Contemporary China: New Socialist Countryside*. London and New York: Routledge.

Ahlers, Anna, and Gunter Schubert. 2013. 'Strategic Modelling: "Building a New Socialist Countryside" in Three Chinese Counties'. *The China Quarterly* 216 (December): 831-849.

Ahlers, Anna, and Gunter Schubert. 2010. '"Building a New Socialist Countryside" – Only a Political Slogan?' *Journal of Current Chinese Affairs* 38(4): 35-62.

Allagui, Ilhem, and Johanne Kuebler. 2011. 'The Arab Spring and the Role of ICTs: Editorial Introduction'. *International Journal of Communications* 5: 1435-1442.

Allsopp, Michelle, Pat Costner, and Paul Johnston. 2001. 'Incineration and Human Health: State of Knowledge of the Impacts of Waste Incinerators on Human Health'. Greenpeace Report. Exeter, UK. http://www.greenpeace.org/international/en/publications/reports/incineration-and-human-health/ (15 October 2016).

Almeida, Paul D. 2003. 'Opportunity Organizations and Threat-Induced Contention: Protest Waves in Authoritarian Settings'. *American Journal of Sociology* 109 (2): 345-400.

Alpermann, Björn. 2013. 'Village Governance Reforms in China: Paradigm Shift or Muddling Through?' In *Towards a New Development Paradigm in Twenty-First Century China: Economy, Society and Politics*, edited by Eric Florence and Pierre Defraigne, 147-164. London and New York: Routledge.

Alpermann, Björn. 2012. 'Qualitative Interviewforschung in China'. In *Qualitative Interviewforschung in und mit Fremden Sprachen: Eine Einführung in Theorie und Praxis*, edited by Jan Kruse et al., 169-189. Weinheim: Beltz Juventa.

Alpermann, Björn. 2010a. ‚Neue Regeln für Dorfwahlen – Revision des Gesetzes der VR China über die Organisation der Dorfkomitees 2009/2010'. *Research Network 'Governance in China' Background Paper* 1. http://regiereninchina.uni-wuerzburg.de/uploads/media/Background_Paper_No.1_2010_03.pdf (16 October 2015).

Alpermann, Björn. 2010b. 'State and Society in China's Environmental Politics'. In *China's Environmental Crisis: Domestic and Global Political Impacts and Responses*, edited by Joel Kassiola and Sujian Guo, 167-209. New York: Palgrave Macmillan.

Alpermann, Björn, and Maria Bondes. forthcoming. 'Networked Contention against Waste Incinerators in China: Brokers, Linkages and Dynamics of Diffusion'. In *Handbook of Dissent and Protest in China*, edited by Teresa Wright. Cheltenham: Edward Elgar Publishing.

Ansfield, Jonathan. 2013. 'Alchemy of a Protest: The Case of Xiamen PX'. In *China and the Environment: The Green Revolution*, edited by Sam Geall, 136-202. London and New York: Zed Books.

ASW villagers microblog. 2013a. '阿苏卫垃圾填埋场 – 被你害惨了阿苏卫垃圾填埋场周边村民亟待解救!' [*Asuwei laji tianmai chang – bei ni haican le Asuwei laji tianmai chang zhoubian cunmin jidai jiejiu!*]. Blog post on Asuwei villagers' Sina microblog, 1 August, 14:20, http://weibo.com/u/3667805421#_rnd1446820509938 (2 August 2013).

ASW villagers microblog. 2013b. '阿苏卫垃圾填埋场 – 被你害惨了阿苏卫垃圾填埋场周边村民亟待解救!' [*Asuwei laji tianmai chang – bei ni haican le Asuwei laji tianmai chang zhoubian cunmin jidai jiejiu!*]. Blog post on Asuwei villagers' Sina microblog, 2 August, 15:26, http://weibo.com/u/3667805421#_rnd1446820509938 (2 August 2013).

ASW villagers microblog. 2013c. '阿苏卫垃圾填埋场 – 被你害惨了阿苏卫垃圾填埋场周边村民亟待解救!' [*Asuwei laji tianmai chang – bei ni haican le Asuwei laji tianmai chang zhoubian cunmin jidai jiejiu!*]. Blog post on Asuwei villagers' Sina microblog, 2 August, 15:33, http://weibo.com/u/3667805421#_rnd1446820509938 (2 August 2013).

ASW villagers microblog. 2013d. '阿苏卫垃圾填埋场 – 被你害惨了阿苏卫垃圾填埋场周边村民亟待解救!' [*Asuwei laji tianmai chang – bei ni haican le Asuwei laji tianmai chang zhoubian cunmin jidai jiejiu!*]. Blog post on Asuwei villagers' Sina microblog, 6 August, 20:15, http://weibo.com/u/3667805421#_rnd1446820509938 (2 August 2013).

Balkan, Elizabeth. 2012. 'The dirty truth about China's incinerators'. *China Dialogue*, 4 July https://www.chinadialogue.net/article/5024-Dirty-truth-about-China-s-incinerators (18 April 2013); also published in *The Guardian*, 4 July, http://www.theguardian.com/environment/2012/jul/04/dirty-truth-chinas-incinerators (18 April 2013).

Bandurski, David. 2019. 'Converging on Control'. *China Media Project,* 16 March, http://chinamediaproject.org/2019/03/16/converging-on-control/ (27 March 2019).

Baxter, Pamela, and Susan Jack. 2008. 'Qualitative Case Study Methodology: Study Design and Implementation for Novice Researchers'. *The Qualitative Report* 13 (4): 544-559.

Beck, Ulrich. 1986. *Risikogesellschaft. Auf dem Weg in eine andere Moderne*. Berlin: Suhrkamp.

Beijing Daily. 2015. ‚阿苏卫循环经济园年中开工'. [*Asuwei xunhuan jingji yuan nianzhong kaigong*]. *Beijing Daily* [北京日报, *Beijing ribao*], 30 April, http://bjrb.bjd.com.cn/html/2015-04/30/content_275848.htm (20 December 2015).

Beijing Evening News. 2011. '住豪宅律师弃业两年践行垃圾分类：驴屎蛋儿和绿房子'. [*Zhu hao zhai lüshi qiye liangnian jianxing laji fenlei: Lüshidaner he Lü fangzi*]. *Beijing Evening News* [北京晚报, *Beijing wanbao*], 17 August, http://www.chinadaily.com.cn/dfpd/shehui/2011-08/17/content_13135212.htm (16 December 2015).

Beijing Morning Post. 2011. ‚北京大工村垃圾发电厂地址公布 将播报排放指标'. [*Beijing Dagong cun laji fadian chang dizhi gongbu jiang bobao paifang zhibiao*]. *Beijing Morning Post* [北京晨报, *Beijing chenbao*], 17 May, http://finance.huanqiu.com/roll/2011-05/1696795.html (15 January 2016).

Beijing Municipal Development and Reform Commission [北京市发展与改革委员会, *Beijing shi fazhan yu gaige weiyuanhui*]. 2006. '北京市 '十一五' 时期固体废弃物处理规划'. [*Beijing shi 'shi yi wu' shiqi guti feiqiwu chuli guihua*]. 12 December, http://law.solidwaste.com.cn/view/id_475 (30 September 2015).

Beijing Municipal People's Government [北京市人民政府, *Beijing renmin zhengfu*]. 2006. '北京市 '十一五' 时期生活垃圾处理设施建设规划实施方案.' [*Beijing shi 'shi yi wu' shiqi shenghuo laji chuli jianshe guihua shishi fangan*]. http://www.bjmac.gov.cn/huanwei/content/browseInfo.jsp?infoId=c00000002803 (30 September 2015).

Beijing Municipality First Intermediate People's Court [北京市第一中级人民法院, *Beijing shi diyi zhongji renmin fayuan*]. 2014. '杨宝华与任再金农村房屋买卖合同纠纷案'. [*Yang Zhuhua yu Ren Zaijin nongcun fangwu maimai hetong jiufen an*]. Civil lawsuit court verdict by the Beijing Municipality First Intermediate People's Court, 4 June, http://www.pkulaw.cn/Case/pfnl_121240424.html?match=Exact (16 January 2016).

Beijing News. 2011a. '多家环保组织联名支持 '绿房子' 垃圾站'. [*Duo jia huanjing zuzhi lianming zhichi 'Lü fangzi' laji zhan*]. The *Beijing News* [新京报, *Xin jingbao*], 12 July, http://news.sina.com.cn/green/news/roll/p/2011-07-12/101822799216.shtml (16 December 2015).

Beijing News. 2011b. '北京海淀大工村循环经济产业园将于明年3月开工'. [*Beijing Haidian Dagong cun xunhuan jingji chanye yuan jiang yu mingnian 3 yue kaigong*]. *The Beijing News* [新京报, *Xin jingbao*], 17 November, http://legal.gmw.cn/2011-12/09/content_3141319.htm (13 December 2012, link no longer available).

Beijing News. 2010a. '广州垃圾焚烧专家意见定稿被改被指'小动作''. [*Guangzhou laji fenshao zhuanjia yijian dinggao beigai beizhi 'xiao dongzuo'*]. *The Beijing News* [新京报, *Xin jingbao*], 17 March, http://news.sohu.com/20100317/n270874496.shtml (15 December 2015).

Beijing News. 2010b. '北京市海淀区苏家坨垃圾焚烧发电项目环评公示'. [*Beijing shi Haidian qu Sujiatuo laji fenshao fadian xiangmu huanping gongshi*]. *The Beijing News* [新京报, *Xin jingbao*], 17 November, http://www.cn-hw.net/html/china/201011/20037.html (14 January 2016).

Beijing News. 2010c. '北京苏家坨拟建垃圾焚烧厂 现有多个自然景区'. [*Beijing Sujiatuo ni jian laji fenshao chang xian you duoge ziran jingqu*]. *The Beijing News* [新京报, *Xin jingbao*], 17 November, http://green.sohu.com/20101117/n277689677.shtml (14 January 2016).

Beijing News. 2010d. '北京苏家坨焚烧厂被指将影响周围景区环境'. [*Beijing Sujiatuo fenshao chang bei zhi jiang yingxiang zhouwei jingqu huanjing*]. *The Beijing News* [新京报, *Xin jingbao*], 18 November, http://news.qq.com/a/20101118/000117.htm (14 January 2016).

Beijing TV. 2013. '垃圾堆里的'驴屎蛋''.[*Laji dui li de 'Lüshidan'*]. TV feature on Beijing TV program 青年公益 [*Qingnian gongyi*], 8 January, http://space.btv.com.cn/video/VIDE1375437105798870 (12 February 2014).

Benford, Robert D., and David A. Snow. 2000. 'Framing Processes and Social Movements: An Overview and Assessment'. *Annual Review of Sociology* 26: 611-639.

Bennett, W. Lance, and Alexandra Segerberg. 2014. 'Three Patterns of Power in Technology-enabled Contention'. *Mobilization: An International Quarterly* 19 (4): 421-439.

Bennett, W. Lance, and Alexandra Segerberg. 2012. 'The Logic of Connective Action'. *Information, Communication & Society* 15 (5): 739-768.

Biao, Teng, and Stacy Mosher. 2012. 'Rights Defence (Weiquan), Microblogs (Weibo), and the Surrounding Gaze (Weiguan): The Rights Defence Movement Online and Offline'. *China Perspectives* 3: 29-42.

Biggs, Michael, and Kenneth T. Andrews. 2010. 'From Protest to Organization: The Impact of the 1960 Sit-Ins on Movement Organizations in the American South'. In *The Diffusion of Social Movements: Actors, Mechanisms, and Political Effects*, edited by Rebecca Kolins Givan, Kenneth M. Roberts, and Sarah A. Soule, 187-203. Cambridge: Cambridge University Press.

Biswas, Asit, and Jingru Zhang. 2014. 'Waste-to-Energy Plants: A Burning Issue in China'. *Shanghai Daily*, 29 May, http://www.shanghaidaily.com/opinion/foreign-perspectives/Wastetoenergy-plants-A-burning-issue-in-China/shdaily.shtml (16 March 2016).

Bondes, Maria, and Thomas Johnson. 2017. 'Beyond Localized Environmental Contention: Horizontal and Vertical Diffusion in a Chinese Anti-Incinerator Campaign'. In *Chinese Environmental Actors Linking Up: Towards a Broader Environmental Movement?*, edited by Thomas Johnson and Maria Bondes. Special Section in the *Journal of Contemporary China*, 26(106): 504-520.

Botetzaglas, Iosif, and John Karamichas. 2009. 'Grassroots Mobilisations against Waste Disposal Sites in Greece'. *Environmental Politics* 18(6): 939-959.

Boudet, Hilary Schaffer, and Leonard Ortolano. 2010. 'A Tale of Two Sitings: Contentious Politics in Liquefied Natural Gas Facility Siting in California'. *Journal of Planning Education and Research* 30(1): 5-21.

Bradsher, Keith. 2012. 'Bolder Protests against Pollution Win Project's Defeat in China'. *The New York Times*, 4 July, http://www.nytimes.com/2012/07/05/world/asia/chinese-officials-cancel-plant-project-amid-protests.html (31 March 2016).

Bradsher, Keith. 2009. 'China's Trash Incinerators Loom as Global Hazard'. *The New York Times*, 11 August, http://www.nytimes.com/2009/08/12/business/energy-environment/12incinerate.html?_r=0 (17 March 2016).

Branigan, Tania. 2008. 'Chinese Figures Show Fivefold Rise in Babies Sick from Contaminated Milk'. *The Guardian*, 2 September, https://www.theguardian.com/world/2008/dec/02/china (26 June 2019).

Broadbent, Jeffrey. 2003. 'Movement in Context: Thick Networks and Japanese Environmental Protest'. In *Social Movements and Networks: Relational Approaches to Collective Action*, edited by Mario Diani and Doug McAdam, 204-229. Oxford: Oxford University Press.

Brown, Phil. 2007. *Toxic Exposures: Contested Illnesses and the Environmental Health Movement*. New York: Columbia University Press.

Burningham, Kate, and Martin O'Brien. 1994. 'Global Environmental Values and Local Contexts of Action'. *Sociology* 28(4): 913-932.

Büsgen, Michael. 2006. 'NGOs and the Search for Chinese Civil Society: Environmental Non-Governmental Organizations in the Nujiang Campaign'. ORPAS Working Paper 422. The Hague: ORPAS Institute of Social Studies.

Cai, Yongshun. 2010. *Collective Resistance in China: Why Popular Protests Succeed Or Fail*. Palo Alto: Stanford University Press.

Cai, Yongshun. 2008. 'Power Structure and Regime Resilience: Contentious Politics in China'. *British Journal of Political Science* 38(3): 411-432.

Cai, Yongshun. 2005. 'China's Moderate Middle Class: The Case of Homeowners' Resistance'. *Asian Survey* 45(5): 777-799.

Cai, Yongshun, and Zhiming Sheng. 2013. 'Homeowners' Activism in Beijing: Leaders with Mixed Motivations'. *The China Quarterly* 215(September): 513-532.

Caijing. 2012. '浙江伟明上市环保核查遭质疑 两位农民提起诉讼'. [*Zhejiang weiming shangshi huanbao hecha zao zhiyi liang wei nongmin tiqi susong*]. *Caijing* [财经, *Caijing*], 19 May, http://finance.china.com.cn/stock/special/wmhb/20120519/744894.shtml (20 July 2015).

CCTV. 2011. "驴屎蛋' 的突围'. ['*Lüshidan' de tuwei*]. TV feature on CCTV 1 program 看见 [*Kanjian*], 25 December, http://tv.cntv.cn/video/C29837/f559f573d9c540a2bc2f2d654fbb3e72 (12 February 2014).

Chen, Liwen [陈立雯]. 2012. '众所期待的村长选举被利益纠葛的垃圾焚烧厂搅乱'. [*Zhong suo qidai de cunzhang xuanju bei liyi jiuge de laji fenshao chang jiaoluan*]. Blogpost on Chen Liwen's personal Sina blog, 30 November, http://blog.sina.com.cn/s/blog_9613153701013xma.html (7 April 2015).

Chen, Xi. 2011. *Social Protest and Contentious Authoritarianism in China*. Cambridge: Cambridge University Press.

Chen, Xudong et al. 2010. 'An Overview of Municipal Solid Waste Management in China'. *Waste Management*, 30(4): 716-24.

Cheng, Yingqi. 2011a. 'Gov't to Burn 40% of Household Waste'. *Xinhua News Agency*, 2 September, http://news.xinhuanet.com/english2010/china/2011-09/02/c_131093710.htm (12 April 2013).

Cheng, Yingqi. 2011b. '"Green House" Plan to Improve Garbage Disposal'. *China Daily*, 26 July, http://www.chinadaily.com.cn/business/greenchina/2011-07/26/content_12986935.htm (16 December 2016).

China Daily. 2012. 'Where There's Muck There's Brass'. *China Daily*, 6 December, http://www.chinadaily.com.cn/bizchina/2010-12/06/content_11656524.htm (16 March 2016).

China Daily. 2011. 'Not Rubbish'. *China Daily European Weekly*, 6 May, http://usa.chinadaily.com.cn/business/2011-05/06/content_12459857.htm (30 September 2015).

China Daily. 2009. 'Officials Firm on Incineration Plants'. *China Daily*, 2 November, http://www.chinadaily.com.cn/cityguide/2009-11/02/content_8881364.htm (22 September 2015).

China Solid Waste Net. 2013. '首届 '邻避运动案例研讨会' 召开'. [*Shoujie 'Linbi yundong anli yantaohui' zhaokai*]. *China Solid Waste Net* [中国固废网, *Zhongguo gufei wang*], 21 May, http://news.solidwaste.com.cn/view/id_47350 (26 August 2015).

China Solid Waste Net. 2011. '北京苏家坨垃圾焚烧项目引争议'. [*Beijing Sujiatuo laji fenshao xiangmu yin zhengyi*]. *China Solid Waste Net* [中国固废网, *Zhongguo gufei wang*], 4 July, http://info.ep.hc360.com/2011/07/040908141852.shtml (14 January 2016).

China Speech. n.d. '赵章元'. [*Zhao Zhangyuan*]. *China Speech*, http://www.chinaspeech.com/SpeechorShow.aspx?sId=502 (5 March 2014, link no longer available).

China Youth Daily. 2013. '虚假环评钻了谁的空子?' [*Xujia huanping zuan le shui de kongzi?*]. *China Youth Daily* [中国青年报, *Zhongguo qingnian bao*], 16 January, http://news.xinhuanet.com/fortune/2013-01/16/c_114381025.htm (3 September 2015).

Chung, Yousun. 2015. ‚Pushing the Envelope for Representation and Participation: The Case of Homeowner Activism in Beijing'. *Journal of Contemporary China* 24(91): 1-20.

CNature. 2011. " 驴屎蛋' 黄小山'. ['*Lüshidan' Huang Xiaoshan*]. *CNature*, 21 April, http://www.cnature.org/a/gongyirensheng/2011/0421/3.html (18 November 2015).

Cody, Edward. 2007. 'Text Messages Giving Voice to Chinese'. *The Washington Post*, 28 June, http://www.zonaeuropa.com/20070601_1.htm (1 October 2015).

Cui, Xiaohuo. 2010a. 'Incinerator Foe on Fact-Finding Japan Trip'. *China Daily*, 22 February, http://www.chinadaily.com.cn/cndy/2010-02/22/content_9481918.htm (14 August 2013).

Cui, Xiaohuo. 2010b. 'Waste Incineration Firm Reports on Pollution Levels'. *China Daily*, 18 March, http://china-wire.org/?p=4877 (22 September 2015).

Cui, Xiaohuo. 2010c. 'Turning Up the Heat on Incineration Plans'. *China Daily*, 13 April, http://www.chinadaily.com.cn/cndy/2010-04/13/content_9719999.htm (14 August 2013).

Cui, Xiaohuo. 2010d. 'Old Mine May House Garbage-Burning Plant'. *China Daily*, 5 February, http://www.chinadaily.com.cn/cndy/2010-02/05/content_9432156.htm (12 April 2013).

Cui, Xiaohuo. 2009. 'Incineration Gets Locals all Fired Up'. *China Daily*, 18 August, http://www.chinadaily.com.cn/china/2009-08/18/content_8581942.htm (18 September 2015).

Cui, Zheng. 2013a. '环保部通报处罚88家环评机构'. [*Huanbao bu tongbao chufa 88 jia huanping jigou*]. *Caixin Magazine* [财新杂志, *Caixin zazhi*], 8 January, http://china.caixin.com/2013-01-08/100480553.html (21 August 2013).

Cui, Zheng. 2013b. '捏造100个环评同意'. [*Niezao 100 ge huanping tongyi*]. *Caixin Magazine* [财新杂志, *Caixin zazhi*], 8 February, http://m.magazine.caixin.com/m/2013-02-08/100490909.html?p1 (21 August 2015).

Cui, Zheng [崔筝]. 2011. '中国城市 '邻避运动' 渐起'. [*Zhongguo chengshi 'linbi yundong' jianqi*]. *China Dialogue* [中外对话, *Zhongwai duihua*], 8 December, https://www.chinadialogue.net/article/show/single/ch/4676-Rise-of-the-Chinese-nimby (26 August 2015).

Della Porta, Donatella, and Sidney G. Tarrow. 2005. *Transnational Protest and Global Activism*. Lanham: Rowman & Littlefield.

Deng, Yanhua, and Guobin Yang. 2013. 'Pollution and Protest in China: Environmental Mobilization in Context'. *The China Quarterly* 214(June): 321-336.

Diamond, Larry. 2010. 'Liberation Technology'. *Journal of Democracy* 21(3): 69-83.

Diamond, Larry, and Marc F. Plattner, eds. 2012. *Liberation Technology: Social Media and the Struggle for Democracy*. Baltimore: Johns Hopkins University Press.

Diani, Mario, and Paolo R. Donati. 1999. 'Organisational Change in Western European Environmental Groups: A Framework for Analysis'. *Environmental Politics* 8(1): 13-34.

Diani, Mario. 2003a. 'Leaders or Brokers? Positions and Influence in Social Movement Networks'. In *Social Movements and Networks: Relational Approaches to Collective Action*, edited by Mario Diani and Doug McAdam, 105-122. Oxford and New York: Oxford University Press.

Diani, Mario. 2003b. 'Networks and Social Movements: A Research Programme'. In *Social Movements and Networks: Relational Approaches to Collective Action*, edited by Mario Diani and Doug McAdam, 295-319. Oxford and New York: Oxford University Press.

Dong, Qingzhang et al. 2010. 'Municipal Solid Waste Management in China: Status, Problems and Challenges'. *Journal of Environmental Management* 91: 1623-1633.

Dorfmann, Paul. 2015. 'China Pushes Ahead with High Risk, High Cost Nuclear'. *China Dialogue,* 5 May, https://www.chinadialogue.net/article/show/single/en/7921-China-pushes-ahead-with-high-risk-high-cost-nuclear (01.08.2016).

Duggan, Jennifer. 2015. 'China targets lawyers in new human rights crackdown'. *The Guardian,* 13 July, https://www.theguardian.com/world/2015/jul/13/china-targets-lawyers-in-new-human-rights-crackdown (27 March 2019).

Earl, Jennifer, and Katrina Kimport. 2011. *Digitally Enabled Social Change: Activism in the Internet Age*. Boston: MIT Press.

Economist. 2015. 'Keep the Fires Burning'. *The Economist,* 25 April, http://www.economist.com/news/china/21649540-waste-incinerators-rile-public-are-much-better-landfill-keep-fires-burning (12 March 2016).

Economy, Elizabeth. 2018. *The Third Revolution: Xi Jinping and the New Chinese State*. New York: Oxford University Press.

Epstein, Steven. 2009. *Impure Science: AIDS, Activism, and the Politics of Knowledge*. Berkeley: University of California Press.

Felstiner, William L.F., Richard L. Abel, and Austin Sarat. 1980. 'The Emergence and Transformation of Disputes: Naming, Blaming, Claiming ...' *Law & Society Review* 15(3-4): 631-654.

Feng, Shu. 2011. 'Surrounded by Garbage'. *Global Times*, 1 December, http://www.globaltimes.cn/content/686699.shtml (30 September 2015).

Fischer, Frank. 2000. *Citizens, Experts, and the Environment: The Politics of Local Knowledge*. Durham: Duke University Press.

FON. 2011a. '环评被指做假 – 五环保组织申请取消气科院资质'. [*Huanping bei zhi zuo jia – wu huanbao zuzhi shenqing quxiao Qikeyuan zizhi*]. Press releases on Friends of Nature website, 21 June, http://www.fon.org.cn/index.php/Index/cate/id/92 (6 August 2015).

FON. 2011b. '环保组织质疑苏家坨垃圾焚烧项目的合法性'. [*Huanbao zuzhi zhiyi Sujiatuo laji fenshao xiangmu de hefaxing*]. In '自然大学通讯总第75期', [*Ziran daxue tongxun zong di 75 qi*], FON newsletter, July 2011, 13-15. Beijing: Friends of Nature.

FON. 2011c. '工作日志 2011年 5-6 月'. [*Gongzuo rishi 2011 nian 5-6 yue*]. In '自然大学通讯总第75期', [*Ziran daxue tongxun zong di 75 qi*], FON newsletter, July 2011, 69-72, Beijing: Friends of Nature.

FON. 2011d. '中国生活垃圾管理: 问题与建议'. [*Zhongguo shenghuo laji guanli: wenti yu jianyi*]. Friends of Nature report, http://www.fon.org.cn/uploads/attachment/27261358328685.pdf (18 March 2016).

Fu, Hualing. 2018. 'The July 9th (709) Crackdown on Human Rights Lawyers: Legal Advocacy in an Authoritarian State'. *Journal of Contemporary China* 27: 554-568.

Fu, Hualing, and Richard Cullen. 2008. 'Weiquan (Rights Protection) Lawyering in an Authoritarian State: Building a Culture of Public-Interest Lawyering'. *The China Journal* 59(January): 111-127.

Fu, Tao [付涛].2014. '邻避运动如何改变中国'. [*Linbi yundong ruhe gaige Zhongguo*]. *China Development Brief*, 26 September, http://www.chinadevelopmentbrief.org.cn/news-16626.html (26 August 2015).

Futrell, Robert. 2003. 'Framing Processes, Cognitive Liberation, and NIMBY Protest in the U.S. Chemical-Weapons Disposal Conflict'. *Sociological Inquiry* 73(3): 359-386.

GAIA. n.d. 'Chinese Environmental Groups Question CDM Application'. http://www.no-burn.org/chinese-environmental-groups-question-cdm-application (6 September 2012).

Gao, Shengke [高胜科]. 2012. '秦皇岛垃圾焚烧项目遭抵制 三年僵局很难打破'. [*Qinhuangdao laji fenshao xiangmu zao dizhi san nian jiangju hen nan dapo*]. *Caijing* [财经, *Caijing*], 2 July, http://news.qq.com/a/20120705/000735.htm (17 April 2015)

Geall, Sam. 2013. *China and the Environment the Green Revolution*. London: Zed Books.

George, Alexander L., and Andrew Bennett. 2005. *Case Studies and Theory Development in the Social Sciences*. Cambridge: MIT Press.

Givan, Rebecca Kolins, Kenneth M. Roberts, and Sarah A. Soule. 2010. 'Introduction: The Dimensions of Diffusion'. In *The Diffusion of Social Movements: Actors, Mechanisms, and Political Effects*, edited by Rebecca Kolins Givan, Kenneth M. Roberts, and Sarah A. Soule, 1-15. Cambridge: Cambridge University Press.

Global Times. 2014. 'Beijing Grapples with Overflowing Garbage Problems'. *Global Times*, 16 December, http://www.globaltimes.cn/content/897061.shtml (28 September 2015).

Global Times. 2011a. 'Trash Incinerator Plans Scrapped after Resident Outcry'. *Global Times*, 21 January, http://www.globaltimes.cn/beijing/society/2011-01/615212.html (14 December 2012).

Global Times. 2011b. '"Greenhouse" Treats Garbage'. *Global Times*, 11 July, http://english.sina.com/china/p/2011/0710/380517.html (16 December 2015).

Global Times. 2009. 'Burning the Trash is an Ugly Necessity'. *Global Times*, 8 September, http://www.globaltimes.cn/content/465588.shtml (12 April 2013, link no longer available).

Global Voices. 2010. 'Waste Incineration and Environmental Protection'. *Global Voices*, 23 February, http://transparency.globalvoicesonline.org/project/waste-incineration-and-environmental-protection-6 (15 December 2015).

Gould, Kenneth A. 1993. 'Pollution and Perception: Social Visibility and Local Environmental Mobilization'. *Qualitative Sociology* 16(2): 157-178.

Grano, Simona A., and Yuheng Zhang. forthcoming. 'Anti-nuclear Protest in China'. In *Handbook of Dissent and Protest in China*, edited by Teresa Wright. Cheltenham: Edward Elgar Publishing.

Grano, Simona A., and Yuheng Zhang. 2016. 'New Channels for Popular Participation in China: The Case of an Environmental Protection Movement in Nanjing'. *China Information* 30(2): 165-187.

Guan, Zhong. 2010. 'Loopholes in Environmental Impact Assessment'. In *Friends of Nature Newsletter: Waste Challenges China*, FON, 20 (spring), 10-15, http://www.fon.org.cn/uploads/attachment/92871363574550.pdf (12 April 2013).

Guangzhou Daily. 2010. '32位专家献策广州垃圾处理 31人主张焚烧'. [*32 wei zhuanjia xiance Guangzhou laji chuli* 31 *ren zhuzhang fenshao.*] *Guangzhou Daily* [广州日报, *Guangzhou ribao*], 25 February, http://news.dayoo.com/guangzhou/201002/25/73437_12114850.htm (15 December 2015).

Guo, Tiehu [郭铁路]. 2011. '苏家坨垃圾焚烧厂址环保组织质疑'. [*Sujiatuo laji fenshao changzhi huanbao zuzhi zhiyi*].*The Beijing News* [新京报, *Xin jing bao*], 28 May, http://news.bjnews.com.cn/2011/0528/119908.shtml (13 August 2013, link no longer available).

Hadden, Jennifer. 2015. *Networks in Contention: The Divisive Politics of Climate Change*. New York: Cambridge University Press.

Haidian District People's Government [海淀区人民政府, *Haidian qu renmin zhengfu*]. 2012. '关于海淀区循环经济产业园再生能源发电厂建设进展情况的报告'. [*Guanyu Haidian qu xunhuan jingji chanye yuan zaisheng fadian chang jianshe jinfa qingkuang de baogao*]. 18 September, http://hdrd.bjhd.gov.cn/rdhy/cwhhy/201303/P020130325615097654258.doc (17 January 2016).

Haidian District People's Government. 2010a. '北京绿海能环保有限责任公司简介'. [*Beijing Lühai neng huanbao youxian zeren gongsi jianjie*]. 11 June, http://hdgzw.bjhd.gov.cn/gzjg/jgqy/201202/t20120229_395875.htm (13 January 2016).

Haidian District People's Government. 2010b. '北京市海淀区循环经济产业园再生能源发电厂项目环评信息公告'. [*Beijing Haidian qu xunhuan jingji chanye yuan zaisheng nengyuan fadian chang xiangmu huanping xinxi gonggao*]. 16 November, http://www.bjhd.gov.cn/zf/zfgg/201011/t20101116_223209.htm (27 November 2012, link no longer available).

Hallman, William, and Abraham Wandersman. 1992. 'Attribution of Responsibility and Individual and Collective Coping with Environmental Threats'. *Journal of Social Issues* 48(4): 101-118.

Han, Nuo [韩娜]. 2011. '北京大工村垃圾发电厂进行2次环评 地址公布'. [*Beijing Sujiatuo laji fadian chang jinxing 2 ci huanping dizhi gongbu.*] *Beijing Morning Post* [北京晨报, *Beijing chenbao*], 17 May, http://news.qq.com/a/20110517/000053.htm (15 January 2016).

Hayes, Ben, Frank Barat, Isabelle Geuskens, Nick Buxton, Fiona Dove, Francesco Martone, Hannah Tworney, and Semanur Karaman. 2017. 'On "Shrinking Space" – A Framing Paper'. *Amsterdam Transnational Institute Issue Brief*, www.tni.org/files/publication-downloads/on_shrinking_space_2.pdf (27 March 2019).

He, Haining [何海宁], and Shi Qiujuan [史秋娟]. 2011. '地方红绿榜6.23：广东一高岭土厂涉嫌非法排污'. [*Difang honglü bang 6.23: Guangdong yi gaolingtu chang shexian feifa paiwu*]. *Southern Weekend* [南方周末, *Nanfang zhoumo*], 24 June, http://news.sina.com.cn/green/news/roll/2011-06-24/115422698894.shtml (14 January 2016).

He, Zuoxiu. 2013. 'Chinese Nuclear Disaster "Highly Probable" by 2030'. *China Dialogue*, 13 March, https://www.chinadialogue.net/article/show/single/en/5808-Chinese-nuclear-disaster-highly-probable-by-2-3- (1 August 2016).

Hedström, Peter, and Richard Swedberg. 1998. *Social Mechanisms: An Analytical Approach to Social Theory*. Cambridge et al.: Cambridge University Press.

Hedström, Peter, and Richard Swedberg. 1996. 'Social Mechanisms'. *Acta Sociologica* 39(3): 281-308.

Hedström, Peter, and Petri Ylikoski. 2010. 'Causal Mechanisms in the Social Sciences'. *Annual Review of Sociology* 36(1): 49-67.

Heiman, Michael. 1990. 'From "Not in My Backyard!" to "Not in Anybody's Backyard!"'. *Journal of the American Planning Association* 56(3): 359-362.

Herriott, Robert E., and William A. Firestone. 1983. 'Multisite Qualitative Policy Research: Optimizing Description and Generalizability'. *Educational Researcher* 12(2): 14-19.

Herrold-Menzies, Melinda. 2010. 'Peasant Resistance against Nature Reserves'. In *Reclaiming Chinese Society: The New Social Activism*, edited by You-Tien Hsing and Ching Kwan Lee, 83-98. London and New York: Routledge.

Ho, Peter. 2007. 'Embedded Activism and Political Change in a Semiauthoritarian Context'. *China Information* 21(2): 187-209.

Ho, Peter, and Richard Louis Edmonds. 2007. 'Perspectives of Time and Change Rethinking Embedded Environmental Activism in China'. *China Information* 21(2): 331-344.

Holbig, Heike. 2002. 'Die Politik Entdeckt die "benachteiligten Gruppen": Die 5. Plenartagung des 9. Nationalen Volkskongresses'. *China Aktuell* 31: 266-274.

Holdaway, Jennifer. 2010. 'Environment and Health in China: An Introduction to an Emerging Research Field'. *Journal of Contemporary China* 19(63): 1-22.

Hook, Leslie. 2012. 'Growing Rubbish Mountain Haunts Beijing'. *Financial Times*, 10 April, http://www.ftchinese.com/story/001044060/en/?print=y (9 October 2015).

Hsing, You-tien, and Ching Kwan Lee. 2010. *Reclaiming Chinese Society: The New Social Activism*. London and New York: Routledge.

Hsu, Carolyn, and Jessica Teets. 2016. 'Is China's New Overseas NGO Management Law Sounding the Death Knell for Civil Society? Maybe Not'. *The Asia-Pacific Journal* 14(4/3): 1-14.

Hu, Hui. 2015. 'A Critical Evaluation of Waste Incineration Plants I Wuhan (China) Based on Site Selection, Environmental Influence, Public Health and Public Participation'. *International Journal of Environmental Research and Public Health* 12: 7593-614.

Huang, Qifei et al. 2006. 'The Current Situation of Solid Waste Management in China'. *Journal of Material Cycles and Waste Management* 8: 63-69.

Huang, Ronggui, and Ngai-ming Yip. 2012. 'Internet and Activism in Urban China: A Case Study of Protests in Xiamen and Panyu'. *The Journal of Comparative Asian Development* 11(2): 201-223.

Hung, Chin-fu. 2010. 'The Politics of China's Weiquan Movement in the Internet Age'. *International Journal of China Studies* 1(2): 331-349.

Hung, Ho-fung. 2011. *Protest with Chinese Characteristics: Demonstrations, Riots, and Petitions in the Mid-Qing Dynasty*. New York: Columbia University Press.

Huo, Weiya [霍伟亚]. 2009. 'Sorting the Rubbish in Beijing'. *China Dialogue*, 11 September, https://www.chinadialogue.net/article/3247-Sorting-the-rubbish-in-Beijing (30 September 2015).

Huo, Weiya. 2013. '邻避运动如何改变中国?' [*Linbi yundong ruhe gaibian Zhongguo?*]. *China Dialogue* [中外对话, *Zhongwai duihua*], 10 June, https://www.chinadialogue.net/article/show/single/en/6080-How-are-nimby-campaigns-changing-China- [English], https://www.chinadialogue.net/article/show/single/ch/6080-How-are-nimby-campaigns-changing-China- [Chinese] (26 August 2015).

Hurst, William. 2010. 'Cases, Questions, and Comparison in Research on Contemporary Chinese Politics'. In *Contemporary Chinese Politics: New Sources, Methods, and Field Strategies*, edited by Allen Carlson, Mary E. Gallagher, Kenneth Lieberthal, and Melanie Manion, 162-180. New York: Cambridge University Press.

Inclán, Mariá. 2009. 'Sliding Doors of Opportunity: Zapatistas and Their Cycle of Protest'. *Mobilization: An International Quarterly* 14(1): 85-106.

IPEN and Green Beagle. 2015. 'China Chemical Safety Case Study: Likeng Incinerator in Guangzhou, Guangdong Province'. February 2015, http://ipen.org/sites/default/files/documents/Case%20Study%20Report%20Likeng%202015r.pdf (29 April 2016).

Jacobs, Andrew, and Chris Buckley. 2015. 'China Targeting Rights Lawyers in a Crackdown'. *The New York Times*, 22 July, https://www.nytimes.com/2015/07/23/world/asia/china-crackdown-human-rights-lawyers.html (27 March 2019).

Jian, Lu, and Chris King-Chi Chan. 2016. 'Collective Identity, Framing and Mobilisation of Environmental Protests in Urban China: A Case Study of Qidong's Protest'. *China: An International Journal* 14(2): 102-122.

Jiang-Wai-Jiang. 2010a. '如本版块出现长时间 '技术故障', 请大家移玉步至乐活奥北论坛'. [*Ru benban kuai chuxian chang shijian 'jishu guzhang,' qing dajia yiyu buzhi lehuo Aobei luntan*]. Post on the *Jiang-Wai-Jiang* [江外江] forum, 15 March, http://www.rg-gd.net/forum.php?mod=viewthread&tid=192685 (17 December 2012).

Jiang-Wai-Jiang. 2010b. '乐活奥北论坛被黑! 又被黑了!' [*Lehuo Aobei luntan bei hei! You bei hei le!*]. Post on the *Jiang-Wai-Jiang* [江外江] forum, 14 June, http://www.rg-gd.net/forum.php?mod=viewthread&tid=204615 (17 December 2012).

Jiang-Wai-Jiang. 2010c. '我们怎么了, 警察为什么抓我们?' [*Women zenme le, jingcha wei shenme zhua women?*]. Post on the *Jiang-Wai-Jiang* [江外江] forum, 16 June, http://www.rg-gd.net/forum.php/forum.php?mod=viewthread&tid=205101 (17 December 2012).

Jin, Yu [金煜]. 2011a. '北京苏家坨垃圾焚烧厂环评报告公布'. [*Beijing Sujiatuo lajifenshao chang huanping baogao gongbu.*] *The Beijing News* [新京报, *Xin jing bao*], 6 June, http://news.ifeng.com/mainland/detail_2011_06/11/6947174_0.shtml (11 January 2016).

Jin, Yu. 2011b. '北京苏家坨垃圾厂公众支持率超9成遭质疑'. [*Beijing Sujiatuo laji chang gongzhong zhichilü chao 9 cheng zao zhiyi*]. *The Beijing News* [新京报, *Xin jing bao*], 17 June, http://news.qq.com/a/20110617/000078.htm (12 August 2015).

Jing, Jun. 2000. 'Environmental Protests in Rural China'. In *Chinese Society: Change, Conflict and Resistance*, edited by Elizabeth Perry and Mark Selden, 143-160. London and New York: Routledge.

Johnson, Thomas. 2013a. 'The Politics of Waste Incineration in Beijing: The Limits of a Top-Down Approach?' *Journal of Environmental Policy & Planning* 15(1): 109-128.

Johnson, Thomas. 2013b. 'The Health Factor in Anti-Waste Incinerator Campaigns in Beijing and Guangzhou'. *The China Quarterly* 214(June): 356-375.

Johnson, Thomas. 2010. 'Environmentalism and NIMBYism in China: Promoting a Rules-Based Approach to Public Participation'. *Environmental Politics* 19(3): 430-448.

Johnson, Thomas, and Maria Bondes, eds. 2017. 'Chinese Environmental Actors Linking Up: Towards a Broader Environmental Movement?' Special Section in the *Journal of Contemporary China*, 26(106): 504-548.

Kasperson, Roger E., et al. 1988. 'The Social Amplification of Risk: A Conceptual Framework'. *Risk Analysis* 8(2): 177-187.

Keck, Margaret, and Kathryn Sikkink. 1998. *Activists beyond Borders.* New York: Cornell University Press.

King, Gary, Robert Owen Keohane, and Sidney Verba. 1994. *Designing Social Inquiry: Scientific Inference in Qualitative Research.* Princeton: Princeton University Press.

Kitschelt, Herbert P. 1986. 'Political Opportunity Structures and Political Protest: Anti-Nuclear Movements in Four Democracies'. *British Journal of Political Science* 16(1): 57-85.

Kong, Lingyu. 2015. 'Residents of Beijing Community Renew Protest against Waste Incineration Plant'. *Caixin,* 22 May, http://english.caixin.com/2015-05-22/100811732.html (20 December 2015).

Lan, Xinzhen. 2010. 'A Burning Question'. *Beijing Review,* 8 July, http://www.bjreview.com/business/txt/2010-07/05/content_283040.htm (16 March 2016).

Lang, Bertram, and Heike Holbig. 2018. 'Civil Society Work in China: Trade-offs and Opportunities for European NGOs'. *GIGA Focus Asia,* 06/2018, https://www.giga-hamburg.de/en/publication/civil-society-work-in-china-trade-offs-and-opportunities-for-european-ngos (27 March 2019).

Lang, Graeme, and Ying Xu. 2013. 'Anti-Incinerator Campaigns and the Evolution of Protest Politics in China'. *Environmental Politics* 22(5): 832-848.

Lau, Mimi. 2010. 'China's New Threat: A Mountain of Garbage'. *South China Morning Post,* 30 April, http://www.scmp.com/article/712956/chinas-new-threat-mountain-waste (16 March 2016).

Leonard, Liam, Honor Fagan, and Peter Doran. 2009. 'A Burning Issue? Governance and Anti-Incinerator Campaigns in Ireland, North and South'. *Environmental Politics* 18(6): 896-916.

Li, He. 2010. 'Asuwei Landfill Strives for Deodorization and Purification'. In *Friends of Nature Newsletter: Waste Challenges China*, FON, 20(spring): 8-9, http://www.fon.org.cn/uploads/attachment/92871363574550.pdf (12 April 2013).

Li, Jing. 2013. 'Protests Predicted at Beijing's Plan to Add Incinerators'. *South China Morning Post,* 13 August, http://www.scmp.com/news/china/article/1296238/activists-warn-protests-over-plans-big-expansion-waste-incinerators (6 May 2016).

Li, Liqiang [李立强]. 2010a. '北京抵制建垃圾焚烧厂网友获政府邀请赴日考察'. [*Beijing dizhi jian laji fenshao chang wangyou huo zhengfu yaoqing fu Ri kaocha*]. *The Beijing News* [新京报, *Xin jing bao*], 21 February, http://society.people.com.cn/GB/41158/10989380.html (14 March 2013); also published in *Xinhua News Agency,* 21 February, http://news.xinhuanet.com/politics/2010-02/21/content_13016353_3.htm (14 March 2013).

Li, Liqiang. 2010b. '北京网友"驴屎蛋"获政府邀请赴日考察垃圾处理'. [*Beijing wangyou 'Lüshidan' huo zhengfu yaoqing fu Ri kaocha laji chuli*]. *The Beijing News* [新京报, *Xin jing bao*], 22 February, http://leaders.people.com.cn/GB/10989500.html# (14 August 2013); also published in *China Daily*, 22 February, www.chinadaily.com.cn/dfpd/2010-02/22/content_9484380_2.htm (14 August 2013).

Li, Liqiang et al. 2010. '北京三大垃圾厂无法按期完工 – 相关领导要求加快建设'. [*Beijing san da laji chang wufa anqi wangong – xiangguan lingdao yaoqiu jiakuai jianshe*]. *The Beijing News* [新京报, *Xin jing bao*], 31 July, http://news.dichan.sina.com.cn/bj/2010/07/31/192730.html (13 October 2015).

Li, Lianjiang, and Kevin O'Brien. 2008. 'Protest Leadership in China'. *The China Quarterly* 193: 1-23.

Li, Shuang. 2011. 'Capital Prepares to Sort out the Trash'. *Global Times*, 18 January, http://www.globaltimes.cn/content/613737.shtml (18 April 2013).

Li, Shuang. 2010a. 'Haidian Plans to Build New Incinerator'. *Global Times*, 18 November, http://www.globaltimes.cn/content/593002.shtml (12 April 2013).

Li, Shuang. 2010b. 'Real-Time Monitor for Incinerator'. *Global Times*, 18 March, http://www.globaltimes.cn/content/513921.shtml (14 August 2013).

Li, Shuang. 2010c. 'Beijing to Build Biggest Incinerator'. *Global Times*, 25 October, http://www.globaltimes.cn/content/585297.shtml (12 April 2013).

Li, Shuang. 2010d. 'Nine Ministries Demand Better Dioxin Supervision'. *Global Times*, 9 November, http://www.globaltimes.cn/content/590522.shtml (16 March 2016).

Li, Shuang, and Jingjing Xuyang. 2010. 'Seeking Garbage Disposal Solution'. *Global Times*, 22 February, http://www.globaltimes.cn/metro-beijing/update/top-news/2010-02/506862.html (14 August 2013).

Li, Sipan [李思磐]. 2010. '反垃圾焚烧维权突围记 北京奥北别墅区业主PK亚洲最大垃圾焚烧'. [*Fan laji fenshao weiquan tuwei ji Beijing Aobei bieshu qu yezhu* PK *Yazhou zui da laji fenshao*]. *Southern Metropolis Daily* [南方都市报, *Nanfang dushi bao*], 3 March, http://news.sina.com.cn/s/2010-03-03/050217155296s.shtml (18 November 2015).

Li, Xiang. 2011. 'It's Waste Not, Want Not in Trash Management'. *China Daily*, 25 March, http://www.chinadaily.com.cn/bizchina/2011-03/25/content_12227624.htm (11 April 2013).

Li, Yuanfang [李远方]. 2012. '上市环保核查制度遭疑'. [*Shangshi huanbao hecha zhidu zao yi*]. *China Business Daily* [中国商报, *Zhongguo shang bao*], 12 June, http://www.zgswcn.com/2012/0612/90907.shtml (3 September 2015).

Li, Zemin. 2012. '律师上书证监会: 上市环保核查制度违反行政许可法'. [*Lüshi shang shuzheng jian hui: Shangshi huanbao hecha zhidu weifan xingzheng xuke fa*]. *National Business Daily* [每日经济新闻, *Mei ri jingji xinwen*], 23 May, http://www.nbd.com.cn/articles/2012-05-23/655563.html (3 September 2015).

Li, Zemin [李泽民]. 2011. '苏家坨垃圾焚烧环评方造假? 环保制度亟待规范'. [*Sujiatuo laji fenshao huanping fang zaojia? Huanbao zhidi jidai guifan*]. *National Business Daily* [每日经济新闻, *Mei ri jingji xinwen*], 8 July, http://stock.sohu.com/20110708/n312781779.shtml (6 August 2015).

Liang, Ruqiao. 2011. 'Beijing Lawyer Makes a Difference by Talking Trash'. *Global Times*, 7 May, http://www.globaltimes.cn/content/652543.shtml (18 April 2013).

Lin, Lihong. 2010. 'The Rise of Rights and Protections for the Disadvantaged'. In *China's Journey Toward the Rule of Law: Legal Reform, 1978-2008*, edited by Dingjian Cai, 301-346. Leiden and Beijing: Brill.

Liu Hongqiao [刘虹桥]. 2010. '苏家坨垃圾焚烧厂环评公告遭质疑'. [*Sujiatuo laji fenshao chang huanping gonggao zao zhiyi*]. *Caixin* [财新, *Caixin*], 24 November, http://china.caixin.com/2010-11-24/100201515.html (15 January 2016).

Liu, Chunru [刘春瑞] et al. 2011. '北京居民生活垃圾拟按体积收费'. [*Beijing jumin shenghuo laji ni an tiji shoufei*]. *The Beijing News* [新京报, *Xin jing bao*], 19 January, http://gb.cri.cn/27824/2011/01/19/2225s3128804.htm (15 January 2016).

Liu, Fei [刘飞]. 2013a. '恶臭侵袭村民封堵垃圾场道路 – 城区爆发 '垃圾危机''. [*Echou qinxi cunmin fengdu laji chang daolu – chengqu baofa 'laji weiji'*]. *China National Radio* [中国广播, *Zhongguo guangbo*], 31 July, http://china.cnr.cn/xwwgf/201307/t20130731_513196501.shtml (13 February 2014).

Liu, Fei. 2013b. '北京垃圾填埋场恶臭袭人 – 村民叫苦堵路垃圾无处清运'. [*Beijing laji tianmai chang echou xi ren – Cunmin jiaoku dulu laji wufa qingyun*]. *China National Radio* [中国广播, *Zhongguo guangbo*], 1 August, http://china.cnr.cn/yaowen/201307/t20130731_513198574.shtml (13 February 2014).

Liu, Haiying. 2011. 'China's Garbage Problem Prompts Soul Searching among NGOs'. *China Development Brief* 49, 28 November, http://chinadevelopmentbrief.cn/articles/chinas-garbage-problem-prompts-soul-searching-among-ngos/ (1 December 2015).

Liu, Jianqiang. 2013. 'China's New "Middle Class" Environmental Protests'. *China Dialogue*, 2 January, https://www.chinadialogue.net/article/show/single/en/5561-China-s-new-middle-class-environmental-protests (30 September 2015).

Liu, Jinsong. 2011. 'Beijing's "Seventh Ring" of Rubbish'. *The Economic Observer*, 22 July, http://www.eeo.com.cn/ens/2011/0722/206842.shtml (30 September 2015).

Liu, Linlin. 2013a. 'Police Tell Election Protesters to Report'. *Global Times*, 8 January, http://china-wire.org/?p=25256 (20 July 2015).

Liu, Linlin. 2013b. 'Defrauding the Environment'. *Global Times*, 17 January, http://www.globaltimes.cn/content/756622.shtml (12 February 2013).

Liu, Linlin. 2012a. 'Villagers Protest Delayed Election'. *Global Times*, 30 November, http://www.globaltimes.cn/content/747434.shtml (5 December 2012); also published in *People's Daily*, 30 November, http://english.peopledaily.com.cn/90882/8040671.html (5 December 2012).

Liu, Linlin. 2012b. 'Stealing Elections'. *Global Times*, 3 December, http://china-wire.org/?p=25256 (20 July 2015).

Liu, Linlin. 2012c. 'Hebei Villagers Disrupt "Illegal" Election'. *Global Times*, 31 December, http://www.globaltimes.cn/content/753122.shtml (7 January 2013).

Lober, Douglas. 1995. 'Why Protest? Public Behavioral and Attitudinal Response to Siting a Waste Disposal Facility'. *Policy Studies Journal* 23(3): 499-518.

Lohmann, Susanne. 1994. 'The Dynamics of Informational Cascades: The Monday Demonstrations in Leipzig, East Germany, 1989-1991'. *World Politics* 47(1): 42-101.

Lora-Wainwright, Anna. 2017. *Resigned Activism: Living with Pollution in Rural China*. Cambridge, MA, and London: The MIT Press.

Lora-Wainwright, Anna. 2013a. 'Plural Forms of Evidence and Uncertainty in Environmental Health: A Comparison of Two Chinese Cases'. *Evidence & Policy: A Journal of Research, Debate and Practice* 9(1): 49-64.

Lora-Wainwright, Anna. 2013b. 'The Inadequate Life: Rural Industrial Pollution and Lay Epidemiology in China'. *The China Quarterly* 214(June): 302-320.

Lora-Wainwright, Anna, ed. 2013c. 'Dying for Development'. Special section in *The China Quarterly* 214 (June).

Lora-Wainwright, Anna. 2013d. 'Introduction. Dying for Development: Pollution, Illness and the Limits of Citizens' Agency in China'. *The China Quarterly* 214(June): 243-254.

Lora-Wainwright, Anna, et al. 2012. 'Learning to Live with Pollution: The Making of Environmental Subjects in a Chinese Industrialized Village'. *The China Journal* 68(July): 106-124.

Lynch, Marc. 2011. 'After Egypt: The Limits and Promise of Online Challenges to the Authoritarian Arab State'. *Perspectives on Politics* 9(2): 301-310.

Ma, Jun. 2009. 'Solving the Incinerator Uproar'. *China Dialogue*, 22 December, https://www.chinadialogue.net/article/show/single/en/3436 (30 September 2015).

Mao, Da. 2013. '毛达：秦皇岛农民的反焚垃圾运动'. [*Mao Da: Qinhuangdao nongmin de fanshao laji yundong*]. *Green Sohu*, 20 May, http://green.sohu.com/20130520/n376486520.shtml (27 November 2013).

Mao, Da. 2012. '见证草根力量 – 走访河北抚宁县反对垃圾焚烧厂建设二村民记'. [*Jianzheng caogen Liliang – Zoufang Hebei Funing xian fandui laji fenshao chang jianshe liang cunmin ji*]. Blogpost on Mao Da's personal Sina blog, 22 October, http://blog.sina.com.cn/s/blog_5fbaf32e01015y2t.html (21 July 2013).

Mao, Da. 2011. '四评 '关于加强二恶英污染防治的指导意见''. [*Si ping 'guanyu jiaqiang ereying wuran fangzhi de zhidao yijian'*]. In 中国垃圾信息工作网络通讯 第1期 [*Zhongguo laji xinxi gongzuo wangluo tongxun di 1 qi*], *China Waste Information Network Newsletter* 1, July: 19-23, http://www.waste-cwin.org/sites/default/files/zhong_guo_la_ji_xin_xi_gong_zuo_wang_luo_ji_kan_zhong__0.pdf (30 March 2016).

Mao, Da [毛达]. 2010. '美国历史上的垃圾焚烧狂热症'. [*Meiguo lishi shang de laji fanshao kuang rezheng*]. In: 通讯总第2期 (2010.03), [*Tongxun zong di 2 qi*], Green Anhui newsletter 2: 8-11, http://www.green-anhui.org/sites/default/files/upfiles/lu_man_jiang_huai_di_2qi_.pdf (30 March 2016).

Mao, Da, and Ang Zhao. 2010. 'On the Issue of Urban Domestic Waste'. In *Friends of Nature Newsletter: Waste Challenges China*, FON, 20 (spring): 5-9, http://www.fon.org.cn/uploads/attachment/92871363574550.pdf (12 April 2013).

Markham, William T. 2005. 'Networking Local Environmental Groups in Germany: The Rise and Fall of the Federal Alliance of Citizens' Initiatives for Environmental Protection (BBU)'. *Environmental Politics* 14(5): 667-685.

Matsuzawa, Setsuko. 2012. 'Citizen Environmental Activism in China: Legitimacy, Alliances, and Rights-Based Discourses'. *ASIANetwork Exchange: A Journal for Asian Studies in the Liberal Arts* 19(2): 81-91.

Mayntz, Renate. 2004. 'Mechanisms in the Analysis of Social Macro-Phenomena'. *Philosophy of the Social Sciences* 34(2): 237-259.

McAdam, Doug. 2003. 'Beyond Structural Analysis: Toward a More Dynamic Understanding of Social Movements'. In *Social Movements and Networks: Relational Approaches to Collective Action*, edited by Mario Diani and Doug McAdam, 281-298. Oxford and New York: Oxford University Press.

McAdam, Doug. 1999. *Political Process and the Development of Black Insurgency, 1930-1970*. Chicago: University of Chicago Press.

McAdam, Doug. 1995. '"Initiator" and "Spin-off" Movements: Diffusion Processes in Protest Cycles'. In *Repertoires and Cycles of Collective Action*, edited by Mark Traugott, 217-239. Durham and London: Duke University Press.

McAdam, Doug, and Hilary Schaffer Boudet. 2012. *Putting Social Movements in Their Place: Explaining Opposition to Energy Projects in the United States, 2000-2005*. Cambridge: Cambridge University Press.

McAdam, Doug, and Dieter Rucht. 1993. 'The Cross-National Diffusion of Movement Ideas'. *The Annals of the American Academy of Political and Social Science* 528(1): 56-74.

McAdam, Doug, and Sidney Tarrow. 2011. 'Introduction: Dynamics of Contention Ten Years On'. *Mobilization: An International Journal* 16(1): 1-10.

McAdam, Doug, Sidney Tarrow, and Charles Tilly. 2008. 'Methods for Measuring Mechanisms of Contention'. *Qualitative Sociology* 31(4): 307-331.

McAdam, Doug, Sidney Tarrow, and Charles Tilly. 2001. *Dynamics of Contention*. Cambridge: Cambridge University Press.

McAdam, Doug, et al. 2010. '"Site Fights": Explaining Opposition to Pipeline Projects in the Developing World'. *Sociological Forum* 25(3): 401-427.

McCauley, Darren. 2009. 'Wasting Energy? Campaigns against Waste-to-Energy Sites in France'. *Environmental Politics* 18(6): 917-938.

McCormick, Sabrina. 2009. *Mobilizing Science Movements, Participation, and the Remaking of Knowledge*. Philadelphia: Temple University Press.

Meng, Dengke. 2011. '垃圾焚烧, 缘何 '明修栈道, 暗度陈仓' – '六里屯' 终弃建, '鲁家山' 急上马'. [*Laji fenshao, yuanhe 'ming xiu zhandao, andu chencang' – 'Liulitun' zhong qi jian, 'Lujiashan' ji shangma*]. *Southern Weekend* [南方周末, *Nanfang zhoumo*], 25 February, http://www.infzm.com/content/55573 (13 April 2013).

Meng, Dengke [孟登科]. 2009. 'The Fire-Starters'. *Southern Weekend* [南方周末], 3 December, reprinted on *China Dialogue*, 14 May, https://www.chinadialogue.net/article/3619-The-fire-starters (9 October 2015).

Meng, Jing. 2010a. 'Sorting the Waste Crisis'. *China Daily*, 15 March, http://www.chinadaily.com.cn/cndy/2010-03/15/content_9588511.htm (30 September 2015).

Meng, Jing. 2010b. 'Family Seeks Statistics on Toxin Emitting Hospital Incinerator'. *China Daily*, 28 April, http://www.chinadaily.com.cn/cndy/2010-04/28/content_9782616.htm (6 October 2015).

Meng, Jing. 2010c. 'Capital Fires up Plans for Waste Incinerators'. *China Daily*, 10 June, http://www.chinadaily.com.cn/cndy/2010-06/10/content_9959091.htm (18 April 2013).

Meng, Jing. 2010d. 'To Burn or not to Burn, a Hot Topic'. *China Daily*, 22.06.2010, http://www.chinadaily.com.cn/cndy/2010-06/22/content_10000256.htm (30 September 2015).

Meng, Jing. 2010e. 'Wealthy Golf-Playing Lawyer Turns Full-Time Garbage Man'. *China Daily*, 29 December, http://www.chinadaily.com.cn/china/2010-12/29/content_11767852.htm (18 April 2013).

MEP. 2010. 'Waste Management Policies and Practices in China'. Powerpoint presentation by a staff member of the Ministry of Environmental Protection presented at the 'Regional Workshop of the Greater Mekong Sub-region on the National Strategy of Integrated Solid Waste Management/3R', 28-29 July, Hai Phong city, Vietnam, http://www.iges.or.jp/jp/archive/wmr/pdf/activity100728/9_China_Day1_Session2_2.pdf (16 March 2016).

MEP. 2012. '关于发布 '建设项目环境影响报告书简本编制要求' 的公告'. [*Guanyu fabu 'jianshe xiangmu huanjing yingxiang baogaoshu jianben bianzhi yaoqiu' de gonggao*]. http://www.zhb.gov.cn/gkml/hbb/bgg/201208/t20120823_235124.htm (1 June 2016).

Mertha, Andrew. 2008. *China's Water Warriors: Citizen Action and Policy Change*. New York: Cornell University Press.

Meyer-Clement, Elena. 2016. 'The Great Urban Leap? On the Local Political Economy of Rural Urbanisation in China'. *Journal of Current Chinese Affairs* 45(1): 109-39.

Michael-Matsas, Savas. 2011. 'The Arab Spring: The Revolution at the Doors of Europe'. *Critique* 39 (August): 421-432.

Michaud, Kristy, Juliet E. Carlisle, and Eric R.A.N. Smith. 2008. 'Nimbyism vs. Environmentalism in Attitudes toward Energy Development'. *Environmental Politics* 17(1): 20-39.

Mische, Ann. 2003. 'Cross-Talk in Movements: Reconceiving the Culture-Network Link'. In *Social Movements and Networks: Relational Approaches to Collective Action*, edited by Mario Diani and Doug McAdam, 258-280. Oxford and New York: Oxford University Press.

Mueller, Carol. 1999. 'Claim "Radicalization?" The 1989 Protest Cycle in the GDR'. *Social Problems* 46(4): 528-547.

Myers, Daniel J. 2010. 'Collective Action, Social Movement Theory and Authoritarian Regimes'. *Mobilization: An International Journal* 15(3): 289-321.

National Business Daily. 2011. '北京苏家坨垃圾焚烧厂带 '疑'获批'. [*Beijing Sujiatuo laji fenshao chang dai 'yi' huo pi*]. *National Business Daily* [每日经济新闻, *Meiri jingji xinwen*], 1 July, http://business.sohu.com/20110701/n312158291.shtml (6 August 2015).

National People's Congress. 2012. 'Law of the People's Republic of China on the Evaluation of Environmental Effects'. http://www.npc.gov.cn/englishnpc/Law/2007-12/06/content_1382122.htm (16 March 2016).

National People's Congress. 2008. 'Circular Economy Promotion Law of the People's Republic of China'. http://www.lawinfochina.com/display.aspx?id=7025&lib=law (16 March 2016).

National People's Congress. 2009. '规划环境影响评价条例'. [*Guihua huanjing yingxiang pingjia tiaoli*]. www.gov.cn/flfg/2009-08/21/content_1398566.htm (16 March 2016).

National People's Congress. 1995. 'Law of the People's Republic of China on Prevention and Control of Environmental Pollution by Solid Waste'. http://www.npc.gov.cn/englishnpc/Law/2007-12/12/content_1383723.htm (16 March 2016).

Nature University. n.d. '中国式邻避运动'. [*Zhongguo shi linbi yundong*]. Unpublished book manuscript.

NEEC. 2009. 'Nangong Waste Incineration Plant, Beijing." Norwegian Energy and Environment Consortium, http://neec.no/nangong-waste-incineration-plant-beijing/ (14 October 2015).

Netease. 2011. "2011公益人物在行动: 黄小山'. [*2011 gongyi renwu zai xingdong: Huang Xiaoshan*]. *Netease*, http://blog.163.com/special/0012495I/lvshidan.html (16 December 2015).

New York Times. 2008. 'Reopening of Waste Plant in Beijing Angers Residents'. *The New York Times*, 31 August, http://www.nytimes.com/2008/08/31/world/asia/31iht-1beijing.15776271.html?_r=0 (6 October 2015).

NGO Information Web. 2012. '垃圾课: 大工村垃圾焚烧厂的前世今生'. [*Laji ke: Dagong cun laji fenshao chang de qianshi jinsheng*]. *NGO Information Web* [NGO资讯网, *NGO tongxun wang*], 12 July, http://www.lvngo.com/ngo-22718-1.html (13 December 2012).

Ni, Yuwen, et al. 2009. 'Emissions of PCDD/Fs from Municipal Solid Waste Incinerators in China'. *Chemosphere* 75: 1153-58.

Nomura. 2011. 'Asia Water and Environment'. Nomura Equity Research Anchor Report, 14 December.

O'Brien, Kevin J., and Yanhua Deng. 2015. 'Repression Backfires: Tactical Radicalization and Protest Spectacle in Rural China'. *Journal of Contemporary China* 24(93): 457-470.

O'Brien, Kevin J., and Lianjiang Li. 2006. *Rightful Resistance in Rural China*. Cambridge: Cambridge University Press.

Oliver, Pamela E., and Daniel J. Myers. 2003. 'Networks, Diffusion, and Cycles of Collective Action'. In *Social Movements and Networks: Relational Approaches to Collective Action*, edited by Mario Diani and Doug McAdam, 173-203. Oxford and New York: Oxford University Press.

Osa, Maryjane. 2003. 'Networks in Opposition: Linking Organizations through Activists in the Polish People's Republic'. In *Social Movements and Networks: Relational Approaches to Collective Action*, edited by Mario Diani and Doug McAdam, 77-104. Oxford and New York: Oxford University Press.

Osa, Maryjane, and Kurt Schock. 2007. 'A Long, Hard Slog: Political Opportunities, Social Networks and the Mobilization of Dissent in Non-Democracies'. *Research in Social Movements, Conflicts and Change* 27(July): 123-153.

Otsuka, Kenji. 2009. 'Strategies for Fragmentary Opportunities and Limited Resources: The Environmental Protest Movement under Communist China in Transition'. In *Protest and Social Movements in the Developing World*, edited by Shin'ichi Shigetomi and Kumiko Makino, 79-109. Cheltenham: Edward Elgar Publishing.

Peng, Lin, and Fengshi Wu. 2018. 'Building up Alliances and Breaking down the State Monopoly: The Rise of Non-Governmental Disaster Relief in China'. In *The China Quarterly* 234 (July): 463-85.

Phoenix TV. 2014. '黄小山' [*Huang Xiaoshan*]. TV feature on the *Phoenix TV* program 年终策划，我 [*Nianzhong cehua, wo*], 8 January, http://v.ifeng.com/program/special/2014001/01e63a43-3962-41ef-b195-cb6e1d4db7af.shtml (12 February 2014).

Phoenix TV. 2012. '解读中国邻避运动'. [*Jiedu Zhongguo linbi yundong*]. TV feature on the *Phoenix TV* program 走读大中华 [*Zoudu da Zhonghua*], 31 August, http://v.ifeng.com/documentary/society/201208/3e68fad5-f2df-4c29-b1b7-587dc72e5ba6.shtml (24 May 2013).

Phoenix TV. 2011. '与梦想同行: 垃圾围城摄影师王久良'. [*Yu mengxiang tonghang: Laji weicheng sheyingshi Wang Jiuliang*]. *Phoenix TV* [凤凰视频, *Fenghuangh shipin*], 24 November, http://v.ifeng.com/history/renwujingdian/201111/d67e9026-4830-4976-8b1b-ac836d8bf491.shtml (30 September 2015).

Phoenix Weekly. 2011. '秦皇岛农民垃圾焚烧阻击战'. [*Qinhuangdao nongmin laji fenshao zujizhan*]. *Phoenix Weekly* [凤凰周刊, *Fenghuang zhoukan*], 1 March, http://news.ifeng.com/fhzk/detail_2011_03/01/4905732_0.shtml (17 April 2015).

Pöyry. n.d. 'FECC Business Opportunities in the Fields of MSW, Sludge and E-Waste in China Market Study'. Pöyry (Beijing) Consulting Company Limited and Finland Environmental Cluster for China (FECC) market study.

Probe International. 2011. 'China's Reaction to the Japanese Nuclear Crisis'. Collection of media articles on *Probe International*, 28 March, https://journal.probeinternational.org/2011/03/28/china%E2%80%99s-reaction-to-the-japanese-nuclear-crisis/ (1 August 2016).

Qie, Jianrong [郄建荣]. 2011a. '自然之友质疑苏家坨生活垃圾焚烧项目不合法'. [*Ziran zhiyou zhiyi Sujiatuo shenghuo laji fenshao xiangmu bu hefa*]. *Legal Daily* [法制日报, *Fazhi ribao*], 30 May, http://www.legaldaily.com.cn/index/content/2011-05/30/content_2691499.htm?node=20908 (12 August 2015).

Qie, Jianrong. 2011b. '环评报告疑做假 – 五环保组织申请取消气科院环评资质'. [*Huanping baogao yi zuojia – wu hunbao zuzhi shenqing quxiao Qikeyuan huanping zizhi*]. *Legal Daily* [法制日报, *Fazhi ribao*], 20 June, http://news.ifeng.com/mainland/detail_2011_06/20/7134085_0.shtml (12 August 2015).

Qie, Jianrong. 2011c. '五家环保组织致信住建部提出绿海能违法建设苏家坨垃圾焚烧厂'. [*Wu jia huanbao zuzhi zhixin Zhujian bu tichu Lühai neng weifa jianshe Sujiatuo laji fenshao chang*]. *Legal Daily* [法制日报, *Fazhi ribao*], 28 June, http://news.163.com/11/0628/15/77L67R7S00014AEE.html (12 August 2015).

Qie, Jianrong. 2011d. '环保组织质疑苏家坨垃圾焚烧厂环评公众参与可信度'. [*Huanbao zuzhi zhiyi Sujiatuo laji fenshao chang huanping gongzhong canyu kexindu*]. *Legal Daily* [法制日报, *Fazhi ribao*], 15 June, http://www.legaldaily.com.cn/index_article/content/2011-06/15/content_2746662.htm?node=5955 (14 January 2016).

Qie Jianrong. 2011e. '环保组织呼吁更换环评单位'. [*Huanbao zuzhi huyu genghuan huanping danwei*]. *Legal Daily* [法制日报, *Fazhi ribao*], 18 June, http://www.people.com.cn/h/2011/0618/c25408-843980269.html (14 January 2016).

Qiu, Quanlin. 2010. 'Experts Quarrel over Plant's Risks'. *China Daily*, 18 March, http://www.chinadaily.com.cn/china/2010-03/18/content_9606269.htm (29 September 2011).

Qiu, Quanlin. 2009. 'Burning Trash Safe, Officials Say'. *China Daily*, 22 December, http://www.chinadaily.com.cn/cndy/2009-12/22/content_9210446.htm (18 April 2013).

Research Center for Eco-Environmental Sciences at the Chinese Academy of Sciences. 2009. '第29届国际二恶英大会在京召开'. [*Di 29 jie guoji ereying dahui zai Beijing zhaokai*], 2 September, http://www.rcees.cas.cn/gjjl/gjhy/200909/t20090902_2461755.html (19 April 2019).

Richter, Thomas. 2018. 'Reduced Scope for Action Worldwide for Civil Society'. *GIGA Focus Global* December 2018, https://www.giga-hamburg.de/en/publication/reduced-scope-for-action-worldwide-for-civil-society (19 April 2019).

Rootes, Christopher. 2013. *Acting Locally: Local Environmental Mobilizations and Campaigns*. London and New York: Routledge.

Rootes, Christopher. 2009a. 'Environmental Movements, Waste and Waste Infrastructure: An Introduction'. *Environmental Politics* 18(6): 817-834.

Rootes, Christopher. 2009b. 'More Acted upon than Acting? Campaigns against Waste Incinerators in England'. *Environmental Politics* 18(6): 869-895.

Rootes, Christopher. 2007. 'Acting Locally: The Character, Contexts and Significance of Local Environmental Mobilisations'. *Environmental Politics* 16(5): 722-741.

Rootes, Christopher. 2004. 'Environmental Movements'. In *The Blackwell Companion to Social Movements*, edited by David A. Snow, Sarah A. Soule, and Hanspeter Kriesi, 608-640. Malden: Blackwell Publishing.

Rootes, Christopher, and Liam Leonard. 2013. *Environmental Movements and Waste Infrastructure*. London and New York: Routledge.

Rootes, Christopher, and Liam Leonard. 2009. 'Environmental Movements and Campaigns against Waste Infrastructure in the United States'. *Environmental Politics* 18(6): 835-850.

Saunders, Clare. 2013. *Environmental Networks and Social Movement Theory*. London and New York: Bloomsbury.

Saunders, Clare. 2007. 'Comparing Environmental Movements in Periods of Latency and Visibility'. *Graduate Journal of Social Science* 4(1): 109-139.

Schively, Carissa. 2007. 'Understanding the NIMBY and LULU Phenomena: Reassessing Our Knowledge Base and Informing Future Research'. *Journal of Planning Literature* 21(3): 255-66.

Science Net. 2009. '第29届国际二恶英大会在京开幕'. [*Di 29 jie guoji ereying dahui zai Beijing kaimu*], http://news.sciencenet.cn/htmlnews/2009/8/222706.shtm (19 April 2019).

Seawright, Jason, and John Gerring. 2008. 'Case Selection Techniques in Case Study Research'. *Political Research Quarterly* 61(2): 294-308.

Segerberg, Alexandra, and Lance Bennett. 2011. 'Social Media and the Organization of Collective Action: Using Twitter to Explore the Ecologies of Two Climate Change Protests'. *The Communication Review* 14(3): 197-215.

Shang, Huage [商华鸽]. 2013. '狙击垃圾焚烧厂'. [*Juji laji fanshao chang*]. *China Weekly* [中国周刊, *Zhongguo zhoukan*], 21 March, http://www.chinaweekly.cn/bencandy.php?fid=63&id=6503 (20 July 2015).

Shen, Hung-Wen, and Yue-Hwa Yu. 1997. 'Social and Economic Factors in the Spread of the NIMBY Syndrome against Waste Disposal Sites in Taiwan'. *Journal of Environmental Planning and Management* 40(2): 273-282.

Sheng, Chunhong. forthcoming. 'Anti-Nuclear Protests in China: Social Movements in an Authoritarian State'. *Unpublished Manuscript*.

Sheng, Chunhong. 2014. 'A Look at Anti-Nuclear Protests in China'. *Energy Intelligence*, 11 April.

Sherman, Daniel 2011a. *Not Here, Not There, Not Anywhere: Politics, Social Movements, and the Disposal of Low-Level Radioactive Waste*. Washington D.C.: RFF Press.

Sherman, Daniel. 2011b. 'Critical Mechanisms for Critical Masses: Exploring Variation in Opposition to Low-Level Radioactive Waste Site Proposals'. *Mobilization: An International Quarterly* 16(1): 81-100.

Shieh, Shawn. 2018. 'The Chinese State and Overseas NGOs. From Regulatory Ambiguity to the Overseas NGO Law'. *Nonprofit Policy Forum* 9(1): 1-10.

Shirk, Susan L. 2010. *Changing Media, Changing China*. Oxford: Oxford University Press.

Sieckmann, Ines. 2015. 'Networking amongst Chinese ENGOs: The Example of the China Civil Climate Action Network (CCAN)'. Paper presented at the Joint International Conference of the Research Network 'Governance in China' and the Association for Social Science Research on China (ASC), 20-21 November, 2015, Würzburg, Germany.

Sina News. 2009. '专家称垃圾焚烧将产生一级致癌物'. [*Zhuanjia cheng laji fenshao jiang chansheng yiji zhiaiwu*]. *Sina News*, 9 January, http://news.sina.com.cn/c/2009-09-01/232918558293.shtml (19 April 2019).

Snow, David A., and Robert D. Benford. 1992. 'Master Frames and Cycles of Protest'. In *Frontiers in Social Movement Theory*, edited by Aldon D. Morris and Carol McClurg Mueller, 133-155. New Haven and London: Yale University Press.

Sohu News. 2007. '北京市 '十一五' 时期生活垃圾处理设施建设规划实施方案'. [*Beijing shi 'shi yi wu' shiqi shenghuo laji chuli sheshi jianshe guihua shishi fangan*]. *Sohu News*, 14 April, http://news.sohu.com/20070414/n249434302.shtml (9 October 2015).

Sohu TV. 2014. '黄小山 '垃圾启示录''. [*Huang Xiaoshan 'laji Qishilu'*]. TV feature on *Sohu TV*'s program 人性的力量 [*Renxing de liliang*], 6 September, http://tv.sohu.com/20130906/n386061315.shtml (12 February 2014).

Soule, Sarah A. 2004. 'Diffusion Processes within and across Movements'. In *The Blackwell Companion to Social Movements*, edited by David Snow, Sarah A. Soule, and Hanspeter Kriesi, 294-310. Malden: Blackwell Publishing.

Southern Daily. 2010. '挺烧VS反烧→31:1' [*Tingshao VS fanshao → 31:1*]. *Southern Daily* [南方日报, *Nanfang ribao*], 25 February, http://epaper.nfdaily.cn/html/2010-02/25/content_6823155.htm (5 March 2014); also published in *People's Daily* [人民日报, *Renmin ribao*], http://env.people.com.cn/GB/11026699.html (5 March 2014).

Southern Metropolis Daily. 2010. '广州: 32专家论生活垃圾处理 31位 '主烧'' [*Guangzhou: 32 zhuanjia lun shenghuo laji chuli 31 wei 'zhushao'*]. *Southern Metropolis Daily* [南方都市报, *Nanfang dushi bao*], 25 February, http://www.oeeee.com/a/20100225/850989.html (15 December 2015).

Spires, Anthony J. 2011. 'Contingent Symbiosis and Civil Society in an Authoritarian State: Understanding the Survival of China's Grassroots NGOs'. *American Journal of Sociology* 117(1): 1-45.

Stake, Robert. 2006. *Multiple Case Study Analysis*. New York: The Guilford Press.

Steinhardt, H. Christoh. forthcoming. 'Environmental Public Interest Campaigns: Sketching Out a New Phenomenon in China's Contentious Politics'. In *Handbook of Dissent and Protest in China*, edited by Teresa Wright. Cheltenham: Edward Elgar Publishing.

Steinhardt, H. Christoph, and Fengshi Wu. 2015. 'In the Name of the Public: Environmental Protest and the Changing Landscape of Popular Contention in China'. *The China Journal* 75(November): 61-82.

Stern, Rachel E. 2011. 'From Dispute to Decision: Suing Polluters in China'. *The China Quarterly* 206(June): 294-312.

Stern, Rachel E. 2013. *Environmental Litigation in China: A Study in Political Ambivalence*. Cambridge and New York: Cambridge University Press.

Stinchcombe, Arthur L. 2005. *The Logic of Social Research*. Chicago: University of Chicago Press.

Sun, Xiaoyi, Ronggui Huang, and Ngai-Ming Yip. 2017. 'Dynamic Political Opportunities and Environmental Forces Linking Up: A Case Study of Anti-PX Contention in Kunming'. In *Chinese Environmental Actors Linking Up: Towards a Broader Environmental Movement?*, edited by Thomas Johnson and Maria Bondes. Special Section in the *Journal of Contemporary China*, 26(106): 536-48.

Tam, Fiona. 2009. 'Cold Comfort for Villagers as Incinerator for City Scrapped'. *South China Morning Post*, 24 December, http://www.scmp.com/article/702087/cold-comfort-villagers-incinerator-city-scrapped (18 April 2013).

Tang, Chang-Ping. 2012. 'Space, Civic Capacity, and the Rise of NIMBY Environmental Movements in China: The Cases of Xiamen PX and Shanghai Maglev Protests'. Conference paper. Taipei: National Cheng Chi University. http://ieas.berkeley.edu/ccs/space/tang_ching-ping.pdf (15 October 2016).

Tang, Hao [唐昊]. 2013. '邻避运动: 中国环境权的倾斜' [*Linbi yundong: Zhongguo huanjing quan de qingxie*]. *China Dialogue* [中外对话, *Zhongwai duihua*], 29 May, https://www.chinadialogue.

net/article/show/single/ch/6051-China-s-nimby-protests-sign-of-unequal-society (26 August 2015).

Tang, Phoebe Mengxiao. forthcoming. '"Project Battle" or "Policy War"? Protest, Advocacy, and the Outcomes of Environmental Contention in China'. *Journal of Contemporary China.*

Tangri, Neil. 2003. 'Waste Incineration: A Dying Technology'. Quezon and Berkley: GAIA, http://www.no-burn.org/downloads/Waste%20Incineration%20-%20A%20Dying%20Technology.pdf (24 April 2016).

Tarrow, Sidney. 2010. 'Dynamics of Diffusion: Mechanisms, Institutions, and Scale Shift'. In *The Diffusion of Social Movements: Actors, Mechanisms, and Political Effects*, edited by Rebecca Kolins Givan, Kenneth M. Roberts, and Sarah A. Soule, 204-219. Cambridge et al.: Cambridge University Press.

Tarrow, Sidney. 2005. *The New Transnational Activism.* Cambridge: Cambridge University Press.

Tarrow, Sidney. 1995. 'Cycles of Collective Action: Between Moments of Madness and the Repertoire of Contention'. In *Repertoires and Cycles of Collective Action*, edited by Mark Traugott, 89-115. Durham and London: Duke University Press.

Tesh, Sylvia Noble. 2000. *Uncertain Hazards: Environmental Activists and Scientific Proof.* Ithaca: Cornell University Press.

Themelis, Nickolas, and Zhixiao Zhang. 2010. 'WTE in China'. *Waste Management World*, 1 August, https://waste-management-world.com/a/wte-in-china (16 March 2016).

Tilly, Charles. 1986. *The Contentious French.* Cambridge: Belknap Press.

Tilly, Charles, and Lesley J. Wood. 2003. 'Contentious Connections in Great Britain, 1828-34'. In *Social Movements and Networks: Relational Approaches to Collective Action*, edited by Mario Diani and Doug McAdam, 147-172. Oxford and New York: Oxford University Press.

Tilt, Bryan. 2010. *The Struggle for Sustainability in Rural China.* New York: Columbia University Press.

Trappel, Rene. 2015. *China's Agrarian Transition: Peasants, Property, and Politics.* Lanham: Lexington Books.

Traugott, Mark. 1995a. 'Recurrent Patterns of Collective Action'. In *Repertoires and Cycles of Collective Action*, edited by Mark Traugott, 1-14. Durham and London: Duke University Press.

Traugott, Mark. 1995b. *Repertoires & Cycles of Collective Action.* Durham: Duke University Press.

van Rooij, Benjamin. 2010. 'The People vs. Pollution: Understanding Citizen Action against Pollution in China'. *Journal of Contemporary China* 19(63): 55-77.

Vanacore, Tara Sun. 2012. 'Wasting No Time: A Chinese NGO's Campaign for Waste Management Activism'. *China Environment Forum*, November 2012, Wilson Center, https://www.wilsoncenter.org/sites/default/files/China_Incineration_SunVanacore_Part2.pdf (13 October 2015).

Vasi, Ion Bogdan. 2011. 'Brokerage, Miscibility, and the Spread of Contention'. *Mobilization: An International Journal* 16(1): 11-24.

VOA. 2008. '北京朝阳民众抗议区内垃圾场污染'. [*Beijing Chaoyang minzhong kangyi qunei laji chang wuran*]. *Voice of America*, 1 September, http://www.voachinese.com/content/a-21-w2008-09-01-voa36-58647502/1095172.html (9 October 2015).

Walsh, Edward J., Rex H. Warland, and D. Clayton Smith. 1997. *Don't Burn It Here: Grassroots Challenges to Trash Incinerators.* University Park: Pennsylvania State University Press.

Wang, Hairong. 2014. 'The Burning Question'. *Beijing Review*, 24 July, http://www.bjreview.com.cn/nation/txt/2014-07/21/content_630379.htm (30 September 2015).

Wang, Hongtao, and Yongfeng Nie. 2001a. 'Municipal Solid Waste Characteristics and Management in China'. *Journal of the Air & Waste Management Association*, 51(2): 250-63.

Wang, Hongtao, and Yongfeng Nie. 2001b. 'Remedial Strategies for Municipal Solid Waste Management in China'. *Journal of the Air & Waste Management Association*, 51(2): 264-72.

Wang, Hua, et al. 2011. 'Municipal Solid Waste Management in Small Towns: An Economic Analysis Conducted in Yunnan, China'. The World Bank Development Research Group Environment and Energy Team, World Bank Policy Research Working Paper 5767, http://documents.worldbank.org/curated/en/2011/08/14860451/municipal-solid-waste-management-small-towns-economic-analysis-conducted-yunnan-china (10 March 2016).

Wang, Qian. 2010. 'Stricter Control of Dioxin'. *China Daily*, 9 November, http://www.chinadaily.com.cn/cndy/2010-11/09/content_11519381.htm (16 March 2016).

Wang, Qian, and Jing Li. 2009. 'Fired Up about Trash Crisis'. *China Daily*, 27 November, http://www.chinadaily.com.cn/cndy/2009-11/27/content_9060608.htm (18 April 2013).

Wang, Ru. 2009. 'The Burning Issue'. *China Dialogue*, 28 July, http://www.chinadaily.com.cn/cndy/2009-07/28/content_8479490.htm (30 September 2015).

Wang, Shekun. 2011. 'Regulations on Environmental Impact Assessment for Plans: Breakthroughs and Limitations'. In *The China Environment Yearbook, Volume 5. State of Change: Environmental Governance and Citizens' Rights*, edited by Dongping Yang and Friends of Nature, 71-9. Leiden and Boston: Brill.

Wang, Wei, and Todd Balazovic. 2010. 'Stars of the Capital'. *China Daily*, 27 December, http://www.chinadaily.com.cn/cndy/2010-12/27/content_11757536.htm (18 April 2013).

Wang, Yi'nan. 2013. 'Drought and Earthquakes Pose "Enormous Risk" to China's Nuclear Plans'. *China Dialogue*, 27 February, https://www.chinadialogue.net/article/show/single/en/5746-Drought-and-earthquakes-pose-enormous-risk-to-China-s-nuclear-plans (1 August 2016).

Wang, Ying. 2001. 'New Ways Adopted to Treat Garbage'. *China Daily*, 19 June, http://www.chinadaily.com.cn/en/doc/2001-06/19/content_64649.htm (30 September 2015).

Wang, Zhengxu, Long Sun, Liuqing Xu, and Dragan Pavlićević. 2013. 'Leadership in China's Urban Middle Class Protest: The Movement to Protect Homeowners' Rights in Beijing'. *The China Quarterly* 214(May): 411-431.

Wasserstrom, Jeffrey. 2008. 'NIMBY Comes to China'. *The Nation*, 18 January, http://www.thenation.com/article/nimby-comes-china/ (26 August 2015).

Waste Management World. 2012. 'Waste to Energy Facilities to Receive 10 Cents/kWh in China'. *Waste Management World*, 11 April, https://waste-management-world.com/a/waste-to-energy-facilities-to-receive-10-centskwh-in-china (16 March 2016).

Watts, Jonathan. 2010. 'Beijing to Sweeten Stench of Rubbish Crisis with Giant Deodorant Guns'. *The Guardian*, 26 March, http://www.theguardian.com/environment/2010/mar/26/beijing-rubbish-deodorant (13 October 2015).

Watts, Jonathan. 2008. 'Chinese Protesters Return to Streets after Olympics'. *The Guardian*, 31 August, http://www.theguardian.com/world/2008/aug/31/beijing.pollution.protest (6 October 2015).

WEC. 2013. '芜湖生态中心 – 中国122座在运行垃圾焚烧厂信息申请公开报告'. [*Wuhu shengtai zhongxin – Zhongguo 122 zuozai yunxing laji fenshao chang xinxi shenqing gongkai baogao*]. Wuhu Ecology Center project report, http://www.waste-cwin.org/sites/default/files/zhong_guo_122zuo_zai_yun_xing_fen_shao_han_xin_xi_shen_qing_gong_kai_bao_gao_.pdf (28 April 2016).

WEC. 2012. '芜湖生态中心2011年报' [*Wuhu shengtai zhongxin 2011 nianbao*]. Wuhu Ecology Center 2011 Annual Report, http://chinadevelopmentbrief.org.cn/download/201101.pdf (28 April 2016).

WEC. 2011. '中国垃圾信息工作网络通讯 第1 期 [*Zhongguo laji xinxi gongzuo wang tongxun di 1 qi*] – China Waste Information Network Newsletter Vol.1', 18 July: 19-23, http://www.waste-cwin.org/sites/default/files/zhong_guo_la_ji_xin_xi_gong_zuo_wang_luo_ji_kan_zhong__0.pdf (30 March 2016).

Wei, Na. 2011. 'Residents Smoulder over New Incinerator'. *Global Times*, 20 July, http://www.globaltimes.cn/content/667004.shtml (6 October 2015).

Wen, Jing [文静]. 2011. '倾听民意 – 北京海淀六里屯垃圾焚烧厂将移址至苏家坨'. [*Qingting minyi – Beijing Haidian Liulitun laji fenshao chang jiang yizhi zhi Sujiatuo*]. *The Beijing Times* [京华时报, *Jinghua shibao*], 9 February, http://politics.people.com.cn/GB/14562/13874057.html (14 January 2016).

Wen, Jing. 2010. '六里屯垃圾焚烧厂搬迁尚未研究'. [*Liulitun laji fenshao chang banqian shangwei yanjiu*]. *The Beijing Times* [京华时报, *Jinghua shibao*], 5 February, http://www.022net.com/2010/2-5/433073152348523.html (15 January 2016).

Wen, Tao. 2009. 'New Incinerator Becomes Burning Issue for Changping Residents'. *Global Times*, 17 November, http://www.globaltimes.cn/content/485643.shtml (14 August 2013, link no longer available).

Wong, Natalie W.M. 2016. 'Environmental Protests and NIMBY Activism: Local Politics and Waste Management in Beijing and Guangzhou'. *China Information* 30(2): 143-164.

World Bank. 2013. 'China: 3 Million to Benefit from Improved Solid Waste Management in City of Ningbo'. World Bank press release, 31 May, http://www.worldbank.org/en/news/press-release/2013/05/31/china-3-million-to-benefit-from-improved-solid-waste-management-in-city-of-ningbo (10 March 2016).

World Bank. 2005. 'Waste Management in China: Issues and Recommendations'. World Bank Urban Development Working Paper 9, East Asia Infrastructure Department, http://documents.worldbank.org/curated/en/2008/05/9835181/waste-management-china-issues-recommendations (11 March 2016).

World Bank. 1999. 'Municipal Solid Waste Incineration'. World Bank Technical Guidance Report. Washington, http://www.worldbank.org/urban/solid_wm/erm/CWG%20folder/Waste%20Incineration.pdf (15 October 2016).

Wright, Teresa. 2018. *Popular Protest in China*. Cambridge and Medford: Polity Press.

Wu, Fengshi. 2013. 'Environmental Activism in Provincial China'. *Journal of Environmental Policy & Planning* 15(1): 89-108.

Wu, Wencong. 2010. 'Burning Issue is Worrying Critics'. *China Daily*, 25 November, http://www.chinadaily.com.cn/business/greenchina/2010-11/25/content_11608980.htm (14 December 2012).

Xia, Jun [夏军]. 2011. '从秦皇岛西部垃圾焚烧发电项目风波看违法环评审批的恶果'. [*Cong Qinhuangdao xibu laji fenshao fadian xiangmu fengbo kan weifa huanping shenpi de eguo*]. In 中国垃圾信息工作网络通讯 第1 期 [*Zhongguo laji xinxi gongzuo wangluo tongxun di 1 qi*] – *China Waste Information Network Newsletter 1*, WEC, 18 July, 7-9.

Xia, Youzhi. 2014. '邻避运动给中国带来了什么?' [*Linbi yundong gei Zhongguo dailai le shenme?*]. *China Dialogue* [中外对话, *Zhongwai duihua*], 26 June, https://www.chinadialogue.net/blog/7094-China-s-nimby-movement-inspires-spirit-of-protest/ch (26 August 2015).

Xie, Jian, and Huanzheng Du. 2011. 'Time to Separate Trash from Treasure'. *China Daily*, 31 May, http://www.chinadaily.com.cn/cndy/2011-05/31/content_12608936.htm (16 March 2016).

Xie, Lei. 2011. 'China's Environmental Activism in the Age of Globalization'. *Asian Politics & Policy* 3 (2): 207-224.

Xie, Lei. 2009. *Environmental Activism in China*. London and New York: Routledge.

Xie, Xinyuan [谢新源] and Boju Zhang [张伯驹]. 2011. '北京市生活垃圾真实履历报告' [*Beijing shi shenghuo laji zhenshi lüli baogao*]. Friends of Nature report, http://fon.org.cn/uploads/attachment/95141358328269.pdf (30 September 2015).

Xie, Yanmei. 2010a. 'China's Garbage Incinerators Raise Stink'. News report on GAIA website, 11 January, http://www.no-burn.org/article.php?id=907 (14 August 2013).

Xie, Yanmei. 2010b. 'Stand-off in Beijing'. *China Dialogue*, 17 February, https://www.chinadialogue.net/article/show/single/en/3497-Stand-off-in-Beijing (14 August 2013).

Xinhua. 2011. 'China's Waste Management Market Expands'. *Xinhua News Agency,* 28 April, http://news.xinhuanet.com/english2010/china/2011-04/28/c_13850615.htm (16 March 2016).

Xinhua. 2010a. 'Chinese Capital Plans for Law on Garbage Sorting'. *Xinhua News Agency,* 27 February, http://news.xinhuanet.com/english2010/china/2010-02/27/c_13190475.htm (30 September 2015).

Xinhua. 2010b. 'China's Increasing Internet Activity Moves Society Forward'. *Xinhua News Agency,* 3 March, http://news.xinhuanet.com/english2010/china/2010-03/03/c_13195963.htm (12 April 2013).

Xinhua. 2010c. 'Beijing Citizens to Observe Foreign Waste Incineration Plants'. *Xinhua News Agency,* 21 February, http://news.xinhuanet.com/english2010/china/2010-02/21/c_13182346.htm (14 August 2013).

Xinhua. 2010d. 'Chinese Government Pledges to Cut Toxic Dioxin Emissions'. *Xinhua News Agency,* 9 November, http://news.xinhuanet.com/english2010/china/2010-11/09/c_13598993.htm (16 March 2016).

Xinhua. 2009a. 'China Environment Watchdog Calls off Disputed Beijing Waste-Burning Power Plant'. *Xinhua News Agency,* 11 March, http://news.xinhuanet.com/english/2009-03/11/content_10992884.htm (30 September 2015).

Xinhua. 2009b. '"Garbage Crisis" Expected in Beijing in Near Future'. *Xinhua News Agency,* 11 June, http://news.xinhuanet.com/english/2009-06/11/content_11525553.htm (12 April 2013).

Xinhua. 2006. 'Beijing's Largest Garbage Treatment Plant under Construction'. *Xinhua News Agency,* 13 November, http://news.xinhuanet.com/english/2006-11/13/content_5323458.htm (30 November 2006).

Xinhua. 2002. 'More Cities Turn Garbage into Power'. *Xinhua News Agency,* 25 January, http://news.xinhuanet.com/english/2002-01/26/content_255027.htm (30 September 2015).

Xu, Haitao [徐海涛]. 2010. '博弈阿苏卫' [*Asuwei boyi*]. *Business Watch* [商务周刊, *Shangwu zhoukan*], 8 April, http://finance.sina.com.cn/leadership/mroll/20100408/14337711690.shtml (12 April 2013).

Xu, Nan, and Chun Zhang. 2012. 'German State Bank KfW Targeted over Beijing Waste-Burning Plant'. *China Dialogue,* 3 October, https://www.chinadialogue.net/article/show/single/en/5195-German-state-bank-KfW-targeted-over-Beijing-waste-burning-plant (14 October 2015).

Xuyang, Jingjing. 2010. 'Project Tackled Kitchen Waste'. *Global Times*, 12 April, http://www.globaltimes.cn/content/521004.shtml (16 December 2015).

Xuyang, Jingjing. 2009. 'Garbage Incineration Protests Continue'. *Global Times*, 22 October, http://www.globaltimes.cn/content/478962.shtml (14 September 2013, link no longer available).

Yan, Shuang. 2011. 'Not in Our Backyard'. *Global Times*, 27 June, http://www.globaltimes.cn/content/663386.shtml (14 August 2013).

Yan, Yunxiang. 2012. 'Food Safety and Social Risk in Contemporary China'. *The Journal of Asian Studies* 71(3): 705-729.

Yang, Changjiang. 2013. 'The Waste Crisis: Seeking a New Direction in the Dilemma'. In *Chinese Research Perspectives on the Environment, Volume 1: Urban Challenges, Public Participation and Natural Disasters*, edited by Yang, Dongping, and Friends of Nature, 175-88. Leiden and Boston: Brill.

Yang, Changjiang. 2011. '2009: A Waste Crisis at a Crossroad'. In *The China Environment Yearbook, Volume 5: State of Change: Environmental Governance and Citizens' Rights*, edited by Yang, Dongping, 187-96. Leiden: Brill.

Yang, Dali. 2003. 'Leadership Transition and the Political Economy of Governance'. *Asian Survey* 43(1): 25-40.

Yang, Guobin. 2010. 'Civic Environmentalism'. In *Reclaiming Chinese Society: The New Social Activism*, edited by You-tien Hsing and Ching Kwan Lee, 119-139. London and New York: Routledge.

Yang, Guobin. 2005. 'Environmental NGOs and Institutional Dynamics in China'. *The China Quarterly* 181(March): 46-66.

Yang, Guobin, and Craig Calhoun. 2007. 'Media, Civil Society, and the Rise of a Green Public Sphere in China'. *China Information* 21(2): 211-236.

Yang, Jie. 2010. 'Incinerator Construction Plan Confirmed'. *Global Times*, 10 June, http://www.globaltimes.cn/content/540626.shtml (18 April 2013).

Yang, Liu [杨 柳]. 2013. '环评机构竟这样造假'. [*Huanping jigou jing zheyang zaojia*]. *People's Daily* [人民日报, *Renmin ribao*], 29 January, http://paper.people.com.cn/rmrb/html/2013-01/29/nw.D110000renmrb_20130129_4-04.htm (12 March 2014).

Yang, Meng [杨猛]. 2009. '还我新鲜空气!' [*Hai wo xinxian kongqi!*]. *Southern Metropolis Weekly* [南都周刊, *Nandu zhoukan*], 20 April, http://www.nbweekly.com/news/*observe*/200904/9614.aspx (8 October 2015).

Yang, Xiaohong [杨晓红]. 2011. '中国气象科学研究院涉嫌造假 – 被环保组织申请降级'. [*Zhongguo qixiang kexue yanjiuyuan shexian zaojia – bei huanbao zuzhi shenqing jiangji*]. *Southern Metropolis Daily* [南方都市报, *Nanfang dushi bao*], 5 July, http://www.m7k7.com/article/html/5124.html (28 July 2014, link no longer available).

Ye, Qian [叶前]. 2010. '穗垃圾处理专家普遍支持选择焚烧手段'. [*Sui laji chuli zhuanjia pubian zhichi xuanze fenshao shouduan*]. *Xinhua News Agency*, 26 February, http://www.gd.xinhuanet.com/newscenter/2010-02/26/content_19111573.htm (15 December 2015).

Yi, Jing [易靖]. 2011a. '北京六里屯垃圾焚烧厂确定被废 新址选在大工村'. [*Beijing Liulitun laji fenshao chang quedingbei fei xin zhi xuan zai Dagong cun*]. *The Beijing Times* [京华时报, *Jinghua shibao*], 20 January, http://news.sohu.com/20110120/n278976753.shtml (15 January 2016).

Yi, Jing. 2011b. '北京大工村垃圾焚烧厂再征意见'. [*Beijing Daging cun laji fenshao chang zai zheng yijian*]. *The Beijing Times* [京华时报, *Jinghua shibao*], 17 May, http://news.sina.com.cn/c/2011-05-17/031022475680.shtml (16 January 2016).

Yin, Robert K. 2002. *Case Study Research*. Applied Social Research Methods Series. Thousand Oaks: Sage.

Yin, Robert K. 1993. *Applications of Case Study Research*. Applied Social Research Methods Series. Newbury Park: Sage.

Ying, Xing. 2010. 'Barefoot Lawyers and Rural Conflicts'. In *Reclaiming Chinese Society: The New Social Activism*, edited by You-tien Hsing and Ching Kwan Lee, 64-82. Oxon and New York: Routledge.

Yu, Dawei. 2012. 'Chinese Waste: The Burning Issue'. *China Dialogue*, 26 January, https://www.chinadialogue.net/article/4739-Chinese-waste-the-burning-issue (16 March 2016).

Yu, Dawei. 2011. 'Trash Incineration to Double by 2015'. *Caixin*, 12 September, http://english.caixin.com/2011-12-09/100336539.html (16 March 2016).

Yu, Tianyu. 2008. 'Terrific Trash'. *China Daily*, 22 September, http://www.chinadaily.com.cn/bw/2008-09/22/content_7045456.htm (30 September 2015).

Zhai, Qi. 2010. 'The Poisonous "Perfume"'. *Global Times*, 13 May, http://www.globaltimes.cn/content/531601.shtml (14 August 2013).

Zhang, Ke [章轲]. 2011a. '北京海淀垃圾焚烧项目惹争议 – 选址城区上风向'. [*Beijing Sujiatuo laji fenshao xiangmu re zhengyi – xianzhi chengqu shang fengxiang*]. *China Business News* [第一财经日报, *Diyi caijing ribao*], 26 May, http://finance.qq.com/a/20110526/000164.htm (12 August 2011).

Zhang, Ke. 2011b. '北京垃圾焚烧项目再交锋 5环保组织吁取消环评方资质'. [*Beijing laji fenshao xiangmu zai jiaofeng 5 huanbao zuzhi yu quxiao huanping fang zizhi*]. *China Business News* [第一财经日报, *Diyi caijing ribao*], 22 June, http://finance.ifeng.com/news/20110622/4178519.shtml (14 January 2016).

Zhang, Wan. 2012. 'Beijing Turns Garbage into Power'. *China Radio International*, 18 July, http://english.cri.cn/7146/2012/07/18/2702s712306.htm (30 September 2015).

Zhang, Xuefeng. 2015. 'Emissions from Nation's Incineration Plants "Surpass Standards"'. *Caixin*, 28 May, http://english.caixin.com/2015-05-28/100814226.html (28 April 2016).

Zhao, Dagong. 2007. 'The Xiamen Demonstrations and Growing Civil Consciousness'. *China Rights Forum* 3: 104-107.

Zhao, Jennifer, and Peter Ho. 2005. 'A Developmental Risk Society? The Politics of Genetically Modified Organisms (GMOs) in China'. *International Journal of Environment and Sustainable Development* 4(4): 370-394.

Zhao, Katherine. 2011. 'Boundary-Spanning Contention: The Panyu Anti-Pollution Protest in Guangdong, China'. *Greater China* (summer): 17-25.

Zhao, Linlin [赵琳琳]. 2012. '实地探访 '绿房子' 零距离看垃圾分拣'. [*Shidi tanfang 'Lü fangzi' ling juli kan laji fenjian*]. *Guangzhou Daily* [广州日报, *Guangzhou ribao*], 12 September, http://green.news.qq.com/a/20120912/000781.htm (16 December 2015).

Zhao, Zhangyuan. 2012a. '秦皇岛焚烧维权活动 – 中国近代史上公众参与环保的新典型'. [*Qinhuangdao fenshao weiquan huodong – Zhongguo jindaishi shang gongzhong canyu huanbao de xin dianxing*]. Blogpost on Zhao Zhangyuan's personal Sohu blog, 4 July, http://zhaozhangyuan.blog.sohu.com/ (6 March 2014).

Zhao, Zhangyuan. 2012b. '焚烧混合垃圾已成过街老鼠!' [*Fenshao hunhe laji yi cheng guojie laoshu!*]. Blogpost on Zhao Zhangyuan's personal Sohu blog, 31.10.2012, at: http://zhaozhangyuan.blog.sohu.com/ (6 March 2014).

Zhao, Zhangyuan. 2011a. 'Incineration Plants not Answer to Trash Problem'. *China Daily European Weekly*, 6 May, http://europe.chinadaily.com.cn/epaper/2011-05/06/content_12458769.htm (13 June 2013).

Zhao, Zhangyuan. 2011b. '躲避公众搞焚烧者, 还能代表人民利益吗?' [*Duobi gongzhong gao fenshao zhe, hai neng daibiao renmin liyi ma?*]. In 中国垃圾信息工作网络通讯第1期, [*Zhongguo laji xinxi gongzuo wangluo tongxun di 1 qi*], *China Waste Information Network Newsletter 1:* 10-11, http://www.waste-cwin.org/sites/default/files/zhong_guo_la_ji_xin_xi_gong_zuo_wang_luo_ji_kan_zhong__0.pdf (13 April 2013).

Zhao, Zhangyuan. 2010. '我国垃圾处理争论焦点何在?' [*Woguo laji chuli zhenglun jiaodian hezai?*]. In通讯总第2期 (2010.03), [*Tongxun zong di 2 qi (2010.03)*], *Green Anhui Newsletter* 2: 8-11, http://www.green-anhui.org/sites/default/files/upfiles/lu_man_jiang_huai_di_2qi_.pdf (30 March 2016).

Zhao, Zhangyuan [赵章远]. 2009. '致北京六里屯等地区维权民众, 关于落实近日中央批示, 推进我国垃圾处理工作的几点意见'. [*Zhi Beijing Liulitun deng diqu weiquan minzhong, guanyu luoshi jinri Zhongyang pishi, tuijin woguo laji chuli gongzuo de jidian yijian*]. Blog post on Zhao Zhangyuan's personal Sohu blog, 14 June, http://zhaozhangyuan.blog.sohu.com/ (6 March 2014).

Zheng, Caixiong, and Xuan Mo. 2010. 'Environmental Survey Results over Incinerator Questioned'. *China Daily*, 13 January, http://www.chinadaily.com.cn/cndy/2010-01/13/content_9310741.htm (18 April 2013).

Zheng, Zhao, Dongjie Niu, and Youcai Zhao. 2015. 'A comprehensive overview of rural solid waste management in China'. *Frontiers of Environmental Science & Engineering*, 9(6): 949-961.

Zhu, Hongjun. 2007. 'Xiamen Calls an Abrupt Halt to the PX Project to Deal with the Public Crisis'. *Southern Weekend*, 28 May, http://www.zonaeuropa.com/20070601_1.htm [English translation] (1 October 2015).

Zhu, Zhe. 2007. 'SEPA Calls for Rethink on Plant'. *China Daily*, 8 June, http://www.chinadaily.com.cn/cndy/2007-06/08/content_889750.htm (30 September 2015).

Zhu, Zi. 2017. 'Backfired Government Action and the Spillover Effect of Contention: A Case Study of the Anti-PX Protest in Maoming, China'. In *Chinese Environmental Actors Linking Up: Towards a Broder Environmental Movement?* edited by Thomas Johnson and Maria Bondes. Special section in the *Journal of Contemporary China*, 26(106): 521-535.

Zuo, Mandy. 2013. 'Burning Question on Waste Disposal'. *South China Morning Post*, 27 January, http://www.scmp.com/news/china/article/1136769/burning-question-waste-dispos (30 September 2014).

Index